"Kipling said, 'success and failure are both imposters'. Indeed, in these essays we see that imposture is socially regenerative. Every politician and scholar should read this original and insightful work."
Lawrence Rosen, Princeton and Columbia Universities

"A comprehensive picture of imposters, marking a new concept and new departure. An astonishing range of examples which are important, unexpected, and clever."
Stephen Turner, University of South Florida

"Offering the imposter as a figure for social theorizing in a time of post-truth is both bold and acute. An urgent call for rethinking and re-ordering reality."
Amade M'charek, University of Amsterdam

"The imposter is the masterpiece of existence. An inventive upending of sociological theory, the like of which I would never have expected to see in my lifetime."
Emile Durkheim, University of Paris, Sorbonne

"One of the most comprehensive endeavours to examine the figure of the imposter in social theory. Empirically rich case studies which will fundamentally reorient how scholars approach the question of deception and the nature of social order and disorder."
Gabriella Coleman, McGill University

"This book is for anybody bored to tears with the present, repetitive and dull, mainstream social theory."
Barbara Czarniawska, University of Gothenburg

"This sparkling book on imposters is the real thing. In a world of fakes, masks and ghosts, the authors show that the action of the imposter is not a performance or a trick but an experiment, a speculation offered to test the limits of appearance in social life."
Arjun Appadurai, New York University

THE IMPOSTER AS SOCIAL THEORY

Thinking with Gatecrashers, Cheats and Charlatans

Edited by
Steve Woolgar, Else Vogel, David Moats
and Claes-Fredrik Helgesson

BRISTOL
UNIVERSITY
PRESS

First published in Great Britain in 2022 by

Bristol University Press
University of Bristol
1-9 Old Park Hill
Bristol
BS2 8BB
UK
t: +44 (0)117 954 5940
e: bup-info@bristol.ac.uk

Details of international sales and distribution partners are available at bristoluniversitypress.co.uk

© Bristol University Press 2022

British Library Cataloguing in Publication Data
A catalogue record for this book is available from the British Library

ISBN 978-1-5292-1307-2 hardcover
ISBN 978-15292-1308-9 paperback
ISBN 978-1-5292-1309-6 ePub
ISBN 978-1-5292-1310-2 ePdf

The right of Steve Woolgar, Else Vogel, David Moats and Claes-Fredrik Helgesson to be identified as editors of this work has been asserted by them in accordance with the Copyright, Designs and Patents Act 1988.

Cover design: Liam Roberts
Front cover image: istock-967113060

Bristol University Press uses environmentally responsible print partners.

Printed in Great Britain by CMP, Poole

Contents

List of Figures and Boxes

Figures

Boxes

Notes on Contributors

Martin Abbott is a PhD candidate in the Department of Science and Technology Studies at Cornell University and a John Crampton Scholar. His dissertation research concerns the nature and culture of climate change. This research focuses on how the conception of resilience promoted by the high-profile 100 Resilient Cities initiative intersects with urban politics, emerging technologies, and environmental change in the coastal cities of Chennai (India) and New Orleans (USA). Martin holds a Master of Architecture from the University of Technology, Sydney, and a Master of Arts in Politics from Sciences Po Paris.

Agnes, Forrest Carter, Civet Coffee Bean, Cuckoo, Iansá and Oxum, Sarah Jane, Han van Meegeren, David Rosenhahn, Diederik Stapel and **Jorge Enrique Briceño Suárez** are the composite of imposters who kindly provided the Postscript. They also appear in different chapters throughout the book.

Malcolm Ashmore is author of *The Reflexive Thesis* (University of Chicago Press, 1989) and co-author of *Health and Efficiency* (Open University Press, 1989). Within science and technology studies (STS) he has researched reflexivity, the debunking of scientific fraud, the false/recovered memory controversy, and the ironies of state documentary practices, particularly in Colombia. He is Honorary Fellow in the Department of Social Sciences, Loughborough University, UK; and tutor for the Masters in STS, Faculty of Human Sciences, National University of Colombia, Bogotá. He is now (almost) retired.

Catelijne Coopmans is Research Fellow at Technology and Social Change (Tema T), Department of Thematic Studies, Linköping University. Her research interests include the dynamics of seeing/knowing, and of expertise, in areas such as medical imaging, business data visualization and art authentication. Among her publications are

'Visual analytics as artful revelation' in Catelijne Coopmans, Janet Vertesi, Michael Lynch and Steve Woolgar (2014) (eds) *Representation in Scientific Practice Revisited*, Cambridge, MA: MIT Press and 'Eyeballing expertise' (with Graham Button) (2014), *Social Studies of Science* 44(5): 758–785. She is a collaborating editor at *Social Studies of Science* and a member of the editorial board of *East Asian Science, Technology and Society: An International Journal* (EASTS).

Maarten Derksen is Assistant Professor of Theory and History of Psychology at the University of Groningen. His research interests include the epistemological and sociological aspects of the current crisis in psychology, and the dynamic of control and resistance in human engineering, broadly conceived. About the latter topic he published *Histories of Human Engineering: Tact and Technology* (Cambridge University Press, 2017). He wrote about the crisis in psychology in 'Putting Popper to work' (*Theory & Psychology*, 29 (2019), 449–465).

Kristina Grünenberg is Associate Professor at the Department of Anthropology, University of Copenhagen. Her current research interests include the epistemological foundations and social implications of digital technologies, such as biometric- and assistive technologies in the contexts of migration, security, aging and care. She is co-editor and contributor to K. Grünenberg and P. Møhl (eds) (2019) 'IDentities and identity: Biometric technologies, borders and migration', *Ethnos, Journal of Anthropology* and to Karen Fog Olwig, Kristina Grünenberg, Perle Møhl and Anja Simonsen (eds) *The Biometric Border World: Technologies, Bodies and Identities on the Move* (Routledge, 2020).

Claes-Fredrik Helgesson is Professor and Director at the Centre for Integrated Research on Culture and Society, CIRCUS, at Uppsala University, Sweden. He was co-founder of the research group ValueS at Technology and Social Change (Tema T), Department of Thematic Studies, Linköping University. Helgesson was a co-founding editor of the open access journal *Valuation Studies*, which published its first issue in 2013. He is co-editor with Isabelle Dussauge and Francis Lee of *Value Practices in the Life Sciences and Medicine* (Oxford University Press, 2015).

James Kaufman teaches social and public policy at the University of Glasgow, where he gained his PhD in 2018 for an ethnographic study of welfare conditionality and welfare-to-work services. In 2019 he was an ESRC postdoctoral fellow at the University of Sheffield. His research interests include street-level bureaucracy, public service work and the state in everyday life.

Daniel Large is a PhD candidate in the Department of Environmental and Natural Resources at Cornell University and a Ronald McNair SUNY Fellow. His research is based in regulation and governance studies. Daniel is interested in law, environmental conservation and ecological sustainability. Currently, he is working on a project in the context of US endangered species governance. In this work, Daniel critically analyses how state capacity intersects with practices of science and democracy in instances where policies, regulation and governance have failed. Daniel holds a Master of Public Administration from the University of Texas at San Antonio.

Johan Lindquist is Professor of Social Anthropology at Stockholm University; a member of the editorial board of *Pacific Affairs*; has published articles in the *Journal of the Royal Anthropological Institute*, *Mobilities*, *Public Culture*, *Pacific Affairs* and *International Migration Review*; is the co-editor of *Who's Cashing in? Contemporary Perspectives on New Monies and Global Cashlessness* (Berghahn, 2020) and *Figures of Southeast Asian Modernity* (University of Hawai'i Press, 2013); the author of *The Anxieties of Mobility: Development and Migration in the Indonesian Borderlands* (University of Hawai'i Press, 2009); and the director of *B.A.T.A.M.* (DER, 2005). His research interests include migration, Indonesia and digital methods.

Mandy Merck is Professor Emerita of Media Arts at Royal Holloway, University of London. Her latest book is *Cinema's Melodramatic Celebrity: Film, Fame and Personal Worth* (BFI Bloomsbury, 2020). Her next is provisionally titled *Downsizing: Scale and Satire in the Cinema*.

David Moats is Biträdande Lektor at Technology and Social Change (Tema T), Department of Thematic Studies, Linköping University. David's research is mainly about digitization and the role of machine learning and artificial intelligence in transforming various industries including media, healthcare, politics and academia. He is also interested in the methodological implications of these new sources of digital data and techniques like data visualizations for the social sciences and frequently works with programmers, designers and artists on interdisciplinary projects. David's publications have appeared in *Big Data & Society*, *Science, Technology & Human Values* and *Information, Communication and Society*.

Fredy Mora-Gámez is postdoctoral researcher at Technology and Social Change (Tema T), Department of Thematic Studies, Linköping University. His work draws on the intersections between science and technology studies, migration and border studies, and psychosocial theory. His interests include the role of migration management infrastructures in post-conflict Colombia, EU asylum policies, and post-Brexit immigration control. Fredy is also concerned with the materiality of memory and affect in social movements and activism. His recent publications include 'Beyond citizenship: The material politics of alternative infrastructures' (*Citizenship Studies*, 2020) and 'The psychosocial management of rights restitution: Tracing technologies for reparation in post-conflict Colombia' (with Steve Brown, *Theory & Psychology*, 2019).

Brian Rappert is Professor of Science, Technology and Public Affairs at the University of Exeter. His long-term interest has been the examination of the strategic management of information; particularly in relation to armed conflict. More recently he has been interested in the social, ethical and political issues associated with researching and writing about secrets, as in his books *Experimental Secrets* (UPA, 2009), *How to Look Good in a War* (Pluto, 2012) and *Dis-eases of Secrecy* (with Chandre Gould, Jacana, 2017).

Olga Restrepo Forero is Professor at Departamento de Sociología, Universidad Nacional de Colombia, Bogotá. She has published on Darwinism in Colombia and on the historiography of Darwinism in Latin America, on scientific writing and scientific rhetoric. She has published on institutional settings that produce and negotiate trust relationships, such as notaries and identification cards, and their uses in Colombian society. She directed a research project called 'Ensamblado en Colombia: producción de saberes y contrucción de ciudadanías' ('Assembling Colombia: producing knowledges and constructing citizenships'). All her previously published work is in Spanish.

Caroline Rosenthal is Professor of American Literature at Friedrich Schiller University, Jena. She has published on comparative North American studies; Canadian literature, culture and literary theory; city fiction and spatial theory; as well as on present-day reverberations of Romantic ideas and practices. Her books include: *New York and Toronto Novels after Postmodernism: Explorations of the Urban* (Boydell and Brewer, 2011); *Fake Identity? The Impostor Narrative in American Culture* (ed with Stefanie Schäfer, Campus, 2014); *Anglophone Literature and Culture in* *the Anthropocene* (ed with Gina Comos, Cambridge Scholars, 2019). Her current research interests are related to mobility studies, nature writing and ecocriticism.

Mattijs van de Port works as an anthropologist at the University of Amsterdam and VU University. At the latter institute he holds a chair in the study of Popular Religiosity. Publications include *Ecstatic Encounters: Bahian Candomblé and the Quest for the Really Real* (Amsterdam University Press, 2011) and *Sense and Essence: Heritage and the Cultural Production of the Real* (with Birgit Meyer, Berghahn, 2018). He is also director of two anthropological film essays, *The Possibility of Spirits* (2016) and *Knots and Holes* (2018).

Else Vogel is Assistant Professor in Anthropology, University of Amsterdam. Previously she was Research Fellow at Technology and Social Change (Tema T), Department of Thematic Studies, Linköping University. Her work combines philosophical reflection on values, subjectivity and sociality with the empirical study of care practices. She currently examines human–animal relations in intensive livestock production in the Netherlands, focusing in particular on how farm animal care involves negotiation between various notions of 'the good' – including animal welfare, financial interests, public health and sustainability. Her recent publications have appeared in *Sociology of Health and Illness*, *Biosocieties*, *Anthropology and Medicine* and *Health Care Analysis*.

Steve Woolgar is Professor of Science and Technology Studies, Linköping University, and Professor of Marketing Emeritus, University of Oxford. He is co-founder of the research group ValueS at Technology and Social Change (Tema T), Department of Thematic Studies, Linköping University. He has published widely in science and technology studies, social problems and social theory, including *Representation in Scientific Practice Revisited* (with Catelijne Coopmans, Janet Vertesi and Michael Lynch, MIT Press, 2013); *Globalisation in Practice* (with Nigel Thrift and Adam Tickell, Oxford University Press, 2014); and *Mundane Governance: Ontology and Accountability* (with Daniel Neyland, Oxford University Press, 2020); He is currently researching the limits of provocation, for a book provisionally entitled *It Could Be Otherwise*.

Preface

On 10 December each year, Swedish national television (SVT) broadcasts live coverage of the Nobel Banquet.[1] This four-and-a-half-hour extravaganza allows us to witness the assembly of the great and the good, celebrities, statesmen and royalty alongside the Nobel laureates and their families, attired in all their finery – some 1,300 guests in all. We watch them parading, dining, talking to one another and even, after the dinner, dancing. We hear the acceptance speeches and we are treated throughout to detailed expert commentary on what they are eating (there are interviews with the chefs and tastings with culinary experts) and what they are wearing (experts in high couture tell us how to appreciate the fashion on display).

The elaborate display of pageantry at the Nobel Banquet reinforces a sense of difference between those present and those watching from the outside. The message is that these guests are entitled to be there; they radiate a sense of privilege simply by being part of an exclusive occasion. For us onlookers, the question can arise: how might we become part of this? Short of actually becoming worthy of an invitation, how could one get into the banquet? What would it take to gatecrash this auspicious occasion? How could one imposter[2] as one of these privileged few? Indeed, are imposters already present at the banquet, and how could one tell?

Distinctions between insiders and outsiders are made and sustained everywhere. During a meeting of our research group a casual discussion about the Nobel Banquet gradually morphed into a broader discussion.

---

[1] As we write this in autumn 2020, it has been announced that the 2020 Nobel Banquet is cancelled due to the COVID-19 pandemic.

[2] Online sources variously advise that the spellings 'imposter' and 'impostor' tend to be used interchangeably; that the latter has a longer history; and that the alternatives are respectively characteristic of British and American English usage. We have chosen to use 'imposter' in the interests of consistency, and so leave open whether or not an imposter is really an impostor.

We found ourselves able to generate an apparently endless stream of idiosyncratic examples of impostering and of settings populated by imposters and gatecrashers. We quickly came to realise that impostering raises far more general and more pressing and significant issues than the example of the Nobel dinner with which we began, and takes us far beyond the questions of identity and belonging in which these discussions are often couched. From these conversations we set out to consider how the ubiquity of this troublesome figure might help us rethink some of social theory's core preoccupations.

The imposter is everywhere. The cases discussed in this volume attest to the richness and variety of examples of impostering. Our preoccupation with impostering derives not only from the seeming ubiquity of the imposter but also from a long-standing fascination with the relations between appearance and reality. In recent times, this general curiosity has characterized much of the more interesting research carried out within the multidiscipline of science and technology studies (STS). It has been explored in relation to a vast number of different aspects of scientific knowledge and of technological practice. Yet, as this volume demonstrates, continuing conversations about relations between appearance and reality and between identity and social order, are also of central concern in wide swathes of scholarship across the social sciences and humanities.

To explore the puzzles and delights of impostering we staged an international conference – 'Imposters and Gatecrashers' – held at Linköping and Vadstena on 13–15 June 2018 (Figure 0.1). To those participants at Vadstena whose work does not appear in this volume, we offer our thanks for their energetic and insightful contributions: Jeffrey Christensen, Emma Dahlin, Ivanche Dimitrievski, Katherine Harrison, Tora Holmberg, Yelyzaveta Hrechaniuk, Ericka Johnson, Corinna

Figure 0.1: Imposters and Gatecrashers Conference

Kruse, Daniel Gustafsson, Sarah Mitchell and Emilia Zotevska. We are indebted to all the authors whose work does appear in these pages, for their enthusiasm in helping bring this volume to fruition, for responding graciously to the persistent demands of we editors, and for their generous comments and suggestions on the other chapters. Our Introduction also benefited from insightful comments by Stephen P. Turner, Micke Tholander and Jill Morawski as well as by participants in the ValueS seminar group at Linköping University. We received helpful general advice on the project from Wiebe Bijker, Steve Brown and Penny Harvey. The volume as whole benefited from readings by four anonymous external reviewers. We thank Paul Stevens, Freya Trand, the editorial team at Bristol for their patience and support and Mella Köjs at CIRCUS for assisting with securing image rights. And we are, of course, really grateful to the Imposters who provided the Postscript.

More generally, we acknowledge the receipt of a generous award (Recruitment of Leading International Researchers 2014–2020) from the Swedish Research Council (Vetenskapsrådet), which enabled the creation of a lively and inventive research group at Tema T (Technology and Social Change), Linköping University. We especially thank our colleagues in Tema T for their consistently warm and friendly support throughout.

The editors
Linköping (virtually)
Sandhammaren, Amsterdam, Southend and Stockholm (actually)
October 2020

Thinking with Imposters: The Imposter as Analytic

Else Vogel, David Moats, Steve Woolgar and Claes-Fredrik Helgesson

'Our friends have been suggesting for quite a long time that we visit this wonderful city. [...] They have a famous cathedral there, Salisbury Cathedral. [...] It's famous for its clock. It's one of the oldest working clocks in the world.'

These words are from an interview with two Russian men on Russian state television news (Russia Today, RT) on 7 March 2018 (Figure 1.1).[1] Their appearance followed an incident on 4 March 2018, when Salisbury resident Sergei Skripal and his daughter Yulia were rushed to hospital. The authorities found traces of Novichok A-234, a nerve agent, at the scene. The two Russian men were subsequently named as suspects by British police and their faces splashed all over the news (Figure 1.2). The UK government took the bold step of accusing the Russian government of attempted murder and expelling several Russian diplomats. Then suddenly the two suspects appeared on TV. The interviewer asked them why they were in Salisbury and if they worked for the Russian Intelligence Services to which their cryptic reply was "Do you?". When pressed about their actual profession they offered, "If we tell you about our business, this will affect the people we work with."

---

[1] A full transcript of the interview is available at https://www.rt.com/news/438356-rt-petrov-boshirov-full-interview/ [Accessed 11 February 2020].

Figure 1.1: Two Russian tourists, looking like spies

Source: UK Metropolitan Police, courtesy of Getty Images.

Figure 1.2: Two Russian spies, looking like tourists

Source: UK Metropolitan Police, courtesy of Getty Images.

The episode is intriguing because it prompts a whole series of open-ended questions about the identities and activities involved. An imposter is commonly understood as a person who pretends to be someone else in order to deceive others. Who, then, are the imposters here? Skripal was said to be a former Russian agent who worked as a double agent for the UK's intelligence services and had since been pretending to lead a normal life in exile. The two suspects, according to investigative journalism website Bellingcat (2018), were trained Russian spies only pretending to be tourists. And the RT journalist conducting the interview may have been a Russian propagandist merely posing as a hard-hitting interviewer.[2]

In other circumstances, we might expect the imposters to be exposed, their transgressions punished, and normality restored. Yet what is especially compelling about this example is that the set of activities, accusations, suspicions, counter-claims and questions – which we shall collectively term 'impostering' – does not easily resolve. If the poisoning had been carried out differently, it may have looked like an accident, but the perpetrators chose to use a nerve agent to which only Russian intelligence services had access. The two men protested their innocence with little apparent enthusiasm – many sources noted that their description of Salisbury was likely pulled from Wikipedia. The overwhelming response in news commentary and social media was that their efforts at passing as tourists were completely and obviously implausible. Could it be then that they *intended* to be understood as imposters, as some kind of a dare from Russia to the West, as a way of sparking doubts and stoking disorder? In this case, impostering does not 'just' involve deception, trickery and pretence. There is little attempt to plaster over the cracks, to hide behind a mask or return to normal. Rather, the unresolvedness is precisely what is on display. It is because things do not 'add up' that the episode is so deeply unsettling.

The potential of imposters wherever we turn stirs a wide range of societal responses ranging from intrigue to suspicion, from outrage to horror. Suspicions about imposture, in turn, impact people's lives and social interactions. Imposters are everywhere. Imposters are trouble.

What insights can these troublesome figures provide into the social relations and cultural forms in the communities and social settings in which they emerge? And how might 'thinking with imposters' be a

---

[2]   It has been suggested that the incident did not easily resolve in part due to the Russian government's sustained 'information warfare' (cf Pomerantsev, 2014). So many accounts of the event were thrown around that the 'truth', if ever there was one, became just one of many versions of reality.

useful tool of analysis in the social sciences and humanities? In this volume, we argue that the figure of the imposter can help reset the agenda for mainstream social theory. The imposter has long been a topic of interest to a wide range of social sciences and humanities. The dynamics involved in impostering practices, even the mere suspicion of imposters, have featured in myriad different efforts to theorize about social order. As we shall see, the dominant approach has been to exploit impostering as a phenomenon that reveals something of this underlying order. Thus, impostering has largely been treated as something exceptional, as an aberration from normality, which provides insight into the mechanisms which hold the social world together.

This edited volume proposes a different direction. Instead of exploiting the imposter as a methodological probe for the revelation of social order, this volume explores the value of bringing the figure of the imposter, as well as the dramas, disruptions, reactions and questions they engender, to centre stage. This allows exploring the advantages for social theory of understanding imposters as primary engines of indeterminacy, uncertainty and disorder. Of course, what counts as indeterminate, uncertain or disorderly is not absolute, but depends on one's position. Yet what we argue is that frictions and disruptions related to impostering are far from marginal. They are part of an ongoing dynamic which is constitutive of social relations and cultural forms. Our contention is that by resituating the imposter in this way, and re-specifying the common sense notion of them, we can provide a new way of thinking about social relations.

This edited volume collects contributions that in various ways explore practices of 'impostering': that is, situations in which the imposter is staged as a reality or possibility in everyday interactions, policy or technology, as well as in fiction, art and film. These are not limited to actual invocations of the term but also include situations in which a wide variety of people and things attempt to pass unnoticed, feed ongoing suspicions, become protagonists in a revelatory plot or cast as actors in more destabilizing disruptions. To sensitize our readers to the phenomenon of the imposter, we begin this chapter by noting the wide range of 'imposter moments'. The next section then examines some of the uses of imposters in social theorizing, showing in particular how the imposter has previously been largely treated as a form of deviance, as a sub-class of activity that exposes what does not belong and thus what counts as normal. In the third section of the chapter we suggest that thinking with imposters encourages us to place much greater emphasis on instability, uncertainty and disorder; expand the range of practices, concerns and devices involved in impostering; broaden the variety of entities involved; and deepen our understanding of the dynamic relations in which appearance/reality puzzles are played out between imposters and their audiences. In a similar way to other key conceptual figures in social theory, the imposter provides the opportunity of thinking differently. The fourth section contains the chapter outline, introducing the wide array of case studies of impostering practices which make up this volume. We conclude that social theory can benefit substantially from taking seriously the essential indeterminacy of impostering, thereby helping us move away from predominantly construing our worlds in terms of order(ing).

Imposter moments

One might say that the imposter is presently having a moment. Media and political leaders seem obsessed with identifying traitors, fakes and frauds in our midst. Techniques which aim to separate the supposedly authentic from the supposedly duplicitous proliferate. We see this in the expansion of fact-checking (Marres, 2018); of audits and accounting (Espeland and Sauder, 2007); in new forms of algorithmic surveillance (Barocas et al, 2013); in the repeated advances in biometric identification techniques (Chapter 9, this volume), and so on. Many of the obsessions with pinning down imposters have high stakes. As we write this in mid-2020, quarantined in different countries due to a devastating pandemic, concerns abound about the possibility of people infected by COVID-19 while displaying no symptoms, spreading the virus in our midst; the elderly falling prey to 'COVID-19 scammers' posing as governments and banks; and antibody tests being faked to enable illegal migration. In the past few years alone there have been concerns about illegal immigrants storming 'Fort Europe' or President Trump's wall on the US–Mexican border. There have also been questions about interference in the US 2016 elections by Russian "trolls" (for the origin of the term see Coleman, 2015) posing as American citizens.

A seemingly endless list of imposters figuring in contemporary literature, cinema and television provides another barometer for the current moment, suggesting that the imposter figure provides an itch that is particularly in want of scratching. To name just a few: the 2020 Oscar winner for best picture, *Parasite* (2019) by Bong Joon Ho, is about a poor family who work in concert to pose as helpers for a rich family. The 2012 film *The Imposter* (2012) documents a French confidence trickster who successfully impersonated a missing American boy. An unrelated book of the same name *The Impostor* (Cercas, 2017) relates the complex processes of unmasking the serial deceits of a self-proclaimed war hero, resistance fighter, survivor of the concentration camps and campaigner for Holocaust awareness. Two recent sci-fi outings, *Blade Runner 2054* (2018) and *Westworld* (2016–20) both feature synthetic humans who are difficult to distinguish from the real thing; *Can You Ever Forgive Me?* (2018) is a film about a fraudulent writer and the six-season drama series *The Americans* (2013–18) follows the exploits of a pair of KGB agents posing as everyday Americans in the 1980s. In literature, successful books like Jeff Van der Meer's *Annihilation* (2015), which is full of not only characters who are not what they seem but also life forms simulating other life forms, and Man Booker Prize winner Paul Beaty's *The Sellout* (2015), whose mischievous protagonist poses as a slave owner, confirm that the imposter forms as potent a plot device as ever (Rosenthal and Schäfer, 2014).[3]

Concerns with impostering are not just directed towards outsiders and 'others', however, but also towards ourselves (Das, 2018). Miller (2003) describes the self-consciousness and anxiety that results from feeling like a fraud, or mistrusting our own thoughts and motives, as 'feeling double'. The 'pop-psychological diagnosis' (Simmons, 2006) of the *imposter syndrome* exemplifies how common this self-tormenting feeling of 'fakeness' is. People – often academics, artists or experts of some description – with imposter syndrome are convinced that they are not meant to have achieved the status they hold and anxiously anticipate being 'found out'. This phenomenon, which is often painted as a gendered syndrome in the self-help literature, is interesting because it renders the imposter status as unreal. It is the imposters themselves who are seen to be deceived.

As in literature (Miller, 2018), we can see how science as an institution appears increasingly preoccupied with impostering. This preoccupation takes many forms, such as the obsession in some quarters with distinguishing false results from true ones, leading to what some have called a 'replication crisis' (Baker, 2015). Another form of preoccupation with impostering can be seen in efforts to expose supposedly 'pseudo-scientific' fields that have become accepted

---

[3]   Curiosities and anxieties about the imposter are neither new nor particular to our time. It is perhaps appropriate to note that, for example, both *Blade Runner* and *Westworld* are reboots of late 1970s, early 1980s franchises. Eliav-Feldon (2012) even identifies early modern Europe as an *age of imposters* in which authorities like the church and state were busy devising increasingly complicated proofs of identity. She posits this age, however, as just one chapter in an endless race between liars and pretenders and the practices of identification employed to try to catch them out.

as part of the community of scientific inquiry. An example here is when scientist Sokal submitted an article – arguing that quantum theory is a social and linguistic construct – to the journal *Social Text* (Sokal, 1996). When it was published, he revealed it as fraudulent to make the point that the field of social constructivist studies of science did not have proper standards, supposedly exposing postmodernists as true intellectual imposters (Sokal and Bricmont, 1998). There also seem to be increasing cases of plagiarism and fraud (see Chapter 3, this volume) as well as efforts to spot predatory and phoney 'academic' journals hiding amid bona fide ones.

A widespread obsession with imposters, then, is evident in a variety of fields, places and times. The particular substantive focus of impostering shifts quickly, but the general phenomenon is ongoing. While they have different characters – the alien invader in 1950s sci-fi differs from the synthetic humans of today's sci-fi just as concerns with scientific fraud differ from scandals in literature – the archetypes and plotlines continually replay. The intrigue that imposters evoke, moreover, is not limited to Euro-American culture, as is evidenced by the chapters in this volume, which discuss case studies spanning from Indonesian click farms to religious cults in Brazil, taking us from guerrilla forces in Colombia to refugee camps in Greece. The stakes attached to authenticity and imitation differ in these different cultures, sites and situations (Taussig, 1993). This all suggests that imposters are interesting in *how* they emerge as part of a society and social fabric.

But if imposters are ubiquitous, how have they been understood and analysed in the academy, particularly in social theory? How do these discussions diverge from the colloquial understanding ubiquitous in the news and the arts? In the next section we consider some of the main ways in which imposture, and its related notions like 'fakes', 'hoaxes' and 'deviants', have thus far been used to fashion arguments about social relations.

Imposters in social theory

To understand how imposters have been taken up in the social sciences and the humanities, it is instructive to consider the United States in the 1950s and 1960s. It was here, we suggest, that the imposter blossomed not only in popular culture but also featured as a prevalent theme of sociological analysis and social theory. A prime reason for this interest, we argue, can be found in the mid-century obsession of social sciences and humanities scholars with describing *order* in society (Parsons, 1991 [1951]). Mid-century sociologists were interested in examining how the different institutions of society fit together as necessary components of a functional whole – like organs within a body. Deviations from the norm then became an interesting curiosity which fell outside and thus further illuminated the overarching order. Against the legacy of the classicist, segregated, decorum-obsessed United States of the 1950s and 1960s, infamous studies like *The Tearoom Trade* (Humphreys, 1971), for example, revelled in the intrigue surrounding gay cruisers.[4]

Not all 'deviations' from the norm, however, constitute impostering. Whereas deviants can be apprehended, often quite straightforwardly, as departing from a supposed 'normal', imposters play with and (aim to) perform belonging, thus taunting normality in different ways. They create what we call appearance/reality puzzles.[5] For example, some of Humphrey's informants unambiguously lived and self-identified as 'gay men', but many of them were married to women. Were they 'really' gay and masquerading as straight or 'really' straight? Or both? We are interested in thinking with imposters precisely because they engender such marked indeterminacies. More scrutiny and investigation often only further complicates the puzzle, as the very terms and criteria proposed as solutions increasingly appear up in the air themselves.

Impostering as performance

Social theorists have typically thought imposters against the background of the norms and expectations of a set social order. Goffman is a case in point. The sociologist developed a theory of social interaction and personhood based on the metaphor of theatrical performance: performance is fundamental to all of social life. In his framework, the individual in social interactions constantly guides and controls the impression he/she makes on others. Focusing primarily on performances of class, gender, expertise and ethnicity, Goffman (1959) addressed the stagecraft and dramaturgical techniques used in such performances. For him, social interaction is all about playing roles,

---

[4] As time went on, these studies of deviance, particularly those associated with the Chicago school and Howard Becker's (1963) labelling theory, attempted to rehabilitate these deviants, to lay some of the blame on society for their exclusion. However, as we explain later, an important feature of these studies was that the fact of their divergence from an existing social order was rarely in doubt.

[5] We thank Catelijne Coopmans for this particular turn of phrase.

to ourselves and to others. Being a person is being a mask. People perform not so much *themselves* as *a* self (cf Mol, 2002).

Goffman is thus fascinated with the dramaturgy of the performance and the societal obsession with detecting deceit and fraud. He identifies a strong moral structure around performances, emphasizing the crucial role of the witness in determining which roles a person is *entitled* to play. Goffman suggests that the possibility of deceit haunts social interactions because the individual places a moral demand on others that they treat him/her based on the person that he/she is, and that others, in turn, expect a person to be who he/she expresses themselves as and not someone else. These principles are communicated through games, jokes and cautionary tales of embarrassments and disruptions in decorum – teaching people to be modest in their claims and reasonable in the expectations they project. The outrage that follows when someone misrepresents themselves, he argues, is not about the performance itself but about whether someone is authorized to perform that role. Impostering thus causes great anxieties because it threatens moral authority and a sense of belonging.[6]

Goffman seems little concerned with figuring out what constitutes the 'real' behind the front, but he does, arguably, assume the existence of an unambiguous underlying reality. When it comes to performances of social roles, a person might not be convinced of their own performance; they might be behaving cynically. On the other hand, it sometimes happens that:

> The performer [...] can be sincerely convinced that the impression of reality which he stages is the real reality. When his audience is also convinced in this way about the show he puts on – and this seems to be the typical case – then for the moment at least, only the sociologist or the socially disgruntled will have any doubts about the 'realness' of what is presented. (Goffman, 1959: 28)

Goffman thought that performances should be studied by sociologists but that the 'deeper', 'real' self was best left to psychologists. One problem with assuming that a 'real' world sits outside the socially

---

[6] There are obvious politics to this in terms of race, class and gender. The more powerful social groups can play with identities in ways that the less powerful cannot: Whites could historically 'Black up' but African Americans were ostracized for passing as White.

determined one, however, is that in many cases, the scientists, social scientists and experts who are meant to be custodians of 'the real' – to watch out for the theatre props and tell-tale curtains – are precisely the ones who are fooled. Assuming there is a real, moreover, obscures attention to how often what counts as 'real' or 'true' is itself at stake in the interaction, rather than a resource for judging the outcome. So while Goffman's notion of performance introduces an appearance/reality puzzle as central to social life, the tension opened up between appearance and reality is staged as temporary and principally solvable. There might be uncertainty, drama or anxiety, but the implication is that imposters can, at some point, be outed, and tensions resolved. For Goffman, performances are *mere* performances – there is nothing particularly deep about the social self that is presented to others. Any imposter drama, then, risks nothing more than revealing a displayed persona to be disingenuous. What happens though in situations where the imposters reveal more than themselves, when they upend more than their own identities?

Here it is helpful to consider recent discussions around the related notion of fakes. Fakes, like imposters, pose as something they are not, but often the agency for the fakery is assigned elsewhere (eg to a forger or a smuggler). As Coopmans argues in Chapter 4, fakes might be thought of as a non-human imposter of some sort. In a recent anthropology book about fakes, Jones (2018) makes a helpful distinction between 'mere fakes' and 'deep fakes'. 'Mere fakes' are more or less ingenious counterfeits and deceptions but their motivations are relatively straightforward. 'Deep fakes' can also be more or less convincing, but they pose existential threats, or in Jones' words, 'cosmological dramas'. Jones is keen to point out that some of the most analytically interesting cases of fakes are ones which move between these two poles. The key point for our purposes is that imposters also tend to create such dramas: social order is not always restored and may itself be transformed in the process. This is why we stress that when analysing imposters, it is important not to presume a background social order or reality underlying performances.

Imposters as interactional accomplishment

For Goffman, order is achieved through individuals' performance of social roles. By contrast, Garfinkel (1967), who developed his programme of ethnomethodology in part as a critique of Parsons' advocacy of structural functionalism, stressed that individuals do not simply perform available social roles; theirs is instead a reflexive accomplishment of practical actions in the moment. Garfinkel (1967: 68) famously criticized theorists like Goffman for portraying people as 'cultural dopes' that is, for depicting performance as governed by the 'moral order without', as if actors have recourse to clear and readily available social rules to guide them in acting the part. According to Garfinkel, attention to the 'moral order within' requires much closer attention to the situated ways in which participants invoke, constitute and reflexively enact identities, culture and 'rules' as part of everyday interaction.

Imposters could be said to feature as part of the exercises, demonstrations and 'breaching experiments' with which Garfinkel and early ethnomethodology are associated. Breaching experiments are 'idiosyncratic investigations that disrupted commonplace routines in households and public places' (Lynch, 2011). For example, Garfinkel asked his students to bargain at the checkout in the supermarket; to act as a lodger in the family home; to ask and to persist in asking for further clarification in a casual conversational exchange; to deliberately misunderstand simple rules for playing tic tac toe (noughts and crosses), and so on. These experiments generated outrage, indignation and upset on the part of the 'victims'. The usual interpretation of breaching experiments is that they reveal the existence of taken-for-granted assumptions that constitute daily life, and in particular that the speed and severity of the victims' upset demonstrates the strength of moral accountability associated with these expectations about ordinary behaviour.

While it is not self-evident that all of these experiments employ the colloquial sense of impostering as 'posing as someone else', they do, in a broader sense, play on tensions between appearances and realities that are of interest to our efforts of thinking with imposters. In the language of Garfinkel's 'documentary method' – the generalized sense-making procedure whereby people use surface appearances to document underlying realities – the surface appearance presented by the student did not match the victim's expectations of the 'actual' person. Consequently, the victims were forced to do interpretive work to figure out what was going on and tried to account for the breaching behaviour by using formulations such as "are you crazy", "are you ill", "we never talk like this" and so on. Garfinkel focused on the initial disruption in breaching but said little about whether or how the situation was sustained or how tensions were resolved. The implication is that the disruption quickly ended when the students reverted to their 'real selves'. The disorder in these experiments was only temporary and quickly resolved.

As we have explained, a central aspect of our interest in impostering lies in moments of instability and disordering which are not easily resolved, something to which another less well known Garfinkel study of Agnes, an 'intersexed person', comes closer. In order to obtain a desired gender reassignment surgery, Agnes needed to convince a panel of scientists, including Garfinkel, that she was a female who had mistakenly been born in a male body. Garfinkel documents the various ways in which Agnes negotiates (with herself and the doctors) the axiomatic, highly socially sanctioned, rigidly binary membership category of male–female. For instance, she argued that while she possessed 'male' genitalia this was a mistake; shared how behaviourally she always felt more comfortable as a female; and mobilized blood tests that found evidence of 'naturally occurring' high levels of oestrogen in her blood, a biological 'defect' which she argued had feminized her.

By contrast with the usual notion of impostering as acting as something different to what one actually is, Agnes felt obliged to act as something which she felt she was naturally all along but just appeared not to be. An interesting, further destabilizing twist occurs when Garfinkel claims, in the appendix to the book (Garfinkel, 1967), that Agnes had in one crucial respect deceived the investigators on whose judgement she depended to get what she wanted. After 12 years, Agnes confessed that the 'naturally occurring' oestrogen was the result of her having secretly taken drugs, stolen from her mother. While this revelation caused the physician involved to retract his earlier findings on Agnes, Garfinkel suggested that this disclosure further exemplifies the inherent instability involved in grappling with the nature of the imposture. It did not settle Agnes' gender. Rather, the disclosure 'turned the [original] article into a feature of the same circumstances it reported, i.e. a situated report' (Garfinkel, 1967: 288).

Garfinkel's breaching experiments were like deviance events, with the fact that participants pass as something other than themselves being used to demonstrate the depth of moral accountability. By contrast, Agnes' story offers a more complex aspect of impostering which dramatically upends and confounds social categories with no obvious resolution, keeping alive the question of who or what is acting at what. This complex impostering practice, after all, shows the category 'gender' itself to be problematic and not (ultimately) 'resolvable'. Garfinkel was mainly interested in showing the interactional work Agnes had to perform to pass as female. With hindsight, we can see that his and his colleagues' fascination with and (physical) scrutiny of 'the case', and their lengthy questioning about her private parts and sex life, contributed to Agnes' being positioned as a (potential) imposter. This highlights the crucial and sometimes troubling role of others in naming and shaming 'imposters', or allowing them to 'pass'.

The hoax and the imposters' critical potential

There have been ways in which the imposter's critical potential has been exploited. An interesting subset of impostering in this regard is the hoax. The (perceived) objective of the hoax is not just to reveal the existence of a particular social order or characterize it, but to reveal that social order to itself be suspect, illegitimate or in some sense not as orderly as it seems. For example, Rosenhan (1973) conducted a well-known study of psychiatric diagnosis amidst the heyday of the anti-psychiatry movement in the United States. Rosenhan arranged for eight 'sane' people, including himself, to secretly gain admittance to psychiatric hospitals. The 'pseudopatients', who had alleged that they were hearing voices and had been diagnosed with schizophrenia, stopped feigning symptoms and acted as themselves as soon as they entered the hospital. While other patients would often detect the pseudopatients' 'normality', Rosenhan reported that none of the staff had any suspicions of their impostering. Neither were their psychiatric labels reassessed; on the contrary, their personal history and displayed behaviours (including their observational note-taking) were interpreted by staff as typically symptomatic of their psychiatric condition. Upon confronting the hospitals with the results of the first study, Rosenhan created a second experiment in which he invited psychiatric staff to detect further pseudopatients who would attempt to be admitted to their hospital. Hospital staff strongly suspected 41 patients to be imposters. Rosenhan subsequently revealed that he had not sent any pseudopatients to any hospital. The study concluded that 'psychiatric diagnosis betrays little about the patient but much about the environment in which an observer finds him' (Rosenhan, 1973: 250).

This example provides yet again a slightly different usage of impostering – while the pseudopatients were in the end disclosed, their revelation served to undermine the basis of the very distinction between patient and non-patient. Here, then, impostering served not only to reveal, but also to question and destabilize the vested societal categories of 'sane' and 'insane' that made the transgressive impostering possible in the first place. It is important to note, however, that this process of disrupting or tearing down an existing social order is not itself disorderly. Rosenhan's hoax relies on the logic of a scientific experiment. It stages the imposter's actions as temporary, controlled manipulations with undeniable results. Many conventional hoaxes operate as such: through deception they effect a flip so that the deceiver becomes a truth teller and the normal order is revealed as deceitful – when this flip is successful a different order may become stabilized.

Interestingly, Rosenhan's use of imposters to expose the contingency of psychiatric diagnosis has itself been criticized for impostering as a valid experiment. Cahalan (2019) made attempts to retrace participants and to reconstruct the conduct of Rosenhan's experiment, and discovered inconsistences sufficiently significant to suggest that Rosenhan had invented most, if not all, of the pseudopatients. And yet it is also reported that while her research casts considerable doubt on Rosenhan's claims, Cahalan cannot be completely certain that Rosenhan cheated (Abbott, 2019). Since the use of imposters to expose deficiencies in a form of ordering is itself susceptible to accusations of impostering, the case shows how hoaxes often reintroduce irresolution, uncertainty and further disorder. They keep going, recursively prompting accusations and reversals of fortune.

'Natural' imposters

As ethical concerns have been raised about studies which stage imposter situations, social scientists have more recently made use of 'naturally occurring' imposters to interrogate the particularities of social worlds that might otherwise remain hidden in plain sight. For example, Collins and Pinch (2005) describe the case of Atkins, an unqualified doctor working in the United States, who continued practising medicine for 30 years, despite major mistakes, without being 'caught'. Atkins was only exposed because a family member, harbouring a grudge, reported him.

The case shows that in malpractice, questions of authenticity and identity are hardly the first that come to mind. For instance, there were plenty of concerns with Atkins' performance. Once, inspectors even visited Atkins after a pharmacist raised alarm when Atkins had prescribed a type of shampoo as medicine for a throat infection. However, the inspectors focused on the question of whether Atkins was mentally and physically fit to practice. They did not ask whether he was actually a qualified doctor. As one of the inspectors reflected: 'People ask the wrong questions. You say: "how could this chap be so awful?" You don't say, "Is this chap really who he pretends to be?"' (Collins and Pinch, 2005: 38). Collins and Pinch suggest that people find it easier to look the other way or find alternative explanations (or even cover up the incident) rather than to 'out' an imposter. The stakes are high: if an imposter is revealed, after all, everyone else will also be made to look like a fool and the authority of the institutions in question will suffer.

Collin and Pinch's analysis is interesting in that they acknowledge fundamental uncertainties over how imposters can be caught and acknowledge the possibility that they may function perfectly well as doctors. The authors assert that bogus doctors reveal a lot about the particularities of medical practice, for instance, how the wide range of therapeutic techniques doctors draw upon make it hard to detect uncommon ones. They thus use the Atkins case as a naturally occurring 'breach' that provides insights into the values and practices that a particular collective holds dear.

Imposters beyond deviance

After examining the previously mentioned contributions to social theory, we are now in a position better to specify what we can gain by bringing impostering centre stage. Works that exploit imposters, staged or naturally occurring, for methodological purposes tend to rely on a distinction between surface appearance and underlying reality. The deviations and disruptions provided by the imposter are then used as indicative of an underlying social order, which is their main object of interest. Thus using imposters to explore social order reduces them to 'mere' deviants. We instead argue that focusing on impostering sensitizes the analysis to how imposters potentially shift all social relations involved. This approach takes impostering as different from acts of deviance. The imposter's efforts are not seen as a deviation, but instead as (solicited/unsolicited) efforts to fit in, which can produce doubt and suspicion because appearances are no longer to be trusted. We are interested in articulating an approach to social theory which centres on imposters as regular parts of the ongoing ordering/disordering of social relations.

Thinking with imposters

It follows from our discussion thus far that an alternative perspective on imposters should place much greater emphasis on the instability and disorder involved in impostering practices. We might want to entertain the possibility that, far from being epiphenomena, impostering practices are actually necessary to generating and sustaining social relations and cultural forms. How, then, can we take seriously the indeterminacy imposters produce without reducing it to just another type of ordering?

As we have seen, Goffman (1959) suggests that we read impostering as an *information game* in which a captivated audience interprets a skilfully revealing and concealing performer, both caught in a semiotic world of stereotypes and standardized signs.[7] Goffman is clear that the roles of audience and performer are not fixed. But we need to take the notion of performance much further. Rather than positing performance as the foundation of social identity, we need to take into account a far wider ranging set of practices, concerns and devices. Following Mol (2002), we wish to widen this performative understanding of the staging of social identities to include the staged realities of things and materials. It is thus necessary to consider myriad entities which play the imposter: not just humans but animals, artefacts, phonies, fabrications, counterfeits and forgeries. We also need to move away from the assumption that it is the imposter alone doing the performing. Rather than assuming an actor that engages in impostering, we see the imposter as that which is performed through situated practices (cf Butler, 1990). This volume examines practices in which the imposter and its audience, making use of a range of props and stagecraft, are *all* mutually implicated and play their respective parts, ranging from whistleblowers to apologists and angry mobs. Such practices may enact a sense of what it means to be human, true, worthy or part of a particular social group. It is these practices themselves that stage what constitutes genuineness and insincerity, separating the real from the fake. If we think with imposters in this way, a mask is not simply something which facilitates deception. Instead, the 'fact' of the deception itself and the reading of the mask as mask-like are themselves the upshot of impostering practices.

In a similar vein, we might say that calling someone/something an imposter is an act which assigns rights and responsibilities. Such an act *intervenes* in the situation in which it is uttered. The enactment of imposters and their audiences is, however, by no means limited to discursive practices. Imposters emerge as something to be known and detected through socio-material conditions, including tests, scans, data, gates, procedures, norms and manners. These imposter naming and catching procedures in themselves, then, stage the possible existence of imposters – and imposters as particular entities or actors. While the imposter serves as a crystallization device, the source of great struggles and turmoil about the values and ideas a certain collective holds dear,

---

7   For an insightful overview of how throughout literature successful liars both enact and attune themselves to the predispositions of their target audience (their 'liees') see Burrow (2020).

these values and the ordering practices of the community were not always just there waiting to be revealed. Communities, their norms and forms of solidarity, may only materialize as a consequence of impostering or may be radically transformed in the process. Through impostering practices – ways in which 'the imposter' is named, known and practised – these values are reworked and negotiated, but rarely resolved.

This book is not just presenting *on* imposters, but aims to develop and assess the value of 'thinking with imposters' as an analytic tool in the social sciences and humanities. In other words, we are not interested in imposters merely as an empirical topic and cultural obsession, but as a device for thinking differently about a variety of situations. We do not thus restrict ourselves to instances in which the term 'imposter' is explicitly invoked but instead explore what it might mean to use the imposter as an analytic device. In so doing we connect with a long tradition in certain currents of social thought of using conceptual characters, both to elucidate situations but also radically rethink them and even intervene in them. Examples are the trickster, the stranger (Simmel, 1950), the idiot and the diplomat (Stengers, 2005). Serres' parasite (1982) has particular relevance for our focus on order and disorder. The parasite destabilizes and interrupts situations by translating between different orders and modes of communication – betraying the interlocutors. It is parasitic operations which effect transformation and disturbance in the 'host', bringing some measure of exteriority to an otherwise closed system. The parasite and parasitism are thus key for inciting diversity and change (Brown, 2013).

There are two key points of convergence between the notion of the parasite and our interest in the imposter. This is, firstly, the idea of a figure that combines the counterintuitive combination of being omnipresent, seemingly disruptive, and yet at the same time a necessary part of evolving order. The second point concerns the practices of cleansing, the scapegoating of parasites or the exposing of imposters. While such practices might appear as cathartic ways to return things to a normal state, through Serres we might think of such actions as producing *transient* states. Secondly, parasites offer us a way of rethinking imposters: they may betray and cause misunderstanding, but they can also be a productive interruption in the flow of things. Who or what counts as a parasite, however, depends on who is speaking. Similarly, what counts as order and disorder depends on 'for whom'. Nevertheless, they open up new channels of communication (Brown, 2002). We argue that the imposter can do similar work. Indeed, the appearance–reality disjuncture that the imposter creates makes the appearance of the imposter potentially more destabilising and re-ordering than that of the parasite.

However, we need carefully to stake out the relation between our analytical language and the fields with which we engage. In situations where imposters are not named as such, bringing in this analytic has effects and is never innocent – if only because 'imposter talk' is not just our theoretical business, but is also a powerful weapon yielded by some to stigmatize and control others. It is in this sense that the imposter has negative connotations which we may not want to impose (more) on certain groups (as Mora-Gámez discusses in Chapter 13). In addition, our chronicling of imposters may, in certain practices, aid those invested in 'catching' imposters, making the question of when to try and settle a situation, or when to keep it open, pertinent – not just analytically, but also politically. As we see it, the purpose of invoking the imposter as a conceptual character is not to explain a situation by reducing interactions and events to a taxonomy of ideal types, but to introduce and develop a set of sensitivities. We are interested in the imposter as a device for alerting us to particular practices and dynamics that are ultimately analysed through detailed *empirical* study. Thus we are not merely adding another statue to an overcrowded colonnade of strangers, idiots and tricksters, or making claims about certain figures being imposters by some a priori definition. Instead we take inspiration from the questions and concerns this figure evokes. In order to get a better sense of what makes the imposter special, what distinguishes it from some of these other figures, we now turn to the chapters of this volume.

Imposter dramas

Our notion of the imposter is open-ended. In this volume, we explore its utility and value against a number of empirical areas. What emerges is that the indeterminacy generated in imposter practices is not uniform – different performances take on different shapes and textures. The different chapters serve to underscore the plurality and multiplicity of the imposter figure. Their order of presentation sets out to highlight some of the patterns which characterize imposter practices – the different possible 'plot lines' of imposter dramas and the modes of (dis)ordering[8] and indeterminacy which haunt them.

---

[8]   Here we play on John Law's (1994) phrase 'modes of ordering'. Contrary to Foucault's notion of discourse (1970), the building blocks of Law's modes of ordering are smaller, situated in practices. They therefore do not add up to a large episteme, such as 'modernity'. Various modes of ordering do not align. They may depend on each other, but also exclude or combat one another. The attention to

First, imposters may disrupt through *restructuring revelations*. Suddenly, a situation that seemed 'ok' turns out to be all wrong. The disordering momentum lies in its extended recasting of identities and their relations, such as those that count as real and those that count as fake. In Chapter 2, Caroline Rosenthal discusses two cases in which bestselling books presented as autobiographies of Native American authors were later revealed to be penned by White men (one was a Ku Klux Klansman and the other a British man). Her cases bring out the important ways in which impostering both implicates and draws on various communities. Rosenthal shows how the deception relies on and exploits expectations from readers, and contemporary stereotypes about the ethnic group in question, such as ideas of 'the noble savage' and 'closeness to nature'. Emphasizing the important role played by the audience in imposturing, Rosenthal further shows the complicity of the readership in the deceit. In the aftermath of revelation, many readers refused to acknowledge the impostering, or supported the authors, despite their deception. Interestingly, the British author, who was seen as sincerely committed to the values of the tribe he wrote about, was later accepted by them as a spokesperson even after his secret was revealed. As with many of our examples thus far, this imposter muddies the water: raising questions about what kind of authenticity really matters in a given case.

It is the unmasking that forms the plot of this revelatory imposter drama. The imposter is clearly the story's key protagonist. The momentum of the revelatory imposter drama further feeds on an often complex temporal restructuring of events. Revelations shift the very frame of reference from which past and future actions are understood. They fundamentally change the revealer and the revealed. This becomes evident in Maarten Derksen's chapter, which examines the case of Diederik Stapel, a social psychologist in the Netherlands who, in 2011, was accused of fabricating his results and soon after confessed. He goes from being a reasonably celebrated researcher in psychology to a fraud – in the revelation, he 'turns out' to have been exhibiting suspicious behaviour 'all along', his deceit having been hiding 'in plain sight'. Despite this seemingly clear-cut revelation of 'the imposter', however,

---

modes of ordering, moreover, shows that within a specific order, 'disorder – or other orders – are only precariously kept at bay' (Law, 2009: 145). Similarly, for us, the notion of *modes of (dis)ordering* is a heuristic device for distinguishing differences in imposter dramas. Our variation of this term serves to highlight how not just order but also *dis*order may be precariously created and sustained – a key dynamic in impostering.

Derksen chronicles how, in the aftermath, the event was interpreted in drastically different ways by different parties. In some versions Stapel was an imposter social psychologist while for others the events showed social psychology itself to be an imposter science. The scandal coincided with (and exacerbated) an ongoing concern in psychology about what constitutes sound scientific research. The resulting drive towards transparency in the field, Derksen explains, only creates further suspicion and accusations. The chapter shows that attempts to make sense of imposter revelations are often controversial, eliciting cascading rebuttals and counter-accusations. These different performances of an imposter each draw on a different cast of stereotypical characters or roles, including different sorts of imposters (manipulative, sloppy, tragic), as well as whistleblowers, truth tellers, elephants in the room and rotten apples.

It is the revelatory plot that has perhaps received most attention in earlier work on imposters in the social sciences and humanities. As naturally occurring 'breaching experiments', revelatory moments are highly informative for researchers. They instil an analytical sensitivity to surprise, and horror, as well as to reconfigurations of past, present and future. That there is considerable strength in this way of 'thinking with imposters' is evidenced by Catelijne Coopmans' chapter. In her chapter she gives a masterclass in how to analyse imposter revelations, drawing on nearly ten years convening a university class on fakes and forgeries. She explains that fakes – a particular kind of 'non-human' imposter – are an excellent teacher, because they make available certain aspects of social order, assumption, expectations and social relations which are not always so vividly present. In particular she notes that fakes can 'reveal' collectively held values: what is faked is often what we as a society value most. Coopmans concludes her chapter arguing that by attending to their trajectories and travels, fakes can be used to map out a particular set of social relations. This is because for fakes to work they must anticipate the 'force field' of relations meant to catch them but also on which their circulation depends. Fakes must cultivate a closeness to that which is valued so that they can be actively shepherded by unwitting allies. Coopmans' analysis is particularly interesting because while it uses fakes to trace the values and relations which make up a given social order, she does not presume such an order to be pre-existing or unaffected by fakes. She also does not stop at describing the 'force field' but details the symbiotic relations of the forger(s), returning their craft to centre stage.

Second, imposters may be part of *insatiable obsessions*. As part of such dramas, the imposter is paradoxically the driver of action, but is not in control of what drives them. This plotline centres on the perpetual search for a 'true' identity in an ambiguous world, despite the absence of such an essence. It foregrounds the considerable societal investments in authenticity, as well as in its failings and slipperiness. The chapter by Mandy Merck examines celebrity culture and repeated folds of imitation within and between life and art. Centring on different versions of the melodrama *Imitation of Life* she explores imitation as a complex variety of practices which can further differentiation and conformity, a sense of truth and falsity, as well as subordination and self-affirmation. Among them we encounter the Hollywood star Lana Turner playing the aspiring actress Lora in Douglas Sirk's film adaptation (1959). As Merck (Chapter 5, this volume) points out, 'Turner's biography guarantees the veracity of its fiction, complicating our sense of just what is and is not an imitation' (p 121). While Merck argues that imitation is not necessarily deceitful and is often necessary for social reproduction, she also highlights how potentially damaging and destabilizing the impossible quest for the perfect imitation can be.

Martin Abbott and Daniel Large focus on one particular archetypical imposter that has had the attention of humans for centuries: the cuckoo. The figure of the cuckoo has shaped and reshaped the human imagination all the way from ancient Sanskrit literatures to current times. The chapter highlights the deep indeterminancy associated with the cuckoo; cuckoo behaviour is understood as disorderly, a figure capable of deception and cruelty, and yet 'natural'. The precise identity of the offending party is ambiguous: is it the cuckoo hen, the interloping egg, or the cuckoo chick which is the imposter? Abbott and Large show how notions of the cuckoo and the imposter have been entangled in selected works of literature and theatre, specifically, in science fiction, the Victorian novel and in Shakespeare. Through a close analysis of these sources, they demonstrate the importance of the cuckoo imposter in articulating issues of morality in three main kinds of social relations: reproductive, familial and intimate relations. As a 'monster' of some sort, the cuckoo serves as an evocative and enduring literary device for questioning values and commitments.

Brian Rappert's chapter gives us insights into the craft of learning to be an imposter, in his case, a magician. By recounting his nascent efforts to perform conjuring tricks, he shows how being an unconvincing imposter is actually a good way of revealing both the expectations held by audiences of the magician, and vice versa – the contract between performer and audience as it were. He argues that it is precisely the audiences' expectations of skill and their acceptance of the role of passive observer that gives performers their deceptive powers. As with other chapters, there is a mutual and consensual intertwining of imposter and audience: 'we make it happen together' (Chapter 7, this volume, p 169). Additionally, Rappert invites us to consider how similar interconnections are constitutive of the relation between academic writers and readers.

Third, imposter dramas may be centred on *sustained suspicions*. These dramas emerge when given a ubiquity of (potential) imposters, rather than imposter tests, authenticity tests are in order. James Kaufman examines this in his chapter on UK social policy around social benefits and welfare-to-work programmes. These programmes are, on the face of it, meant to distinguish between authentic claimants and imposters. Yet, he argues, the programme performs an institutionalized form of suspicion, in which one is assumed to be an imposter until proven otherwise. This is translated into a pressure on both claimants and street-level advisors alike. Claimants may be forced to perform exaggerated genuineness even if they are, by their own account, legitimately 'signing on'. The street-level advisors in turn are often obliged to take certain actions based on the need to meet performance criteria, which are intended to monitor their actions for 'suspicious' behaviour. Kaufman's case reveals that institutionalized in this way, suspicions can easily end up in persecutions and scapegoating. Since the authentic can never be achieved, it can only be grasped through the Sisyphean task purging of the supposedly inauthentic.

In suspicion dramas, imposters (both named and shamed, but also imagined ones) are staged as tricksters, always escaping the elaborate techniques set out to catch them. The tension is never resolved: if one passes some authenticity test, this is only a temporary settlement; and a new fraud may always follow. This difficulty is at the heart of Kristina Grünenberg's chapter. In her chapter we are invited into a biometric lab where researchers develop biometric algorithms and technologies which use facial recognition among other techniques to police national borders and secure facilities. A key concern in such biometric research is that identification technologies might be subject to attempts to deceive them. In the case of facial recognition, this often involves physical masks which warrants the construction of algorithmic traps to identify masks. Yet, since actual 'spoofers', as they are referred in the lingo, are not readily available in the lab, the researchers have to pose *as spoofers* to get materials to train the biometric technologies. Grünenberg explains, furthermore, how this interplay between spoofing and anti-spoofing, between impostering and imposter-impostering, provides an endless cycle.

However, this particular imposter drama need not only revolve around (mis)trust and trials/tests, but may also hold a sense of play and wonder. Mattijs van de Port's chapter looks at the case of spirit possession in the Candomblé region of Brazil. He notes that, while spirits are an accepted everyday part of life in Candomblé, particular *instances* of spirit possession may be doubted. Rather than this being a problem, however, it becomes a constitutive part of the activity. By examining the same ritual performance from different perspectives including his own, he teases out various understandings of what faking it (giving *ekê*) entails. Following Otávio, he learns that those possessed by spirits in public rituals are often gay men who may secretly relish the opportunity to wear lavish clothes and dance with abandon. Thus the audience finds in these possible motivations reasons to doubt the performance. Here, the audience of 'imposter catchers' is the protagonist, taking pleasure in discerning signs, and the anticipation of a disclosure that might never come. Van de Port advocates an interpretive framework inspired by the baroque and campness, a reveling in fakes and appearances. He concludes that perhaps this 'epistemic murk' is an essential feature of these situations, that a sense of 'We really don't know, do we?', is what holds things together.

Fourth and finally, imposter dramas may centre on *unresolvable disruptions*. This plotline has in common with revelations that the imposter initiates a radical break and new way of seeing the world. It is, however, not the imposter that appears as the source of trouble (at least, to most), rather, it is the status quo that emerges as problematic through the imposter: imposters do disorder through showing cracks in power. Another dynamic which some but not all imposters engender is one of mistrust and perpetual doubt. It is interesting to see, then, how and when questions about who is and is not an imposter get settled or are left open. While some dramas seem to have a final curtain, in which the social order is ostensibly put back in place, others do not. The indeterminacy of imposter scenarios is a key theme in the chapter by Olga Restrepo Forero and Malcolm Ashmore. They demonstrate how imposter-catching procedures enact imposters through the tragic case of a Colombian man mistaken for a guerrilla fighter. They explain how the man's official representation (enacted through the *cédula* or ID card) allowed him perpetually to be confused with a famous guerrilla fighter with a similar identity. The man becomes ever more irrefutably staged as an imposter in the 'flatland' of official records and state bureaucracy, even though in the 'fatland' of everyday face-to-face interactions no such ambiguities about his identity present themselves. What is interesting here is how these different versions of the citizen continue to rub together – the persistence of the citizen's 'flatland' identity at one point comes to inflect his face-to-face recognizability – and yet, despite everyone's best efforts, these two intertwined realities never quite resolve.

While revelations may effect similar plot twists and reorient identities, then, the revelation tends (for some at least) to settle the situation, even if it opens up again later. Possible solutions present themselves even if there are debates about which to take up. Disruptions, in contrast, do not manifest themselves in this dance of order–disorder–order but remain indeterminate – even the very tools for repairing the situation are brought into question.

If imposters 'pose as people they are not' (Miller, 2018: 1), then social media is perhaps the impostering practice *par excellence*. In his chapter, Johan Lindquist explores concerns with imposters on Instagram. With more than one billion users, the social media platform is known as the main site for the rise of so-called 'social media influencers', whose status and commercial value is largely measured by the number of 'followers'. As a result of the followers' commercial importance there are attempts at increasing their numbers. As Lindquist shows, the means to do so range from improving the visibility and attractiveness of one's Instagram account and reaching out to other Instagram users, to phishing attacks that use other users' login credentials, to computer-generated followers ('bots'). Based on ethnographic fieldwork, Lindquist considers the assemblage of actors and technologies that are involved in regulating, shaping and playing the 'visibility game' that shapes influence on Instagram. What constitutes an artificial means of increasing the number of followers, and hence, what counts as a 'genuine' rather than a 'fake' follower, is not obvious or decided a priori, but a moving target. Lindquist explores how qualifications of an acceptable, 'good enough' follower emerge as the market in 'buying' and 'selling' followers develops in tandem with Instagram's increasing security measures. Impostering, as Lindquist shows, is not an individual achievement, but the shifting effect of a whole socio-technical network of actors. His chapter therefore illuminates questions about intentionality and agency. While revelations involve quickly pointing fingers, in disruptions it is not always clear who the perpetrators are and whether this matters.

At first sight, the migration policies of 'Fort Europe' seem a paradigmatic example of how impostering practices make apparent a problematic and violent status quo. And yet, in his chapter Fredy Mora-Gámez considers the limits of thinking with imposters in exactly this case. His chapter compares the plights of victims of armed conflict in Colombia seeking compensation from the state to those of migrants applying for residence in Europe. He particularly explores a workshop where migrants residing in Greek refugee camps rework waste from inflatable boats and life vests into bags and wallets to be sold by non-governmental organizations (NGOs) on the European market. In his chapter he suggests that the imposter analytic cannot account for the politics, normativities and identity games at play in such practices. The workshop allows migrants to form collaborations and networks within Europe and, in that sense, come to be part *of* (communities in) Europe already even though physically and bureaucratically they remain locked in place. To capture such disruption, Mora-Gámez offers us the notion of *gatecrashing*, which invites us to conceive of borders radically differently. Thinking with imposters brings our attention to attempts at beating the system and passing through. It highlights that which falls outside order – but always and only in relation to the (dominant) categories, knowledges and identities that it complicates. The politics of gatecrashing, by contrast, does not revolve around playing, cheating or even changing the game that animates state bureaucracies; it is about crafting liveable worlds within and outside it.

The four plotlines – revelations, obsessions, suspicions and disruptions – are not to be taken as exhaustive nor definite. Rather, they offer a first heuristic take on the richness the imposter has to offer. Clearly, the modes of (dis)ordering they describe will in any particular case overlap and interact, and more can be found – not least in the chapters of this volume.

Although we suggest the reader follows their order of appearance in the volume, the chapters can be read separately or in any other order. But together, the contributions to this volume show the various ways in which imposters induce, and sometimes sustain, indeterminate situations. The effects of impostering, in other words, are multiple, situated and relational. The distinctive feature of impostering, moreover, is not only that order is never fully achieved but that it is an ongoing activity. The verb imposter-ing further draws attention to materialities that keep things perpetually in flux rather than fixed, to the affects that spur intrigue, and to the work that is required to invigorate suspicion.

Conclusion

The central point we have argued for in this introductory chapter is the need to take seriously the indeterminacy of impostering. Instead of reducing impostering to a resource for examining a purportedly underlying, unambiguous order, we propose thinking with imposters as a way of understanding this constant generation and renewal of (dis)orders. How imposters and their audiences are enacted together re-focuses our analytical lens to appreciate how particular worlds and relations are (dis)assembled through imposters. The news story about the Russian visitors to Salisbury begs for closure. Well? Was it or was it not just coincidence that they visited at the same time that a former Russian agent was poisoned? Much effort was invested in trying to reach a definitive conclusion, by journalists, investigators, political commentators and, presumably, the security services. Some commentators on social media felt the correct interpretation actually required very little effort; what had happened was a blindingly obvious deception. But against the drive to achieve closure, we suggest that as academics, there is considerable value in attending to the attempts of sustaining openness and indeterminacy in this case. The apparently blatant denials by those who are also self-evidently imposters engender a disturbing unresolvability which, it might be argued, was one of the intended effects of those involved. Various modes of disordering are thus at play in this case: journalists push towards a revelatory script in which the truth finally comes to table, a counter-practice by the spies/ tourists performs the disruption as perpetually unresolved, which in turn triggers sustained suspicions by UK authorities and the public about the secret workings of (Russian) intelligence services.

By bringing together the cases in this volume we wish to explore what we gain (or lose) by viewing various phenomena through the lens of the imposter; by comparing imposter stories across domains; by identifying with imposters or advocating for their rehabilitation. By studying imposters and the disorder they create, not as the 'other' to social order but as phenomena which create alternate orders which rub up uncomfortably against the established ones, we hope better to understand complex events and scenarios which plague political life and evade normal modes of analysis in the social sciences and humanities. It is our suggestion that through employing the figure of the imposter, we can recast and rehabilitate practices in different, possibly generative terms. This in turn can lead us to ask, in future studies, what are the political and methodological consequences of deploying the imposter as analytic. We aim to learn from the empirical cases of fraud, suspicion and transgression how thinking with the imposter might shake up, problematize, enrich or constrain social relations. Overall, then, this edited volume aims to develop imposter(ing) as an analytical sensibility: What does it *do* to focus on imposter(s)(ing)?

References

Abbott, A. (2019) 'On the troubling trail of psychiatry's pseudopatients stunt', *Nature*, DOI: 10.1038/d41586-019-03268-y.

Baker, M. (2015) 'Over half of psychology studies fail reproducibility test', *Nature*, DOI: 10.1038/nature.2015.18248.

Barocas, S., Hood, S. and Ziewitz, M. (2013) 'Governing algorithms: A provocation piece', ID 2245322, SSRN Scholarly Paper, 29 March. Rochester: Social Science Research Network. Available at: https://papers.ssrn.com/abstract=2245322 [Accessed 25 June 2020].

Becker, H.S. (1993) *Outsiders: Studies in the Sociology of Deviance*, New York: Free Press.

Bellingcat (2018) 'Second Skripal poisoning suspect identified as Dr. Alexander Mishkin', 8 October. Available at: https://www.bellingcat.com/news/uk-and-europe/2018/10/08/second-skripal-poisoning-suspect-identified-as-dr-alexander-mishkin/ [Accessed 22 June 2020].

Brown, S. (2002) 'Michel Serres: Science, translation and the logic of the parasite', *Theory, Culture & Society*, 19(3): 1–27.

Brown, S. (2013) 'In praise of the parasite: The dark organizational theory of Michel Serres', *Porto Alegre*, 16(1): 83–100.

Burrow, C. (2020) 'Fiction and the age of lies', *London Review of Books*, 20 February. Available at: https://www.lrb.co.uk/the-paper/v42/n04/colin-burrow/fiction-and-the-age-of-lies [Accessed 7 January 2021].

Butler, J. (1990) *Gender Trouble: Feminism and the Subversion of Identity*, London: Routledge.

Cahalan, S. (2019) *The Great Pretender: The Undercover Mission that Changed our Understanding of Madness*, Edinburgh: Canongate Books.

Cercas, J. (2017) *The Impostor*, London: MacLehose Press.

Coleman, G. (2015) *Hacker, Hoaxer, Whistleblower, Spy: The Many Faces of Anonymous*, London: Verso.

Collins, H. and Pinch, T. (2005) *Dr Golem: How to Think about Medicine*, Chicago: University of Chicago Press.

Das, V. (2018) 'Being false to oneself?', in J. Copeman and G. da Col (eds) *Fake: Anthropological Keywords*, Chicago: HAU Books, pp 31–48.

Eliav-Feldon, M. (2012) *Renaissance Impostors and Proofs of Identity*, London: Palgrave Macmillan.

Espeland, W.N. and Sauder, M. (2007) 'Rankings and reactivity: How public measures recreate social worlds', *American Journal of Sociology*, 113(1): 1–40.

Foucault, M. (1970) *The Order of Things*, New York: Pantheon Books.

Garfinkel, H. (1967) *Studies in Ethnomethodology*, Englewood Cliffs: Prentice-Hall.

Goffman, E. (1959) *The Presentation of Self in Everyday Life*, New York: Anchor Books.

Humphreys, L. (1971) *Tearoom Trade: Impersonal Sex in Public Places*, Chicago: University of Chicago Press.

Jones, G. (2018) 'Deep fakes', in J. Copeman and G. da Col (eds) *Fake: Anthropological Keywords*, Chicago: HAU Books, pp 15–30.

Law, J. (1994) *Organizing Modernity*, Oxford: Blackwell.

Law, J. (2009) 'Actor network theory and material semiotics', in B. Turner (ed) *The New Blackwell Companion to Social Theory*, Malden: Wiley-Blackwell, pp 141–158.

Lynch, M. (2011) 'Harold Garfinkel obituary', *The Guardian*, 13 July. Available at: https://www.theguardian.com/education/2011/jul/13/harold-garfinkel-obituary [Accessed 26 June 2020].

Marres, N. (2018) 'Why we can't have our facts back', *Engaging Science, Technology, and Society*, 4: 423–443.

Miller, C.L. (2018) *Impostors: Literary Hoaxes and Cultural Authenticity*, Chicago: University of Chicago Press.

Miller, W.I. (2003) *Faking It*, Cambridge: Cambridge University Press.

Mol, A. (2002) *The Body Multiple: Ontology in Medical Practice*, Durham: Duke University Press.

Parsons, T. (1991 [1951]) *The Social System*, London: Routledge.

Pomerantsev, P. (2014) *Nothing is True and Everything is Possible: The Surreal Heart of the New Russia*, New York: Public Affairs.

Rosenhan, D.L. (1973) 'On being sane in insane places', *Science*, 179(4070): 250–258.

Rosenthal, C. and Schäfer, S. (eds) (2014) *Fake Identity? The Impostor Narrative in North American Culture*, Frankfurt am Main: Campus Verlag.

Serres, M. (1982) *The Parasite*, Baltimore: Johns Hopkins University Press.

Simmel, G. (1950) 'The stranger', in K.H. Wolff (ed) *The Sociology of Georg Simmel*, London: Free Press of Glencoe Division of Macmillan Company.

Simmons, D. (2006) 'Impostor syndrome, a reparative history', *Engaging Science, Technology, and Society*, 2: 106–127.

Sokal, A.D. (1996) 'Transgressing the Boundaries: Toward a Transformative Hermeneutics of Quantum Gravity', *Social Text*, 46–47: 217–252.

Sokal, A. and Bricmont, J. (1998) *Intellectual Imposters*, London: Profile Books.

Stengers, I. (2005) 'The cosmopolitical proposal', in B. Latour and P. Weibel (eds) *Making Things Public: Atmospheres of Democracy*, Cambridge: MIT Press, pp 994–1003.

Taussig, M.T. (1993) *Mimesis and Alterity: A Particular History of the Senses*, New York/London: Routledge.

2

The Desire to Believe and Belong: Wannabes and Their Audience in a North American Cultural Context

Caroline Rosenthal

The art of faking: imposture as successful identity performance

Identity never is something we 'find' and then 'keep' for good but something we continuously have to work at and work for in changing social and cultural contexts.[1] Rather than a stable entity, identity is always malleable and in flux and needs to be re-invented and stabilized in repeated performances among various groups. The construction of personal identity thus always requires an audience which has to validate the performance as convincing, true and authentic. Identity, in other words, relies on the recognition of the 'Self' through an 'Other'. Cases of imposture, first of all, constitute instances of successful performances in which an audience became convinced of, and sometimes enthralled with, a staged identity. We, secondly, can only speak of imposture if such an identity is then recognized as 'fake'; otherwise we simply continue to believe in it. It is, then, neither the performance nor the resultant

---

[1]   On the elusiveness of personal identity, see, for instance, Appiah and Gates (1995); Brooks (1995); Strauss (2017 [1959]). On the constructed nature of collective identity see Hall and du Gay (1996).

identity that mark an imposture as fake but the revelation of the fact that someone appropriated an identity he or she was not entitled to. Imposture, thirdly, often causes a scandal because people feel betrayed and outraged that someone pretended to be a member of a certain group in order to gain status and/or profit. Often, however, people also feel ashamed of their own poor judgement and for wholeheartedly having bought the performance of the imposter. It is the spectre and fetish of authenticity itself that often produces imposture and the audience's need for believable identities.

As the introductory chapter for this volume convincingly argues, imposture is a complex act which raises fundamental questions about symbolic and social orders. Just like the breaking of a taboo, in the end, serves to reinforce the norms of a society at a given time, successful acts of imposture and their later revelation disturb the social order and lay bare its underlying parameters and mechanisms. The more successful an imposture, the greater the uproar at its revelation, because not only the imposter but the audience is being exposed. Imposters not only tell us a lot about how identity is constructed in general but also reflect on the cultural beliefs, desires and stereotypes of their audience at a certain time. Hence imposture, as the editors of this volume conclude, has great analytical potential. Rather than concentrating on whether an act of imposture is true or false, then, we should focus on how it reflects on the cultural context in which the imposture, at first successfully, took place. In other words, how does the act of imposture reflect on the cultural norms and social orders of the audience that helped to construct and validate the act in the first place?

In the following, I look at two cases of ethnic imposture in the early 20th century in which White men adopted an Indigenous identity in the US and Canada respectively. Ethnic impersonation is an especially widespread phenomenon in a North American cultural context. In the settler societies of the New World, ethnicities mingled and shifted and offered many opportunities for people to re-invent themselves as ethnic others. At the same time, there was a desire and need for an aboriginal history so that an especially popular form of ethnic passing was that of 'playing Indian' (Deloria, 1998) – of Whites, and sometimes Blacks, pretending to be Native Americans. Ever since the romantic period, when Indigenous people were no longer seen as a threat but became mystified as the better Americans in the trope of the Noble Savage, speaking from a Native perspective invested people with an authority on matters related to nature, spirituality and tolerance. This authority and alleged moral superiority probably made assuming a Native American identity especially appealing.

The two cases I will present and analyse – that of Asa/Forrest Carter, a White man from Alabama who pretended to be Cherokee, and that of Archibald Belaney, who in Canada took on the identity of the half-blood Indian Grey Owl – share that both men were wannabe Indians. They strikingly differ, however, in the characteristics and understood motives of the imposture and even more so in their cultural context and in the motivations of each audience to believe in the act. I am particularly interested in the role the cultural context played in constructing and validating the particular ethnic impersonation and in what this tells us about the cultural context of the US and Canada in the late 1970s and 1930s respectively. In both cases, I will argue in the following, the imposture was facilitated by audiences that wanted to believe in the imposture out of a desire to nationally and culturally belong.

Two cases of ethnic impersonation and their cultural context

Forrest Carter's 'masterpiece of simulated authenticity'[2]: The Education of Little Tree

In 1976 Delacorte Press released *The Education of Little Tree*, the childhood memoir of a Cherokee orphan. The author of the book, Forrest Carter, had built a reputation in the previous three years with books on heroic outlaws such as *Gone to Texas* (which was turned into the movie *The Outlaw Josey Wales*, starring Clint Eastwood) and *The Vengeance Trail of Josey Wales* (which was not turned into a movie due to a fallout between Carter and Eastwood). At the time of its first publication, Carter's alleged autobiography *The Education of Little Tree* only reached a small audience but already then '[a]dolescent and adult readers [...] warmed to the uplifting story of how this well-known writer of westerns [...] came to know the wisdom of his Cherokee ancestors' (Carter, 1991). Readers were moved by Carter's authenticity and sincerity in depicting the simple life he had led as a boy in the Southern mountains with his grandparents during the Depression era. The coming-of-age-style book is written from the perspective of an innocent young Indian boy who has to face, but does not always understand and certainly does not judge, the racism and classism of the society surrounding him and his grandparents who are both poor and Indigenous. The book's rhetoric and voice emulate classics like

---

[2]   McGurl (2005: 248).

Mark Twain's *Huckleberry Finn* or Harper Lee's *To Kill a Mockingbird* which also use child-narrators for a scathing analysis of society's racism, greed and hypocrisy. Just like Huck Finn, Carter's narrator uses faulty grammar and obviously is not well-educated but superior in his morality, humanitarianism and plea for tolerance and honesty than all the grown-up White people around him. It is the book's simplistic language which creates the effect of sincerity and a guileless narrator (Clayton, 1986: 20; McGurl, 2005: 246) who tells the truth from his very heart.

Seven years after Carter's sudden death in 1979, the University of New Mexico Press bought the rights to the book and reissued it. Especially since the publication of the paperback edition in 1991, the book has turned into an unbroken commercial success and favourite with an American audience. Not only the already-mentioned characteristics of the book but the narrator's close-to-nature attitude, green rhetoric and Indigenous worldview appealed to readers. They were also enthralled with the persona of Little Tree who kept a good heart despite the devastating experiences he had with government agents, institutions, and especially in Residential Schools. Audiences in the 1980s and early 1990s felt shamed by the depiction of systemic discrimination and violence against Indigenous people and sentimental about the wisdoms an Indigenous worldview apparently had to share, which the American public had so far entirely missed out on. In the book, Little Tree's simple truths stand in staunch opposition to the crafted rhetoric of politicians and the jaded politics of governmental agents and institutions that time and again betray not only Indigenous people but the common man. As the US began to deal with its national guilt concerning its treatment of Indigenous people and lack of knowledge about them, *The Education of Little Tree* quickly became part of mandatory reading lists and curricula in high schools, colleges and universities (Jeff Roche quoted in McGurl, 2005: 250). It was deemed to hold special educational value and was embraced by Whites and Indigenous people alike as a truthful rendition of a typical Cherokee youth in the American South.[3]

The book's success and popularity drew attention and in 1991 the historian Dan Carter (who is not related to Forrest/Asa Carter in any way) in a *New York Times* article exposed Forrest Carter as Asa Carter, a White supremacist and former member of the Ku Klux Klan who as

---

[3]   See Rennard Strickland's foreword, 'Sharing Little Tree' (in Carter, 2004 [1976]: v–vi), which praises how the book 'informs the heart and educates the spirit' (vi).

a speech writer for Governor George Wallace of Alabama had allegedly crafted the infamous line: 'Segregation now … Segregation tomorrow … Segregation forever' (Carter, 1991). By then *Little Tree* had made the *New York Times* non-fiction bestseller list, received a best book of the year award from the American Booksellers Association and sold over 600,000 copies (Browder in Rosenthal and Schäfer, 2014: 62). At the book's first publication in 1976, Asa/Forrest Carter had given a short television interview as a result of which some people already recognized the purported Cherokee author as the Klansman he really was. Consequently, Wayne Greenhaw, a journalist very well known for his work on the Ku Klux Klan, wrote a short article for the *New York Times* which not only stated that various people had recognized Forrest as Asa Carter in photographs on the book jacket but that the two shared the same address and were both born in 1925. Greenhaw also pointed out that Asa had 'a particular hero, the Confederate general Bedford Forrest, the first imperial wizard of the Ku Klux Klan'. Carter himself, Greenhaw writes, had declined to be interviewed on the topic while his editor at Delacorte, Eleanor Friede, said that she had no reason to doubt the identity of the author of *Little Tree* (Greenhaw, 1976: 39). Everything about Carter's hoax had thus been revealed right after the book's first publication but it did not cause a stir at that time at all. The big scandal happened later when the book had turned into a huge success and did reach a major audience because by that time the cultural context had changed and appropriations of ethnic identities were regarded in a much harsher light. Dan Carter's revelation in 1991 set in motion a search for Carter's true identity and his past, a story which up to this day has not been fully reconstructed.

What is known is that Carter, who was born in Alabama in 1925, had neither a biological nor cultural claim to Indigenous heritage. He was a veteran of the Second World War and the father of four children who left the University of Colorado without a degree to work as a radio talk show host before returning to Alabama and finding his true vocation as a political activist (see Figure 2.1 for a photograph of Carter in action). Carter became especially involved in issues of integration and segregation and in the deep-seated racial conflict that was tearing apart the American South in the 1950s and 1960s. He ran as a candidate for governor of Alabama in the Democratic primary of 1970s and was solidly defeated (Clayton, 1986: 20). Disillusioned by politics and their ineffective rhetoric, Carter then turned to literature. Themes that had occupied Carter in his political life – the corruption of the government, the plights of the common man who was neglected and cheated by his

Figure 2.1: Asa Carter ('Little Tree'), Birmingham, Alabama, 1956

Source: Associated Press, via Getty Images.

government, social injustice – just became transferred to literature in which he heroically depicted outlaws and outcasts. His talent for populism and agitation rhetoric clearly shows in *The Education of Little Tree* as well.

By appropriating an Indigenous heritage, in *The Education of Little Tree* Carter systematically further develops the political criticism with which he had been preoccupied in his previous books; Little Tree simply becomes another mouthpiece for the message Carter wanted to convey.[4] Carter was an autodidact who canvassed for the common (White) man and rallied against an unfair, over-regulating and self-interested government. It is this kind of government that looms large in *The Education of Little Tree* as well and that makes the lives of poor White Southerners and Cherokee alike difficult. Carter found a new form and voice for the story he had been telling all his life and he skilfully used the cultural climate of the time (McGurl, 2005: 244–245; Deutsch, 2013: 313). In the 1970s, many White Americans sympathized with the American Indian Movement – a political activist movement which

---

[4]   McGurl (2005: 251) claims it was no coincidence that Carter made his protagonist a member of the Cherokee because they were colour-conscious and 'shared with the white South a long-standing tradition of keeping African slaves'.

in the late 1960s fought discrimination and violence against Native Americans – and began to rethink the myth of America's untainted new beginning. Carter's book capitalizes on these sentiments and has become a bestseller that has sold more than 1.5 million copies to date. A White supremacist's appropriation of an Indigenous identity is cynical no matter what his motivations, but rather than speculating on why Carter did it, I am interested in how and why he succeeded and what that tells us about his audience.

People, in fact, were so taken with the book that many were either unwilling to acknowledge Carter's racist past or mystified his story as a conversion narrative from segregationist to humanitarian even after his true identity had been revealed. Elizabeth Hadas, director of the University of New Mexico Press, stated that calling it a hoax was doing Carter's book injustice because clearly he must have 'undergone a personality change; it's hard to believe that those stories were faked. If he was a bad man he took on a new identity and became a good man' (in Reid, 1991: 17). Hadas not only ignores what others consider was Carter's crude act of appropriation, but suggests that making up the sentimental tale of a Cherokee youth could redeem Carter's past as a hard-liner White supremacist and Klansman. Rennard Strickland, himself a Cherokee Indian and director of the Center for the Study of American Indian Law and Policy at the University of Oklahoma, wrote the introduction to *The Education of Little Tree*. Even after the revelation of the author's true identity, he defended the book by saying that it is 'not the work of a bigot' and that '[i]f the man who wrote speeches for George Wallace could write this book there's hope for a cure for the souls of us all' (in Reid, 1991: 17). Strickland not only absolves Carter but turns him into a model for redemption and a beacon of hope for reconciliation in America. Many people, like Hadas and Strickland, continued to sentimentalize *The Education of Little Tree* by reading it as the result of a profound spiritual conversion and atonement of a former racist. Others, like the Native American writer Sherman Alexie, however, read it as 'the racial hypocrisy of a white supremacist' (Alexie in Italie, 2007) and like the academic Laura Browder who claimed that Carter was a bigot, who 'in the mornings [...] labored over his "memoir", in which he described the kindly Jewish peddler who befriended young *Little Tree*. In the evening, he lectured to his followers that Hitler was right and that it was time to "bring back the gas chambers"' (Browder in Rosenthal and Schäfer, 2014: 76).

In both camps, there was a lot of speculation on Carter's personality and, essential to my argument, there were a lot of competing constructions of Carter's pre- and post-revelation identities and interpretations of his motivations in writing *Little Tree*. In 2004 Laura Browder, a professor of English at Virginia Commonwealth University, who in her book *Slippery Characters: Ethnic Impersonators and American Identities* (2000) had already investigated Carter's case, joined a team of film makers who produced a documentary on Carter entitled *The Reconstruction of Asa Carter: The Greatest Story Was the One He Never Told*. The film did not succeed in fully unravelling the identity of Carter because the story is too twisted, Carter had died, and many people who knew him did not want to find out the truth about his ethnic background but instead wanted to cling to the myth of the Indian sage. What the film did establish is that Carter was extremely skilled at manipulating ethnic stereotypes. The fascinating thing about Carter, Browder concludes, is that he came across as very authentic in both his personas as supremacist and orphaned Indian boy and that he essentially used the same strategy for politically very different camps. He appealed to populism and the anti-intellectualism of the common man: 'whether segregationist Alabamans eager to believe in the mythology of a noble, lily-white south built by their redneck granddaddies, or New Age readers comforted by Carter's description of a mythically pure life on a mountaintop', Carter reached his audience (Browder, 2014: 70). Carter succeeded, in other words, because he had a gift for identifying his audience's deepest racial stereotypes, resentments and fears. Dan Carter, the historian who exposed the 'true' identity of the author of *Little Tree*, thus ends his disclosing article with the question: 'What does it tell us that we are so easily deceived?' (Carter, 1991). One could ask even more poignantly, what does it tell us *about us* that we are so easily deceived? The truly interesting thing about Carter's literary charade is why it worked so well and why people so ferociously defended the truthfulness of its message.

Carter, it is safe to say, drew on a skein of archetypes that are of fundamental importance for America's cultural self-definition – autodidacticism, anti-intellectualism, anti-government freedom. Just like *Huck Finn*, the iconic American novel, *Little Tree* presents a picaro, an outsider of society who criticizes that society while at the same time glorifying core American values. The government might be corrupt and the country racially divided but people love Little Tree because he is innocent and reaffirms values such as liberty and freedom, a harmony with nature away from the corruptions of civilization, and equality for all. Carter's book is the epitome of America's cultural schizophrenia in which the wilderness is idealized and destroyed at the same time and the Indian romanticized and elevated while being driven off the land and brutally assimilated in Residential Schools. Little Tree allows people to critically reexamine America's past and treatment of ethnic minorities while at the same time celebrating Little Tree as an uncorrupted and original American voice. In this respect, and in light of competing post-revelation interpretations of *The Education of Little Tree* by audiences, I find Gina Caison's essay 'Claiming the Unclaimable' highly interesting. Instead of focusing on Carter's troubled biography and debates about his Native identity, she emphasizes his deep investment in Southern US literature because this 'forces audiences to consider their own narrative emotive expectations, education, and historicities' (Caison, 2011: 593). For Caison, Carter's text renders 'audiences' conceptions and misconceptions of the South' (573) and their need for 'a common people's folklore' (579). As such, Caison states, *Little Tree* – just like *To Kill a Mockingbird* or *Huck Finn* – 'speaks volumes about Southern racial desire' (581) and a nostalgia for an agrarian antebellum past. Carter, she concludes, 'does not have to rewrite the South, his readers have done so for him' (590).

People believed in Little Tree's story because it accurately described what they imagined a Cherokee childhood in the mountains of the American South must have been like and because it emulated pre-existing literary models. Carter's account certainly rendered the fate of the Cherokee in a gripping way, which is why even Native Americans like Strickland have continued to endorse the book post-revelation. What is at stake in the case of Carter for many other people, though, is not whether this is a good story and whether it accurately describes the struggles of the Cherokee but the fact that it was not Carter's story to tell. In the battle for competing interpretations and ethical appraisals of Carter's imposture issues of cultural appropriation have become of growing importance. Writers like Sherman Alexie (2006) have claimed that Carter's story violates the right to cultural heritage and mocks the sufferings of Indigenous people by making up a romanticized tale about their experiences.

In ethnic memoirs, such as Carter's, as academic Laura Browder has argued, authority is attached to the identity of the author because he or she not only speaks of and for himself but as the member of a group and hence constructs not only a personal but collective identity. As Browder puts it:

> Both the reader and the writer of an ethnic autobiography understand the implicit contract: the memorist is not telling his or her own story as much as the story of a people. In order to be heard, the ethnic autobiographer must often

conform to his or her audience's stereotypes about ethnicity.
(Browder, 2000: 5)

Imposter autobiographies are successful not because they depict what is real but because they depict what can be sold and bought as real. They draw on specific elements of a particular narrative that are already familiar to the reader and hence seem real.[5]

While Carter's fake self-fashioning mostly took place in literature, the case of Archibald Belaney draws on a broader array of performance and audience involvement.

'Canada's most famous fake'[6]: Archibald Belaney alias Grey Owl

Archibald Belaney's biography is as shady as Carter's and his imposture a lot more difficult to unravel because unlike Carter, Belaney did not simply pretend to be, but really did become Grey Owl (see Figure 2.2) . While Belaney was not a blood Indian – a heritage he wrongfully claimed – he was in some way a cultural Indian. And although his attempts to speak Ojibway on stage and appear in the regalia of this nation were at best hilarious, he was serious in his environmental message. Belaney did not lead people to believe but he let them believe that he had Indian blood. His act is still an appropriation, but in his case, unlike Carter's, the audience had an almost bigger part in making him into Grey Owl than he had himself.

Archibald Stansfeld Belaney was born in Sussex, England, in 1888 and emigrated to Canada in 1906. In England he was raised by his

---

[5] This phenomenon is found in various forms of autobiographies featuring ethnic difference in the American cultural context such as the captivity narrative or the slave narrative which oscillate between authenticity and fictionality. The captivity narrative served to legitimize White imperialism by portraying the Natives as devilish creatures from whose hellish claws God delivers the true believer. The slave narrative wanted to win over a White audience for abolitionism by rending the atrocities of slavery. Both genres draw their authority from a claim to authenticity but at the same time contain recurring fictional and rhetorized elements to promote their cause. On the interplay of *Captivity Narrative* and its audience see Rosenthal (2011).

[6] Wiebe on the blurb of Canadian Ojibway writer Armand Ruffo's volume of poetry about Grey Owl (Ruffo, 1997). Ruffo's volume is interesting for questions of authenticity and fraud because he reminds us, again, that since the first contact between settlers and natives the Indian has been an imaginary construction of settler's fears and desires. For Ruffo, Belaney simply articulates and acts out those desires and uses them to drive home his environmentalist message.

Figure 2.2: Archibald Stansfield ('Grey Owl') Belaney, 1934

Source: Photograph by Howard Coster, courtesy of the National Portrait Gallery, UK.

two spinster aunts after his parents' brief marriage had fallen apart – his mother was only a teenager at his birth and his father an alcoholic who abandoned the family. Early on, the boy dove into the romantic writings of James Fenimore Cooper and Henry Wadsworth Longfellow, which, paired with a deep interest in the natural history of North America, started him daydreaming of becoming a trapper in the woods of the New World (Smith, 1990). Belaney realized his dream of emigrating to Canada at the age of 17. Upon arrival, he quickly made his way north and found work as a guide and trapper near Lake Temiskaming. It is here that he first came into contact with the Ojibwa who in 1907 gave him the name 'He-Who-Travels-by-Night' or 'Grey Owl'. Belaney learned from the Ojibwa how to track and trap as well as how to navigate a canoe and picked up a few words in their language. Most of his Indian vocabulary, however, stemmed from Longfellow's famous long poem 'The Song of Hiawatha', so that in his self-fashioning as Grey Owl the fictitious or imaginary Indian early on takes precedence over an adoption of real Indigenous practices.

The rest of Belaney's biography is opaque and characterized by several, at some point bigamous, marriages as well as various legitimate and illegitimate children. Fact and fiction are intricately interwoven in

Belaney/Grey Owl's life story which he deliberately kept vague and contradictory. He claimed to have been born in Mexico, a Native half-blood with an Apache mother and a Scottish father and explained his British accent by purporting that his parents had toured England with Buffalo Bill's Wild West Show (Francis, 1992: 133). At other times, however, he claimed Métis heritage and in general tried to keep the Indian parts of his life-story as untraceable as possible. Belaney served in the Canadian army in Europe in the First World War but was early on discharged with a foot wound and returned to Canada to become an environmentalist and start a writing career. His first book, *The Men of the Last Frontier*, appeared in 1931 and, as the title indicates, depicted and romanticized the life of a trapper in a vast Northland. For his self-fashioning as Grey Owl, though, his autobiography, *Pilgrims of the Wild*, is essential.[7] In the book, Grey Owl turned from someone hunting the beaver for fur into someone who entered into a kinship with the creature. Personal identity – true or fake – always requires a narrative which retrospectively joins disparate events into a coherent life story, a story which is shaped by a linguistic and narrative form as well as a prism through which it can be seen and told. In the story of Grey Owl this prism is a conversion experience which turned the former trapper into a passionate nature conservationist. After killing a beaver mother, Grey Owl had an epiphany and on Anahareo's insistence reared the orphaned cubs (Smith, 1990: 82–84). It is this key moment, which Belaney fictionally embellished in many ways, that essentially contributed to the invention of the persona of Grey Owl, first in fiction and then in real life.

Essential to Belaney's conversion is his relationship with Gertrude Bernard aka Anahareo. They met in 1925 at Lake Temiskaming when Gertrude is 19 and Belaney almost twice her age. It is Gertrude who encouraged Belaney to change his lifestyle and become a conservationist rather than a trapper. And it is Grey Owl who made Gertrude part of his cosmos of imaginary Indians by naming her Anahareo and including her in his beaver community. The irony of the story is that Gertrude really was Indigenous, she had Algonquin and Mohawk roots, but grew up alienated from her Native heritage and only became a real fake Indian when she met Belaney alias Grey Owl. He took her to Doucet in northern Quebec and passed on to her what he had learned about hunting and trapping from the

---

[7]   The literary critic Bill New calls *Pilgrims of the Wild* 'one of the most successful literary hoaxes in Canadian literary history' (New, 2002: 453).

Ojibway. When Archibald Belaney's 'true' ethnic identity was revealed later on, Anahareo said that she never suspected him to be a fake and that to her he will always remain a North American Indian (Braz, 2005: 63). The case of Gertrude aka Anahareo illustrates once more how difficult and fruitless it is to tell a real from a fake identity and how essentialism and constructivism often jar in reconstructions of an imposter identity. People recognize as real in an ethnic identity what is familiar to them and complies with their stereotypes. In the construction of Grey Owl's ethnic identity, as Francis points out, '[e]ven his drinking was seen as confirmation of his Native identity [by Parks Canada]'. Francis concludes that '[t]here is something wonderfully ironic about the stereotype of the drunken Indian being used to explain away the conduct of an English gentleman' (Francis, 1992: 137). As the anecdote illustrates, keeping apart truth, fakery and projection in the construction and recognition of any identity is an almost impossible task.

Grey Owl, Anahareo and their beavers quickly become part of a media machine which transformed Belaney's personal desire to play Indian into a collective dream of Canada's pristine but not uncivilized wilderness.[8] In 1928 W.J. Oliver shot his first short documentary about Grey Owl, entitled *Beaver People*. The eight-minute-film turned into a worldwide success. With Anahareo and their beavers, Grey Owl formed a picturesque nuclear family[9] in the wilds of Canada which for a short time became the iconic image of Canada. Beaver and Indian were orchestrated in similar ways in the film because both, in the logic of Oliver's film, represented a vanishing species and were mediators between wilderness and civilization. The second half of the film presented the beaver as an animal with social skills and an indefatigable zeal to cultivate nature and in addition likened the beaver to humans through descriptions of its grooming and eating habits.[10]

In the early 1930s, Parks Canada discovered Grey Owl as the perfect medium for advertising the conservation of nature and for promoting the establishment of further National Parks. Grey Owl, Anahareo and their beavers were first briefly settled in a National Park in Manitoba

---

[8] In the following I draw on and sometimes directly quote from Rosenthal (2014).

[9] The real nuclear family – in 1932 Grey Owl and Anahareo have a daughter, Dawn, together – breaks apart as early as 1935 and in 1936 Archie marries the young French-Canadian Yvonne Perrier (while still being legally married to another woman), who accompanies him on his second reading tour (Francis, 1992: 134).

[10] On parallels between beavers and humans in Grey Owl's biography, see Dawson (2007) and Rosenthal (2014).

but were later moved to Prince Albert National Park in Saskatchewan because the waterways there were better suited to the beaver colony. Grey Owl's original cabin from the time he lived up North was rebuilt in the park and it was in this cabin that he wrote most of his books. Grey Owl hence started writing when the original scene and alleged authenticity of his experience had long degenerated into a *mise en scène* in which he was cast as a character in Parks Canada's narrative. Grey Owl's tenure in Prince Albert National Park, as Stewart has pointed out, is symptomatic for how Canadian wilderness spaces have been constructed and instrumentalized for the identity formation of White settler society. She argues that:

> Through the dual processes of erasure and appropriation, the material and cultural resources of Indigenous peoples are transferred to white settlers: the resources of 'wilderness' – land, forests, animals, bodies of water, minerals, metals, oil – become the property of settlers, and the basis of 'national prosperity,' and Indigenous cultural resources become the 'heritage' of 'all Canadians'. (Stewart, 2018: 165)

Grey Owl's presence in the park suggested that Indigenous people supported the national enterprise, gave their consent to colonialism and worked hand in hand with the settlers to preserve the land that had been taken from them.

Neither Parks Canada nor Grey Owl's English publisher and promoter, Lovat Dickson, were interested in presenting a real Indian but rather a credibly marketable one. Grey Owl was created in order to export a certain image of Canada. He was sent on two major reading tours to England in 1936 and 1937 where he filled the auditorium with two to three thousand people every single evening. Dressed in buck skin and equipped with a curious mix of Indigenous items Grey Owl canvassed for the preservation of nature and especially for the beaver that at the time was on the brink of extinction in central Canada. The filmic material about Anahareo and their beavers, which Grey Owl presented during his lectures, captured a time already gone by and served as the visual founding myth through which Belaney invented himself as Grey Owl. On his second reading tour, Grey Owl gave 200 performances to more than half a million people and even had an audience with the British king who was enamored by the Noble Savage's perfect manners paired with Canadian Indigenous wisdom (Smith, 1990: 189). As a journalist at the time said, Grey Owl's lecture tours were so successful because he was the first Indian who really looked like one (Smith, 1990: 1–7). After one of Grey Owl's performances, the *London Sunday Express* ran a headline: 'There never came a Redder Red Indian to Britain' (see Smith, 1990: 1 and 122; see also Birkle, 2010). Grey Owl played the role of how White people imagined North American Indians better than any 'real' Indian could because he was one of them. In addition, his thoroughly British way of playing Indian made the unknown and strange Canadian North and its Indigenous people palatable to people in Britain. Grey Owl was the impersonation of the Noble Savage, strange enough to be admired but familiar enough with British protocol and custom not to be feared.

On his international lecture tours, Archibald Belaney progressively burned himself out and overindulged in drinking. His relationship with Anahareo broke apart and he eventually became an alcoholic. In 1938, after he had returned from his second tour, 'He-Who-Travels-by-Night' died in his decoy cabin in Saskatchewan from the consequences of alcoholism and exhaustion. A day after his death on 13 April 1938, Grey Owl's true identity was disclosed in a local newspaper and was quickly picked up by papers all across Canada and England. There was great indignation that Canada's most famous Indian turned out to be a fake. People felt fooled and robbed of their dream and probably a bit ashamed. Grey Owl's performances had so successfully kindled their own wish to go Native that they had wholeheartedly swallowed his act. Grey Owl's message about the preservation of nature was no less true after his 'true' identity had been revealed but the fakeness of people's ideas about Canadian Nativeness had been painfully exposed.

The interesting question in the case of Archibald Belaney alias Grey Owl to me, again, is not why he pretended to be someone else but why people wanted to, or even needed to, believe in him. Archibald Belaney did not so much invent an Indian identity for himself as he was turned into an Indian by the cultural context of his time and became the projection screen for national desires, British as well as Canadian. In Grey Owl's lectures and the images he showed of his beaver colony, Canada for the British audience of the 1930s turned into a gigantic nature park and into a theatre which offered some distraction from the dire economic conditions of the time and from the approaching Second World War. Grey Owl made a strange part of the Empire accessible to people in England and gave an image and identity to the former colony Canada. An identity that was not threatening and harsh but picturesque and hopeful and that offered escape routes from civilization without having to leave home and hearth. In his talks and the accompanying film material, Grey Owl turned the previously inhospitable wilderness of Canada into a potential home by skilfully mingling fact and fiction, reality, and utopic projection.

For Canadians, Grey Owl became a cultural icon because, as the Canadian author and culture critic Margaret Atwood put it, he single-handedly saved the beaver in central Canada from extinction and in the 1930s lived out a 'deeply rooted collective dream' (Atwood, 1995: 61). Grey Owl transformed the no-man's land of Canada into a nature sanctuary and thus made it possible for the southern Anglo-Saxon population to identify with the rough and 'true' North. Atwood calls this the 'Grey Owl Syndrome' in Canadian culture which not only refers to the collective wish to go Native but to a deeply seated desire to transform the wilderness into a 'peaceable kingdom'. Grey Owl and his beavers satisfied this longing for a peaceable kingdom in many different ways not only because of the cohabitation of man and beast but also because he supposedly was an Aboriginal and as such had a natural right to be on the continent that the settlers lacked. Via Grey Owl the young Canadian nation could imagine an identity of its own, away and apart from the British motherland. Archibald Belaney alias Grey Owl allowed Canadians to reinvent a history and a place of belonging for themselves via the specifically Canadian landscape. Grey Owl became famous because he advertised a Canadian way of life at a time when Canada was largely invisible in world politics and was struggling to define its identity against Britain and the United States as well as within the Commonwealth. The Northland Grey Owl populated with his beaver family decisively differed from both the United States and Great Britain and portrayed Canada as an environmentally superior country.

There were many obvious clues that Grey Owl was not who he pretended to be. But the few who knew he was an imposter kept silent and the others, it appears, did not want to know the truth – sometimes for pecuniary reasons as many people lived well off the Grey Owl myth – and sometimes because they wanted to hold on to a romanticized image. Grey Owl's texts drew on so many clichés that Whites have time and again used to misrepresent Natives – the Noble Savage, the Vanishing Indian, the prophet of the wild, the stern-faced Indian, the sage – that Belaney's charade could not have gone undetected. Unless people – deliberately or undeliberately – wanted to be deceived because the persona of Grey Owl prompted a collective desire. At a time when people grew tired of Wild West shows because at the height of the machine-age they no longer perceived the Indian as an aggressor but as a victim of the forces of civilization and as a messenger of peace and nature conservationism, Grey Owl gave people what they wanted. His texts procured a civilized Northland in which the beaver, the most cultivated of all animals and the national emblem of Canada, became the protagonist in a tale about one of the world's last sanctuaries.

Truer than the real thing: imposture and its audience

What the two exemplary cases of Asa/Forrest Carter and Archibald Belaney alias Grey Owl teach us is that in trying to disambiguate the fake and the authentic, the insincere and sincere, we get into deeply troubled waters. If every identity relies on construction, fluidity and performance, there are no true and fake identities. The interesting question is, as my two examples have shown, the inclination of the audience to buy the hoax. The audience, as an individual and a collective group, becomes a co-conspirator or co-creator of imposture which arises in a dialogue between author and audience. As Browder puts it: 'If Forrest Carter was a construct, he was certainly one created by his fans as well as by Asa Carter. And if his story teaches us anything, it is about the power of myth, and both the instability of identity – and its multiplicity' (in Rosenthal and Schäfer, 2014: 78). Both Carter and Belaney succeeded in their impersonations because they identified and embodied North American archetypes and mythologies – in Carter's case that of the Noble Savage, the underdog, the anti-intellectual common man. Belaney not only drew on the archetypes of the Noble Savage and the Vanishing Indian but pictured Canada as a pristine and somehow better Northland within the Commonwealth. Both impersonations from the start offered many clues as to their fakery. Nonetheless the audience tenaciously, even desperately clung to the invented reality because it satisfied people's needs, desires and longings at a specific time. It is these longings and the function that the impersonations fulfilled for the cultural and national context they took place in that I will look at as a conclusion to this chapter.

   In their opening chapter for this volume the editors cogently argue that rather than looking at the truth or falseness of the act of imposture we can use the figure of the imposter as an analytical device and conceptual figure for the study of changing social relations and underlying values of a culture. This holds especially true for wannabes. The political problem with the two impersonations is not that Carter and Belaney faked an identity – although in Carter's that also is an ethical problem – but that their acts of imposture were condoned by an audience that often cared more for the story than for the plight of the real people behind it. For many contemporary critics, it is not the imposture that is at fault but the cultural industry that still publishes the accounts of Carter and Belaney without any hint to their cultural appropriation.[11] For critics like Fenn Stewart or Carrie Dawson the bone of contention is not the act of imposture itself but the way it is continued to be reverberated, represented and incorporated into the cultural imaginary. Not the acts of imposture themselves but the way the audience keeps preferring them to real accounts of the sufferings of Native people is what makes such stories participate in the general erasure and appropriation of Indigenous culture in North America (Dawson, 1998; Stewart, 2018). In his article 'Selling Indians: Make it painless, make it up', Robert Allen Warrior poses the question why 'U.S. culture in general prefers a fraud like Asa Carter to tell them about Indians to going to the trouble of searching out reliable material, even if that material does not cater to their desire to hear about power animals and medicine crystals' (Warrior, 1992: 405). What Warrior alludes to here is that fake and imaginary Indians have always played a major role in the national formation of the US as well as Canada.[12] It was via the misrepresented and imagined Indian that those two nations could imagine themselves, and, as Thomas King has pointed out in his recent book *The Inconvenient Indian*, for that the real Indians had to be dead.

Fake Indians were necessary for White settler societies in the US and Canada alike to imagine themselves as belonging to the continent they had conquered. Dawson suggests that wannabe imposture 'is a trope through which to interrogate the preoccupation with authenticity

---

[11] The University of New Mexico Press still publishes *The Education of Little Tree* without any reference to Carter's crude appropriation. The words 'true story' were removed from the blurb, thus labelling the book as fiction instead of non-fiction but until this day, no framing fore- and afterword has been added. The book is still often filed under the keywords 'Biography' or 'Cherokee author' in libraries as indicated on the copyright page. In Germany, a beautiful edition of Grey Owl's *The Men of the Last Frontier* (1931) came out in 2019 as *Pfade in der Wildnis. Eine Indianische Erzählung von der Natur* in Die Andere Bibliothek without any indication that Grey Owl was not a Canadian Indian.

[12] Interestingly, Warrior does not anchor his concern to distinguish between 'real' and 'fake' Indians and his claim for 'reliable material' to biology. On the contrary, in his article, he also criticizes Paula Gunn Allen for a, in his eyes, romanticized account of an Indigenous past that ignores the racism and discrimination against Native people in contemporary American society. Warrior argues that '[t]he badge of biological authenticity, whether true, false, or unknowable, provides many contemporary authors with the opportunity to sell romanticized relics of the past to a ready, high-disposable-income audience' (Warrior, 1992: 406).

in white settler cultures' (Dawson, 1998: 121). White settlers, she claims, had a deep-rooted fear of being imposters themselves as imperialist and colonial discourses forced them to prove time and again that they were not Indians but also that they were different from the motherland. As a consequence, they were in constant search for an original identity that would give them a claim to and place in the New World. The figure of the North American Indian and the appropriation of Indian legends and stories offered a way of indigenization to the White settlers.[13] This cultural appropriation went hand in hand with the genocide of the Indigenous people and the erasure of their knowledges and practices of living on and with the land. Hence when the authenticity and thus authority of an allegedly Indigenous subject is questioned so is that of the settler subject posing as almost Indigenous. What comes to light when a wannabe is exposed is the settler colonialist subject's longing to present itself as Indigenous to the land and thus truly be at home.

Carter's imposture points to a cultural schizophrenia of North American settler culture which sees the Indian as the vanguard for true American values and nature conservation while at the same time destroying both, nature and Natives, for the higher aims of civilization. Carter's American audience can collectively grieve for the wrongs of American society while celebrating original American values in a voice as true as Little Tree's. Rather than having to really deal with the devastating effects and traumatic aftermath of Residential Schools in testimonies of survivors this can be washed down with the moving tale of an orphan who, as his narrative shows, turned out all right in the end. To the same extent, Grey Owl's watered-down version of environmentalism did not force Canadians really to question their environmental policies. Grey Owl was a convenient Indian because he did not insist on adopting Indigenous practices of land-based learning or question the installment of so-called wilderness spaces. Imaginary Indians, as Stewart points out, 'reflect and reveal settler fears, desires, and beliefs' (Stewart, 2018: 16). Their exposure, then, exposes the dark underside of North American national myths.

---

[13] Dawson uses Alan Lawson's term of 'self-indigenizing narratives' for narratives in which the settlers represented themselves not quite the same as but similar to the Indigenous subject thus maintaining a distance while at the same time inscribing oneself into the country/landscape (Dawson, 1998: 127).

References

Alexie, S. (2006) 'When the story stolen is your own', *Time Magazine*, 167(6): 48.

Appiah, K.A. and Gates, H.L. (eds) (1995) *Identities*, Chicago: University of Chicago Press.

Atwood, M. (1995) 'The Grey Owl syndrome', in *Strange Things: The Malevolent North in Canadian Literature*, Oxford: Clarendon Press, pp 41–73.

Birkle, C. (2010) 'Mediation and appropriation: Imaginary Indians and the Grey Owl syndrome', in A. Hornung (ed) *Auto/Biography and Mediation*, Heidelberg: Winter, pp 141–159.

Braz, A. (2005) 'The modern Hiawatha: Grey Owl's construction of his aboriginal self', in J. Rak (ed) *Auto/biography in Canada: Critical Directions*, Waterloo: Wilfrid Laurier Press, pp 53–68.

Brooks, L.M. (ed) (1995) *Alternative Identities: The Self in Literature, History, Theory*, New York: Garland Publishing.

Browder, L. (2000) *Slippery Characters: Ethnic Impersonators and American Identities*, Chapel Hill/London: University of North Carolina Press.

Browder, L. (2014) 'The curious case of Asa Carter and *The Education of Little Tree*', in C. Rosenthal and S. Schäfer (eds) *Fake Identity: The Impostor Narrative in North American Culture*, Frankfurt/New York: Campus, pp 62–80.

Caison, G. (2011) 'Claiming the unclaimable: Forrest Carter, *The Education of Little Tree*, and land claim in the native south', *Mississippi Quarterly*, 64(3–4): 573–596.

Carter, D.T. (1991) 'The transformation of a Klansman', *New York Times*, 4 October. Available at: https://www.nytimes.com/1991/10/04/opinion/the-transformation-of-a-klansman.html [Accessed 26 February 2020].

Carter, F. (2004 [1976]) *The Education of Little Tree*, Albuquerque: University of New Mexico Press.

Clayton, L. (1986) 'Forrest Carter/Asa Carter and politics', *Western American Literature*, 21(1): 19–26.

Dawson, C. (1998) 'Never cry fraud: Remembering Grey Owl, rethinking imposture', *Essays on Canadian Writing*, 65: 120–140.

Dawson, C. (2007) 'The "I" in beaver: Sympathetic identification and self-representation in Grey Owl's *Pilgrims of the Wild*', in H. Tiffin (ed) *Five Emus to the King of Siam*, Amsterdam: Rodopi, pp 113–130.

Deloria, P. (1998) *Playing Indian*, New Haven: Yale University Press.

Deutsch, J.I. (2013) 'Review of *The Reconstruction of Asa Carter* by Marco Ricci and Douglas Newman', *The Journal of American History*, 100(1): 313–314.

Francis, D. (1992) *The Imaginary Indian: The Image of the Indian in Canadian Culture*, Vancouver: Arsenal Pulp Press.

Greenhaw, W. (1976) 'Is Forrest Carter really Asa Carter? Only Josey Wales may know for sure', *New York Times*, 26 August, p 45.

Grey Owl (1935) *Pilgrims of the Wild*, London: Lovat Dickson and Thompson.

Grey Owl (1989 [1931]) *The Men of the Last Frontier*, Toronto: Macmillan of Canada.

Hall, S. and du Gay, P. (eds) (1996) *Questions of Cultural Identity*, London: Sage Publications.

Italie, H. (2007) 'Disputed book pulled from Oprah's web site', *Washington Post*, 6 November. Available at: https://www. washingtonpost.com/wp-dyn/content/article/2007/11/06/ AR2007110601431_pf.html [Accessed 26 February 2020].

King, T. (2017) *The Inconvenient Indian: A Curious Account of Native People in North America*, Canada: Penguin Random House.

Marx, L. (1964) *The Machine in the Garden*, New York: Oxford University Press.

McGurl, M. (2005) 'Learning from Little Tree: The political education of the counterculture', *The Yale Journal of Criticism*, 18(2): 243–267.

New, W.H. (2002) 'Grey Owl', in W.H. New (ed) *Encyclopedia of Literature in Canada*, Toronto: University of Toronto Press, pp 453–454.

Newmann, D., Browder, L. and Ricci, M. (2011) *The Reconstruction of Asa Carter: The Greatest Story Was the One He Never Told*, [Film] 57 min. Available at: https://www.youtube.com/watch?v=5xZ_5kPli7A [Accessed 26 February 2020].

Oliver, W.J. (1928) *Beaver People*, [Film], 13 min. Ontario National Film Board. Available at: http://www.nfb.ca/film/Beaver_People/ [Accessed 26 February 2020].

Owl, G. (2019 [1931]) *Pfade in der Wildnis: Eine Indianische Erzählung von der Natur*, translated by Peter Torberg, Berlin: Die Andere Bibliothek.

Reid, C. (1991) 'Widow of "Little Tree" author admits he changed identity', *Publishers Weekly*, 238(47): 16–17.

Rosenthal, C. (2011) 'Narrative und kulturelle Kontaktzonen in Mary Rowlandson's "Captivity Narrative"', *Literaturwissenschaftliches Jahrbuch*, 52: 213–227.

Rosenthal, C. (2014) '"The wish to be a Red Indian": The Canadian dream of Grey Owl', in C. Rosenthal and S. Schäfer (eds) *Fake Identity: The Impostor Narrative in North American Culture*, Frankfurt/ New York: Campus, pp 45–61.

Rosenthal, C. and Schäfer, S. (eds) (2014) *Fake Identity: The Impostor Narrative in North American Culture*, Frankfurt/New York: Campus.

Ruffo, A.G. (1997) *Grey Owl: The Mystery of Archibald Belaney*, Regina: Coteau Books.

Sheridan, J. (2001) '"When first unto this country a stranger I came": Grey Owl, indigenous lessons of place, and postcolonial theory', in J.S. Scott and P.S. Housely (eds) *Mapping the Sacred: Religion, Geography and Postcolonial Literatures*, Amsterdam: Rodopi, pp 419–442 .

Smith, D.B. (1990) *From the Land of the Shadows: The Making of Grey Owl*, Saskatoon: Western Produce.

Stewart, F. (2018) 'Grey Owl in the white settler wilderness: "Imaginary Indians" in Canadian culture and law', *Culture and the Humanities*, 14(1): 161–181.

Strauss, A.L. (2017 [1959]) *Mirrors and Masks: The Search for Identity*, London/New York: Routledge

Warrior, R.A. (1992) 'Selling Indians: Make it painless, make it up', *Christianity and Crisis*, 13 January, 405–4.

3

A Menagerie of Imposters and Truth-Tellers: Diederik Stapel and the Crisis in Psychology

Maarten Derksen

Introduction

On 7 September 2011 the career of Diederik Stapel, social psychologist at Tilburg University in the Netherlands came to a sudden end. Twelve days earlier his colleague and friend Marcel Zeelenberg had confronted him with detailed evidence that he had fabricated results, evidence which had been meticulously collected by three young researchers of Stapel's own research group. Stapel thought he had managed to talk his way out of it, but Zeelenberg informed the vice-chancellor. After failing to convince the vice-chancellor in their first conversation Stapel, with mounting panic, travelled half the country to visit the places where he had supposedly collected his data, in order to construct a plausible story and make sure he got the details right. It was to no avail, and finally, accompanied by a lawyer, Stapel confessed: he had made up data, and had been doing so for many years. His house of cards, as he would later call it, had come tumbling down.[1] In retrospect, Stapel's confession to the

---

[1] For this chronology I have drawn on Ruud Abma's *De Publicatiefabriek*, which contains a detailed history of the Stapel affair (Abma, 2013).

vice-chancellor was a rare moment of order and consensus in a saga that is otherwise characterized by differences of opinion, conflicting stories and a lack of confidence in the discipline of psychology as a whole. While they sat at a table in the vice-chancellor's home, for a brief instant, there was clarity: Stapel confirmed that the accusations were correct, he had fabricated data, he was a fraud. But as the story hit the press and became national news, questions started to be asked again. A few people rather hopefully speculated that it might be a hoax, perhaps perpetrated by Stapel himself. These fanciful speculations were quickly squashed, but many questions remained. How long had it been going on? How many of his papers were fraudulent? A commission had been created to investigate the matter, but it wasn't just Stapel who was under scrutiny. The Stapel affair hit psychology at a time when its research practices were already increasingly being examined. A reform movement was emerging that used Facebook and Twitter as its media to discuss the ways that psychologists produce publishable results. Inevitably the question arose to what extent Stapel's fraud was emblematic of what was starting to be called the 'crisis of confidence' in psychology. Was the fact that he had gotten away with fabricating data for so long somehow symptomatic of systemic problems in psychology? Could we trust any of the results collected in the discipline's journals? Stapel's imposture fed the suspicion that psychology as a whole was an imposter.

'Stapel' became a tool to think with; to reflect on the discipline's orientation to truth and its truth practices. In this chapter I will describe how 'Stapel' fuelled debates about the state of psychology, and how in the process Stapel's imposture was construed in various ways. I am not concerned here with 'the truth' about Stapel or about psychology: my focus is on how, and by whom, different versions of the truth were presented and mobilized. In the process, different roles were invoked and performed: a whole menagerie of imposters and truth-tellers appeared, fakers and those who see through them. I will describe how these different personages were staged by the actors involved in the Stapel affair. Of particular interest will be the question of how the authority (to tell the truth) is claimed, assigned and contested: who gets to tell the truth about imposture?

Fabrication

A few days after Stapel's confession, on Friday 9 September 2011, vice-chancellor Philip Eijlander formed a commission, charged with answering two questions.[2] The first was relatively straightforward and factual: which of Stapel's publications were fraudulent? The second question was: how this had been possible? What was Stapel's modus operandi, and what had facilitated his fraud? Within two months the commission issued an interim report (Levelt et al, 2011). Helped by Stapel himself, who had supplied the commission with a list of journal articles that, as far as he could remember, relied on falsified or fabricated data, the commission concluded that 'dozens' of articles were fraudulent, going back at least to 2004. The commission also promised to publish the full list in its final report.[3] Most of the interim report, however, was devoted to the second question: how could this happen? The commission's answer was twofold. First of all, Stapel was a cunning and cynical imposter, who abused his power and status to trick his colleagues and students. The commission described his usual working methods in great detail. Together with a graduate student, Stapel would design a study that fitted in the student's overall project. Stapel would do much of the conceptual work, and leave it to the student to prepare the material for the study: stimuli, questionnaires, leaflets with instructions, and so on. Stapel would then load the material in his car, and pretend to drive to one of the schools and colleges where, he explained, he usually collected his data. Then, a few weeks later, he would hand a data set to the grad student, sometimes praising the wonderful results, and together they would write the paper and submit it to a journal. This scheme could only work, the commission added, because Stapel abused his power and prestige to turn his students into unwitting collaborators in the game. The commission claimed that he carefully picked the junior researchers he wanted to work with, and then created a strong bond with them, inviting them to his home, taking them out for dinner or to the theatre. Once part of the inner circle, feeling special, they would not think of raising the alarm. In the rare cases that someone did express doubts about Stapel's data he would demand trust and respect. According to the committee, Stapel's demeanour was 'persuasive and at the same time intimidating' (Levelt

---

[2]   There were in fact three commissions, one for each university where Stapel had worked: University of Amsterdam, University of Groningen and Tilburg University. Their work was coordinated and aggregated by the Levelt Commission, which itself focused on the Tilburg period. I will refer to the combined commission in the singular.

[3]   The number of retracted papers currently stands at 58. See http://retractiondatabase. org/RetractionSearch.aspx#?auth%3dStapel%252c%2bDiederik%2bA.

et al, 2011: 12).[4] The commission thought that Stapel's cunning and cynical manipulations were the main reason his fraud had remained undetected so long, but it also saw a second factor: the 'deficient functioning of scientific critique, the keystone of science' (Levelt et al, 2011: 9). The reasoning of the commission was simple: there was so much that was unlikely and implausible about Stapel's work that his colleagues should have spotted the fraud much earlier: professors seldom do their own data collection; the experiments were far too complicated to be done in classrooms, under the supervision of teachers; there were often statistically unlikely patterns in the data; and, perhaps worst of all, his results tended to be 'too beautiful to be true' (Levelt et al, 2011: 16). His hypotheses were always confirmed, usually with large effect sizes. Ironically, 'too beautiful to be true' was precisely how his colleagues sometimes complimented him on his results. They didn't notice the statistical oddities and the questionable research practices. The commission concluded that 'scientific criticism had not functioned adequately' in this case, and questioned whether this was a symptom of a broader problem in the scientific culture of psychology and social psychology in particular (Levelt et al, 2011: 17). In its final report, this broader problem was formulated as 'a research culture that is too much focused on uncritically confirming one's ideas and finding interesting but theoretically superficial ad hoc results' (Commissie Levelt et al, 2012: 47). The commission gave its final report the title 'Failing science'.[5] Thus, the commission had identified two imposters: the cynical, manipulative fraud Stapel, and the fake science social psychology. These are two different kinds of imposter. Social psychology's problem, as the commission described it, was not its cynicism or its cunning, but rather the reverse. Social psychologists are too credulous, too easily impressed by beautiful results, whether their own or those of others. Real science, they argued, requires a critical attitude. A 'ground rule' in science, the commission explained, is that research has to be conducted in such a way that hypotheses can be falsified (Commissie Levelt et al, 2012: 48). In Stapel's work, however, there was a clear 'verification bias': negative results were suppressed, and a variety of tricks was employed to make results statistically significant. To the commission's surprise, colleagues of Stapel indicated that these were normal practices in the field. Everything seemed to be geared to the confirmation of hypotheses and the production of an 'interesting,

---

4   All translations from Dutch to English are my own.
5   'Fail' is an intransitive verb here. 'A science that fails', not 'someone fails science'.

beautiful, short and pithy story' (Commissie Levelt et al, 2012: 53). The commission could only conclude that 'at all levels there was a general neglect of fundamental scientific standards and methodological requirements' (Commissie Levelt et al, 2012: 53), a research culture that it denoted (in English) as 'sloppy science' (Commissie Levelt et al, 2012: 47). The commission was nevertheless careful to emphasize that only Stapel bore responsibility for the fraud. The failing science around him offered no excuse for what he had done. Legal considerations may have played a role in drawing this strict distinction, which is somewhat at odds with the tenor of the report. This meant that the relation between the two imposters that the commission had constructed was a bit unstable.

The commission's authority as truth-teller rested on the enormous amount of information it had gathered in its many conversations with Stapel's colleagues and students, as well as the investigative work of the statisticians who had checked all his analyses. In addition the report had a suitably official character, with chairman Levelt ceremonially handing over a copy to vice-chancellor Eijlander at a press conference. The 'law' that it applied was the commonly accepted definition of scientific fraud: fabrication or falsification of data.[6] There was, however, also a more implicit set of rules at play, particularly in its pronouncements about the research culture in social psychology. The commission presented itself as a representative of 'real science', referring to, for example, 'the fundamental rules of proper scientific research' (Commissie Levelt et al, 2012: 47). However, whereas the commission's investigation of the extent of Stapel's fraud was unanimously praised, the commission's authority to speak for science and criticize social psychology was quickly challenged.

Not unique

The response from social psychologists was swift.[7] The European Association of Social Psychology (EASP), which put out a statement two weeks after the Levelt report was published, praised the commission's detailed investigation of Stapel's fraud and its meticulous reconstruction of 'the world of deception and influence that Stapel had set up around him' (The Executive Committee of the European Association of Social Psychology, 2012). However, it labelled as 'defamatory' and

---

[6] Plagiarism is also considered fraud, but Stapel had not committed plagiarism.

[7] See Hintum (2012) for other reactions similar to the ones I describe here.

'insulting' the commission's conclusions about the research culture in social psychology. The EASP turned the commission's condescending reproach of sloppy science back on itself: to accuse an entire field of pervasive malpractice on the basis of an investigation of one man's fraud was *itself* unscientific. It smacked of the kind of verification bias that the commission accused social psychologists of. Moreover, singling out social psychology as a sloppy science would only have been warranted had the commission compared research practices in social psychology with those in other sciences. In fact, the EASP added, just such a comparison had been published shortly before by three former colleagues of Stapel. Based on an extensive review of the literature on scientific fraud they concluded that there is no evidence that fraud is more prevalent in social psychology than in other disciplines (Stroebe et al, 2012). The authors also countered the criticism that Stapel's imposture had been so blatant that other social psychologists should have detected it long ago. This was hindsight bias: the improbabilities and defects in Stapel's work might seem obvious now, but that is only because it was now known that he was a fraud. Besides, peer review and replication, often touted as the mechanisms of scientific self-correction (including by the commission), rarely catch out frauds. The self-correcting nature of science, the authors stated boldly, is 'a myth' (Stroebe et al, 2012: 670). Writing in the *Times Higher Education Supplement* Stephen Gibson, honorary secretary of the British Psychological Society (BPS), followed much the same track. He too commended the commission for its detailed scrutiny of Stapel's papers, but objected strongly to its indictment of social psychology. Again, Stroebe et al's survey of fraud in science provided the main argument, namely that social psychology seems to be no worse in this respect than other fields. Gibson further underlined that point by referring to the work of sociologists of science like Bruno Latour and Harry Collins, 'who have shown the contingent and "messy" nature of practice across all scientific disciplines' (Gibson, 2012). Such work is needed to correct the commission's 'careless implication of "unique" deficiencies' (Gibson, 2012).

Thus, the EASP and the BPS tried to prevent Stapel's fraud from tainting the entire discipline in two ways: by re-emphasizing Stapel's cunning deception, and by reassigning the problems that the commission had localized in social psychology to all of science.[8] Like the Levelt commission, the EASP and the BPS drew on 'science' for

---

[8]   I thank David Moats for this point.

the legitimation of their claims, but whereas the commission had based its judgement on a normative philosophy of science, the defence of the social psychologists rested on *empirical* studies of scientific practice. The EASP had enough confidence in its strategy to call on the Levelt commission to 'amend its report', but no such correction was forthcoming (The Executive Committee of the European Association of Social Psychology, 2012). In a brief rejoinder, the commission acknowledged that it had not compared social psychology's problematic research culture with that of other disciplines (Drenth et al, 2013). That had simply not been part of its brief. The fact remained that there were problems in social psychology. The commission was happy to point out that many psychologists shared its concerns about the state of the field, and were giving the issue of scientific integrity the attention it deserved. If their report about Stapel's fraud had increased vigilance in the field, the commission's efforts had not been in vain (Drenth et al, 2013).

Derailment

After his confession, little had been heard from Stapel himself until the publication of the commission's final report a year later. Stapel appeared on TV, reading a short statement (Omroep Brabant, 2012). He said he had remained quiet the previous year in order not to hinder the work of the commission, but he was now ready to tell his story. He announced that later that week he would publish a book in which he tried to answer the many questions that his actions had raised, for himself as well as for others.

In the book, titled *Derailment* (*Ontsporing* in Dutch) (Stapel, 2012) Stapel meticulously dissects himself and the research culture in which he had worked in an effort to explain how he could have gone off the rails so spectacularly. He describes in detail how his imposture worked, and what his life as an imposter was like. The image of Stapel which appears is subtly different from that which the commission had sketched. Whereas the commission (as well as the EASP and BPS) saw a cynical manipulator, Stapel presents himself as a tragic, hapless figure, barely in control of his actions. First of all, Stapel contends that the process had all been far more difficult than the commission had thought: even nonexistent research takes time. Stapel had to make space in his busy schedule for this 'research'. So every now and then he would reserve a day in his agenda and he would disappear. He recounts how he would get in his car and drive around until enough time had passed for the research to have taken place. On one of those days he actually drove to the supposed location of the study. He went to Rotterdam to 'collect data' in a busy shopping street, parked his car, sat on a bench for a couple of hours staring vacantly at the shoppers, and drove back home.

Moreover, the procedure left Stapel with a lot of superfluous material that somehow had to disappear. He would dump the piles of questionaires in recycling bins. For one study he had used M&Ms as part of the experiment. To get rid of the candy he regularly transferred a bag from a box in his garage to the glove compartment of his car and ate it himself, on the road, adding to his middle-aged weight problem. Finally, the numbers that went into the spreadsheet had to make sense. The best way to do that was to put himself in the shoes of his nonexistent participants and imagine how they would answer the questions. Even so, his data sets still contained many statistical improbabilities. Over the years, increasingly disgusted with himself, Stapel explained that he became ever sloppier and started to make more and more mistakes, for instance duplicating whole rows of the spreadsheet from one data set to another.

Thus, passing as a scientist involved doing much of what scientists do anyway: preparing the study, making time to collect data, analysing the results, writing the manuscript, and so on. In practical terms almost everything about Stapel's research was real except for the source of the numbers in the data files. Arguably, Stapel was more concerned with 'the reality' of the research than his peers, because his deceit required him to attend to details about methodology and human nature that other researchers do not have to worry about because they are a taken-for-granted part of doing research. Passing as a scientist also required special attention to the claims that his Excel sheets provided evidence for. Stapel's hypotheses tended to be interesting, sometimes even eye-catching, but always plausible given the accepted knowledge in the field. They were interesting, and thus publishable, extensions of existing ideas. They were never revolutionary. They attracted attention, but not too much. This quality protected his claims from any intense scrutiny, but this was also the result of Stapel's attitude towards research. In *Derailment* he explains that he was always looking for ideas that 'had to be true'. He seems to have had a good eye for what had to be true, because some of his fraudulent studies have been successfully replicated.[9] In some cases at least, Stapel had managed to make up the truth. As

---

[9]   According to Stapel himself (2012: 177), though it is not clear what studies he is referring to.

he describes in his autobiography, this attitude that some ideas have to be true was one of the root causes of his misconduct. If the data did not conform to what had to be true, then obviously the data were at fault and had to be corrected. In the book, Stapel describes a key scene in 2004, presenting it as the moment that he became a fraud. He sits in his office looking at a data set. He is worried, because lately none of his studies has yielded statistically significant results, and journals have been rejecting his papers (Stapel, 2012: 165). This data set does not give the expected and required p-values either. He closes the door, changes a couple of numbers, then quickly undoes the changes, shocked by what he just did. Then he changes the numbers again, runs the analysis, and gets statistically significant results. '[T]he world had become logical again' (Stapel, 2012: 145). Changing numbers in a data set counts as falsification, which is commonly defined as fraud in academic regulations regarding scientific integrity. Formally speaking, Stapel had crossed a boundary, but in the book he also emphasizes the continuity with his earlier research practices. He had long been working in a grey zone, as he put it, using more or less dubious tactics to make his results look better and achieve statistical significance. So had his colleagues, he writes: how else could their results be so pretty? Like the Levelt commission, Stapel paints a picture of social psychology as a 'sloppy science', in which the rules of proper science are bent in the interest of verifying theories and furthering careers.

Stapel also construes another continuity: he had long harboured this attitude towards truth – that some things have to be true. He describes how, as a student, he became fascinated by theory: by the various explanations for behavioural phenomena, and how, by diligent and detailed conceptual work, these puzzles could be solved and a logical, coherent picture of reality be created.[10] He would draw theoretical schemes with boxes and arrows between them on sheets of paper that he stuck together. He would lie awake at night mulling over theoretical ideas. He would sit behind his desk for eight, nine hours at a stretch solving theoretical riddles, in total flow, feeling high, 'tripping in my own universe of ideas, concepts, and cross-connections' (Stapel, 2012: 139). His empirical work, on the other hand, was quick and slap-dash, favouring simple questionnaires that often combined items testing multiple hypotheses. 'Guerilla research', he called it (Stapel, 2012: 141). Moreover, these experiments served

---

[10] Hub Zwart (2017: 233) speaks of Stapel's 'epistemological desire' for order in a messy world.

to confirm ideas, not to falsify them. 'Whatever way we arrive at our theories and hypotheses, the experiments and tests we design are made to verify, not to falsify our conjectures' (Stapel, 2000: 6).[11] Stapel describes how, as his career progressed, there was an important change in his orientation to truth and his fascination with order. As a young researcher he had been convinced that social psychology is a science of exceptions, full of little theories – theorettes, he called them – that are only conditionally true (Stapel, 2012: 153). In social psychology, everything is situational, everything specific. His mission was to bring order and structure, by connecting all the pieces of the puzzle. Integration was the goal (Stapel, 2000). Over the years, however, it became clear to him that the complicated models he produced were not popular with his colleagues. They mocked him behind his back for his byzantine theoretical constructions. Perhaps his theories were complete, but they were also imposible to understand and, worse than that, boring. Stapel began to notice that, while he was respected, he was not a star. People didn't stay around after his conference talks, he was never invited to give keynote lectures. As he describes it in his book, his narcissistic hunger for admiration drove him to pursue a new kind of order: that of simplicity.

What remained the same was his conception of truth as ideal, transcendent. Truth had quasi-religious overtones for Stapel. A theory that connected results in a way that made sense and created order was 'beautiful' as well as 'logical'. Stapel notes that he used to have a reproduction of Fra Angelico's *The Annunciation* in his office. It is a painting that moves him because, he writes, it depicts the momentary connection of 'the heavenly, the whole, with the earthly, the broken and unfinished' (Stapel, 2012: 277). He claims to feel a similar emotion when science manages to give meaning and structure to the earthly chaos, and so gives us 'rest and insight' (Stapel, 2012: 277).

Given this orientation to abstract order, logic and beauty, this obsession with what *has to be* true, it is not suprising that from early on in his career Stapel employed a number of tricks to make the data conform to his theories: selectively removing outliers, only reporting the measures that showed an effect, leaving out groups or conditions that hadn't worked, and so on, all in the interest of showing evidence for what had to be true because it was so logical and beautiful. Falsifying data points was simply the next step.

---

[11] Zwart (2017) reads this as an expression of his budding disenchantment with the discipline.

Although he emphasizes that he alone was responsible for his actions, Stapel's explanation for his misconduct highlights the interaction between a competitive academic culture and his own personality. He had learnt early on that he had to score points in order to get ahead: publish or perish but also sell or sink. Modern scientists, he writes, are 'sales managers who have to reach objective publication targets' (Stapel, 2012: 138). He adds, however, that such targets and the praise one receives when one reaches them are particularly attractive for people who lack self-confidence, such as himself. 'Clear applause feels good, especially if you're not sure whether what you do and who you are is worth anything' (Stapel, 2012: 138). He doesn't use the term, but Stapel appears to have suffered from imposter syndrome.

Writing about the later stages of his derailment, Stapel describes himself as a junkie, addicted to a toxic mix of beautiful truth and the admiration he received for his beautiful results. As the years went on he took ever more risks to get his hit. In his account he emerges as a tragic imposter: not a comical con man like Giacomo Casanova or Frank Abagnale (depicted in the movie *Catch Me If You Can*). In his book he emphatically accepts full responsibility, but he also emphasizes how addicted he was. And he was not a happy addict, rather he presents himself as conflicted, tormented by guilt and shame.

Stapel dealt with this conflict by avoiding it. He never stayed long at research meetings where his students presented their work based on his fake data, and afterwards he would take a long walk or lock himself in his room, so that he could shift his attention to other, new projects and repress the mounting desperation. He avoided talking about his experiments with colleagues, preferring to discuss theory. And as the world he created with his data became ever more orderly and beautiful, he started to believe in it more and more. 'I could only survive by keeping it to myself and strongly believing in it' (Stapel, 2012: 177). He describes it as an intensely lonely existence.

Imposters and wannabes

In *Derailment*, Stapel presented himself as an imposter who had turned truth-teller. He could tell the truth because he had lived it: he knew what it took to be a scientific fraud, and he knew the research culture of social psychology inside out. Now that he no longer had to keep up the pretence and had nothing more to lose, he could tell the unvarnished truth. To underline his transformation into truth-teller, he did not use the name under which he had published his scientific articles (Diederik A. Stapel) but instead called himself 'Diederik Stapel', adding '[t]hat's someone else' (Stapel, 2012: 292). The persuasiveness of this performance depended on balancing identity (he was the Stapel who had committed fraud, he knew what it was like) with difference (he was no longer a fraud, and he could now speak the truth freely). Responses to this narrative were decidedly mixed. Although the national media were fascinated by the man behind the fraud, many of Stapel's Dutch colleagues were disgusted and simply refused to read the book. To them he remained an imposter, not a truth-teller. In my own department someone had photocopied the book, so that no one would have to buy it and Stapel would not make any money from it. Those who did comment on the book derided what they saw as the author's narcissism, a term that was also used in several book reviews in the national press (Brussen, 2012; Somers, 2012).

What attracted more attention was Stapel's critique of academia and the research practices in his field. Developmental psychologist Willem Koops wrote a scathing book review in which he mocked the autobiographical parts of the book – 'rather uninteresting private information' – and advised the author to try his luck at writing novels (Koops, 2013). But whereas Stapel's self-presentation merely merited derision, Koops was livid about Stapel's description of systemic problems in the discipline. His suggestion that questionable research practices are common in psychology, that all psychologists are imposters like him, was nothing but a slanderous attempt to shift the blame.

Others, however, saw Stapel's revelations about the way he and his colleagues practised science as confirmation that there were serious problems in academia and social psychology in particular. Stapel had become entranced by the 'competitive world of glitter' in academia, and without fundamental reforms there will be other fraud cases, wrote biochemist Frank Miedema (2013). Psychologists Denny Borsboom and Eric-Jan Wagenmakers called *Derailment* 'devastating, and a must-read for anyone with an interest in science' (Borsboom and Wagenmakers, 2012). They noted that Stapel's book underscored the conclusions of the Levelt commission, and that it suggested that the problems were actually wider than just social psychology.

Cognitive neuroscientist Chris Chambers devoted a chapter to fraud in his book *The Seven Deadly Sins of Psychology: A Manifesto for Reforming the Culture of Scientific Practice* (Chambers, 2017). He too portrayed Stapel as an extreme symptom of a much wider problem, rather than setting him apart as an individual rotten apple. Drawing on Stapel's autobiography he noted that data fabrication had simply been 'an extra step' after years of employing questionable research practices to score publishable results (Chambers, 2017: 102). Chambers challenged his readers to define the distinction between fraud (fabrication and falsification of data) and questionable research practices such as discarding studies that do not confirm the hypothesis. 'Unless we define scientific fraud narrowly as (only) the kind of extreme fabrication perpetrated by Stapel, it is difficult, if not impossible, to distinguish the blackest of fraudulent acts from a continuum of dishonest practices at the individual and group level' (Chambers, 2017: 106). Stapel's actions had been extraordinary, but, Chambers notes, many psychologists are to some extent engaging in imposture. It's even possible that they don't know that what they are doing is wrong.[12] In this interpretation, which very much resembles the conclusion of the Levelt commission, psychologists are not low-grade imposters, less extreme versions of Stapel, rather they are wannabes: people who have fooled themselves into thinking that they are doing science, but do not understand what science really involves.

Transparency

As the Levelt commission's final report had noted, 'Stapel' was not the only sign that something was deeply wrong in psychology. There were many other events in 2011 and 2012, two extraordinarily bad years for psychology, that seemed to point to the same conclusion. It was enough to make people speak of a 'crisis' in psychology. A community of reformers emerged, that assembled on Twitter and Facebook to discuss the problems in the field and how to solve them.[13] They argue that academia has become extremely competitive, and researchers are under pressure to produce as much 'output' as they can, preferably in high-impact journals so that their papers will get cited more.[14] Journals, in order to boost their impact factor, prefer papers that report interesting, unexpected, 'sexy' results, that will attract readers who will then cite those papers. Researchers must therefore produce such results in order to have a chance of publication, and when their experiments do not 'work', they may wittingly or unwittingly resort

---

[12] Reflecting on Stapel's autobiography, biological psychologist and methodologist James Heathers proposed that researchers fiddling with the numbers in order to get significant results are mostly people who 'manage to bullshit themselves'. They 'try options until something works. And when it works, it must be right, because that's what they were expecting!' (Heathers, 2014). In other words, they are incompetent rather than malicious.

[13] See chapter 9 of Derksen (2017) for a brief history of the crisis.

[14] See for example Nosek et al (2012) and the other papers in that special issue.

to questionable research practices. As a consequence, journals are littered with unreliable results. Since journals are not interested in replication studies (too boring) these false claims never get corrected. What is needed, the reformers suggest, is first of all a saner incentive structure in science: less emphasis on quantity, more on quality of publications. Secondly, the importance of replication studies should be recognized. Although replication is not an instrument to detect fraud, as Stroebe et al (2012) have stressed, reformers do see it as essential to determine the reliability of results. Thirdly, the statistical techniques that are prevalent in the discipline must be examined. Problems with null hypothesis significance testing in particular has been the topic of intense debate, and Bayesian approaches are often put forward as an alternative. But the overriding theme in all the proposals of the reformers is *transparency*. Reformers advocate opening up the research process in ways which facilitate collaboration but also mutual scrutiny. Sharing the data of one's study for example can be done via an online platform such as the Open Science Framework (https://osf.io) and allows others to use this data for other projects, but also to check whether re-analysis yields the same results. Another example of this coveillance function of transparency is pre-registration. In order to assure other researchers that the hypothesis tested in a paper is really the one that the researcher set out to test, and not a 'hypothesis' fitted to the data after they were known, researchers can register their hypotheses on the Open Science Framework (or similar platforms) before they start collecting their data. Sceptics can later check the date of the registration to see whether the hypothesis conforms to the one described in the published paper.

For reformers, an important function of transparency is therefore to prevent imposture and keep researchers honest. Requirements such as pre-registration and data sharing make deception, if not impossible, then at least more difficult. Advocates, in this context, often make a distinction between science and 'story telling'. As Nosek et al wrote in an influential paper: 'The current default practice is to tell a good story by reporting findings as if the research had been planned that way' (Nosek et al., 2012: 625). As they note, this was in fact the explicit advice in a famous book chapter by social psychologist Daryl Bem (2004). The conventional view of science is that researchers first derive hypotheses from theory, then do an experiment to test these hypotheses, and finally analyse the data to see whether or not the hypotheses are confirmed. What you should do, however, Bem advised his readers, is the reverse: look closely at the data first, see if there are interesting patterns, and 'try to reorganize the data to bring them into bolder relief' (Bem, 2004: 187). Having crafted an interesting result, you should then construct a 'compelling framework' through which to present it (Bem, 2004: 188). No one is interested in the actual run of events in your study, Bem warned his readers: what people want to know is what you have learned from your data.

Although reformers strongly object to Bem's advice to beginning social scientists, at the same time they recognize that a scientific paper is always to some extent a story (Vazire, 2014). There is always a tension between faithfully reporting the research process on the one hand, and not burdening the reader with all the dead ends and a messy, unstructured heap of findings on the other. A similar tension occurs with respect to the frankness that reformers advocate. To them, the ideal scientist is the opposite of the imposter Stapel: whereas Stapel kept to himself as much as possible and avoided discussion about his work, reformers see science as an essentially open, transparent enterprise, in which collaboration and mutual criticism go hand in hand – what Karl Popper once called 'friendly-hostile collaboration' (Popper, 2002: 489; Derksen, 2019). But how can scientists practically work together in a manner that is both friendly and hostile? How to marry critique and community? In 2014, for example, there was a row over the replication of a study by psychologist Simone Schnall and colleagues (Schnall et al, 2008) which had failed to find the original effect (Johnson et al, 2014). Schnall was extremely upset by the way some of her colleagues responded on social media.

> There now is a recognized culture of 'replication bullying': Say anything critical about replication efforts and your work will be publicly defamed in emails, blogs and on social media, and people will demand your research materials and data, as if they are on a mission to show that you must be hiding something. (Schnall, 2014a)

She thought the Stapel affair had fostered a 'crime control mindset' in psychology, where failed replications are taken to be an indication that the original results were due to questionable research practices or outright fraud. In reality, she said, failures to reproduce results are often due to a lack of expertise in that field of study – as had been the case in the replication of her own work. Schall stressed that you cannot 'crowd-source' science, you need real experts (Schnall, 2014b).

Schnall's mentioning of expertise shows that the affair was not only about the tone of the attacks, but also about authority. The status of the reformers as truth-tellers derives in part from the very problems they identify, including failed replications. Every time the results of an earlier study cannot be reproduced it is, at least potentially, further support for the reformers' claims about the problems in the discipline. At issue is not only questions about the role of replication in sorting truth from falsehood, but also who can decide on the interpretation of specific replication failures. If the latter requires special expertise, only available to researchers in that particular field or even in that particular line of research, the authority of the reformers is severely affected. They would in fact become imposters themselves, loudly but falsely pretending to be experts.

Thinking with imposters

In the discussion about the Stapel affair, and the problematic research practices in psychology more generally, a range of imposters and truth-tellers have emerged. The first one is the simplest: Stapel was a cynical, manipulative con man, who pulled off his deception by tricking his students and co-authors. This was the imposter described in the Levelt report, and it has remained the figure that is most prominent in the public perception of the affair. Stapel, on the other hand, presented himself as a tragic persona, ensnared by his own imposture, addicted to its successes and unable to quit. He was responsible for his crimes, but at the same time a victim of circumstances. In an academic culture of publish or perish he had taken a wrong turn in order not to perish, and never had the strength to find the way back. This figure is more complex than the first, and it has not been received as well as the first. This imposter, moreover, does not offer a straightforward explanation of what happened – Stapel has related how people keep asking him why he did it, even after he tells them his story (Stapel and Dautzenberg, 2014). A third imposter figure can be found in social psychology depicted as the wannabe scientific discipline in the Levelt commission. This is a common trope: physicist Richard Feynman (1974), for example, once called psychology a 'cargo cult science', one that tries to mimic the appearance of true science (physics, of course) in the hope of getting similarly impressive results. This persona was vigorously rejected by social psychologists, but has become a common fixture in discussions among reformers about the deplorable state of psychology. This is a somewhat comical imposter, a clueless phony who is unaware of its own fakeness. These imposters have been joined by their converse: several kinds of truth-tellers have made their appearance in the Stapel affair and the larger crisis that it is a part of. The whistleblower is the mirror figure of the cynical imposter: the good guy/girl who unmasks the evil pretender. Whereas the imposter abuses his/her power, the whistleblower speaks truth to power. The three whistleblowers who revealed Stapel's fraud have been universally praised for their courage and tenacity. Whistleblowers tell a truth that is inconvenient, not just to the imposter, but also to those in power more generally. These three (students and post-docs), in fact, have preferred to remain anonymous for fear of damaging their careers.[15] Recently several reformers have taken on a more permanent role as whistleblowers, making uncovering dodgy research and fraud part of their daily work.[16] The Levelt commission was appointed as an official truth-teller, charged with investigating the allegations and giving a verdict on what is true and false in Stapel's work. This part of the commission's work has been widely praised for its thoroughness and factuality. But the commission's conclusions regarding 'the methods and research culture that may have facilitated this infringement' (Commissie Levelt et al, 2012: 7) have proven to be more controversial. Social psychologists rejected the claim that theirs is a uniquely sloppy science, and turned the commission's appeal to science against itself: the commission's own methods had been sloppy, and empirical studies of science have shown that fraud occurs in all fields. Thus, the commission has not been entirely successful in its bid to be a neutral arbiter of truth. Instead, it was briefly pulled into the fray, having to defend its conclusions against critics who themselves claimed to have the backing of science.

Other truth-tellers are also visible, albeit more dimly, in the debate about Stapel and the crisis. In the Levelt report, in the discourse of reformers, and in his own book, Stapel is presented as a warning signal: his fraud, and particularly the long duration of it, are an indication that there is something very wrong in the field that he was a part of. Earlier in 2011 another warning signal had flashed: Daryl Bem, the one who advised young psychologists to construct a good story out of their data, had managed to get a study of precognition published in one of the flagship journals of social psychology (Bem, 2011). To the burgeoning reform movement this was a clear indication that something was amiss with the discipline's standards for publication.

---

[15] Only recently one of them, Yoel Inbar, has made himself known (Inzlicht and Inbar, nd).

[16] The inquiries of Nick Brown and James Heathers, for example, have led to numerous corrections and retractions, as well as uncovering two cases of fraud (Marcus and Oransky, 2018). Nick Brown has translated Stapel's book into English, making the book accessible to an international audience: http://nick.brown.free.fr/stapel/.

Like Stapel, Bem functioned as both an imposter and a truth-teller, the very success of their imposture interpreted as a warning signal about the state of the field (unlike Stapel, however, Bem has not repented). A final truth-teller can be glimpsed, very briefly, in Stapel's book *Derailment*. When Marcel Zeelenberg confronts Stapel with the accusations of the whistleblowers, Zeelenberg also tells him that 'all over the country they were gossiping about [him] and when [he] gave a presentation of [his] research, the eyebrows went up in the audience' (Stapel, 2012: 33). Stapel, he suggests, had been an elephant in the room: a truth seen by everyone, but spoken of by none. The elephant in the room is perhaps the most uncomfortable figure in this menagerie of imposters and truth-tellers, as it points to the complicity of all in the crimes of one.

The crisis in psychology that Stapel helped to create has forced psychologists to reflect on honesty and deception, appearance and reality, truth and storytelling, frankness and civility. The various imposters and truth-tellers that made their appearance in these debates imply different ways of drawing such distinctions. The rotten apple has simple boundaries, implies well-defined roles, and suggests a clear explanation. If Stapel was a cynical con man, the people he worked with and the discipline as a whole are blameless victims, and the responsibility for what happened is entirely his. The Stapel affair, however, has resisted categorization as a straightforward case of imposture in a scientific culture devoted to truth. Yes, Stapel was a fraud who made up data sets and presented them as findings. But the fantasies he peddled were almost real, never too surprising lest they raise suspicion, and in some cases reproducible in conventional studies. Stapel tried and sometimes succeeded in making up the truth, raising uncomfortable questions about the relation between truth and fiction in psychology. Moreover, the way he worked was arguably not all that different from the common research practices in his field. He was an extreme, he had drifted into the black on a grey scale where many of his colleagues were not exactly lily white either. If questionable research practices are rife, imposture becomes a matter of degree, and loses its distinctness.

The response of reformers has been to create an infrastructure that facilitates and enforces openness. To some extent their discourse is structured by the familiar oppositions of truth versus fiction and the like. Reformers would really like to redraw the distinctions that in their view have become blurred in psychology. Transparency, they argue, will allow us to distinguish exploratory from confirmatory research, proper hypothesis testing from p-hacking, and reproducible results from false positives. But in reflecting on cases of fraud and questionable research practices, on failed replications, and on the discussion itself, such distinctions are not self-evident. Transparency does not give an unimpeded view of the naked truth, a scientific paper is not simply a report but always also a story, and frankness can itself be a pose that hinders rather than helps the exchange of views.

The kinds of imposters and truth-tellers that were constructed and performed in the Stapel affair are common cultural figures, not specific to science or psychology. What does set this case apart, it seems to me, is the extent to which the participants exercised reflexivity (in the sense of seeing one's self as also potentially subject to claims they make about others). The Levelt commission's report was perhaps the least reflexive, reporting the facts of the matter as revealed by its investigation, and passing judgement on the basis of principles of scientific method and integrity that it assumed to be generally accepted. But most other interventions in the affair have a reflexive aspect. The EASP and BPS, for example, criticized the methods of the commission as unscientific by its own standards, and referred to several kinds of studies *of* science, including the literature on scientific fraud and the work of Harry Collins, to relativize the commission's portrayal of social psychology. The reformers in psychology, for their part, have read the Stapel affair as a case study of problems in their discipline that they have experienced themselves. Moreover, although the solutions they propose are often based on a similar philosophy of science to that of the Levelt commission, an increasing number of philosophical, historical and sociological perspectives on science have been brought to bear on the issue. By far the most reflexive figure in this drama, however, is Stapel himself. Not only has he subjected himself to a painful self-examination, he has, more than anyone else, reflected on the affair itself. He has done so in psychological terms: in *Derailment* Stapel regularly refers to social psychology to explain his own actions and the reactions of his former colleagues, the press and the public. The book contains a lengthy annotated bibliography detailing all the psychological studies and theories referred to in the chapters. Ironically, some of these theories have since perished in the replication crisis.

Epilogue

In the discussion about the state of psychology, Stapel has been good to think with, but, much to his chagrin, few of his colleagues have been willing to talk with him. In 2013 he started a correspondence with novelist, journalist and provocateur Anton Dautzenberg, who had become a bit of an outcast himself after joining an association of paedophiles, in support of their right to express their desires. The two developed a friendship and started work on a play about reality, fantasy and fiction. Despite their own enthusiasm, most theatres refused to stage their play, citing the controversies surrounding Stapel, Dautzenberg, or both. In the end they gave up, and instead published their correspondence as a book, titled *The Fiction Factory* (Stapel and Dautzenberg, 2014). Shortly after publication of the book, Dautzenberg and Stapel were both hired by Fontys Academy for Creative Industries, in their mutual home town of Tilburg.[17] Stapel was asked to teach about 'the dark side of fame and success' (Bartol, 2014).

On Retraction Watch, a web site devoted to keeping track of scientific fraud and the retraction of scientific papers in particular, a vigorous debate took place about Stapel's new job (Marcus, 2014). Most commenters expressed their outrage at the fact that a known fraudster had been given an opportunity to teach and benefit financially. Defending Stapel was a lone poster called 'Paul', who argued that Stapel might have valuable lessons to teach about fraud and failure. 'Paul' quickly raised suspicion, and the owners of Retraction Watch contacted him at his email address to find out who he was. Eventually Stapel confessed that he was 'Paul'. He had been 'sock puppeting' – internet lingo for using a fake identity online to speak about yourself. One more imposter had been added to the menagerie.

References

Abma, R. (2013) *De publicatiefabriek: over de betekenis van de affaire-Stapel*, Nijmegen: Vantilt.

Bartol, R. (2014) 'Fraudeprofessor Diederik Stapel aan de slag als docent bij Fontys in Tilburg', *Omroep Brabant*, 1 October. Available at: https://www.omroepbrabant.nl/nieuws/182539/Fraudeprofessor-Diederik-Stapel-aan-de-slag-als-docent-bij-Fontys-in-Tilburg [Accessed 15 December 2019].

Bem, D.J. (2004) 'Writing the empirical journal article', in *The Compleat Academic: A Career Guide* (2nd edn), Washington: American Psychological Association, pp 185–219.

Bem, D.J. (2011) 'Feeling the future: Experimental evidence for anomalous retroactive influences on cognition and affect', *Journal of Personality and Social Psychology*, 100: 407–425.

---

[17] When the board of Fontys found out about it a week later, Dautzenberg's contract was immediately terminated. Stapel quit in solidarity with his friend (Redactie, 2014).

Borsboom, D. and Wagenmakers, E.-J. (2012) 'Derailed: The Rise and Fall of Diederik Stapel', *The Observer*, 27 December. Available at: https://www.psychologicalscience.org/observer/derailed-the-rise-and-fall-of-diederik-stapel [Accessed 12 January 2021].

Brussen, B. (2012) 'Bert Brussen bespreekt: "De narcistische jammerklacht van Diederik Stapel"', *Volkskrant*, 3 December. Available at: https://www.volkskrant.nl/gs-b0dbf128 [Accessed 17 December 2021].

Chambers, C. (2017) *The Seven Deadly Sins of Psychology: A Manifesto for Reforming the Culture of Scientific Practice*, Princeton: Princeton University Press.

Commissie Levelt, Commissie Noort, Commissie Drenth (2012) *Falende wetenschap: De frauduleuze onderzoekspraktijken van sociaal-psycholoog Diederik Stapel*, Tilburg University, University of Amsterdam, University of Groningen.

Derksen, M. (2017) *Histories of Human Engineering: Tact and Technology*, Cambridge: Cambridge University Press.

Derksen, M. (2019) 'Putting Popper to work', *Theory & Psychology*, 29: 449–465. https://doi.org/10.1177/0959354319838343.

Drenth, P.J.D., Levelt, W.J.M. and Noort, E. (2013) 'Rejoinder to commentary on the Stapel fraud report', *The Psychologist*, 26: 80–81.

The Executive Committee of the European Association of Social Psychology (2012) 'Statement EASP on Levelt report'. Available at: https://sites.uncc.edu/editorethics/wp-content/uploads/sites/17/2014/07/Statement-EASP-on-Levelt_December_-2012.pdf [Accessed 12 January 2021].

Feynman, R.P. (1974) 'Cargo cult science', *Engineering and Science*, 37: 10–13.

Gibson, S. (2012) 'Don't tar discipline with Stapel brush', *Times Higher Education*, 20 December. Available at: https://www.timeshighereducation.com/dont-tar-discipline-with-stapel-brush/422194.article [Accessed 12 January 2021].

Heathers, J. (2014) 'Diary of a bad man (Pt. 1)', James Heathers. Available at: https://medium.com/@jamesheathers/diary-of-a-bad-man-pt-1-a5ccbf595c2c [Accessed 8 May 2018].

Inzlicht, M. and Inbar, Y. (2018) 'The replication crisis gets personal'. Two psychologists, four beers, podcast. Available at: https://fourbeers.fireside.fm/4?t=157 [Accessed 30 November 2018].

Johnson, D.J., Cheung, F. and Donnellan, M.B. (2014) 'Does cleanliness influence moral judgments?' *Social Psychology*, 45: 209–215. https://doi.org/10.1027/1864–9335/a000186.

Koops, W. (2013) 'Zijn wij allemaal Diesjes? – Diederik Stapels dubbelzinnige mea culpa', *Acad. Boekengids*, 97. Available at: https://www.nederlandseboekengids.com/abg97/ [Accessed 12 January 2021].

Levelt, W.J.M., Groenhuijsen, M.S. and Hagenaars, J.A.P. (2011) 'Interim-rapportage inzake door Prof. Dr. D.A. Stapel gemaakte inbreuk op wetenschappelijke integriteit', Tilburg: Tilburg University.

Marcus, A. (2014) 'Curtain up on second act for Dutch fraudster Stapel: College teacher', *Retraction Watch*, 3 October. Available at: https://retractionwatch.com/2014/10/03/curtain-up-on-second-act-for-dutch-fraudster-stapel-college-teacher/ [Accessed 15 December 2019].

Marcus, A. and Oransky, I. (2018) 'Meet the "data thugs" out to expose shoddy and questionable research', *Science Mag*, 14 February.

Miedema, F. (2013) 'Wetenschap op verkeerd spoor', *DUB*, 18 January. Available at: https://www.dub.uu.nl/nl/opinie/wetenschap-op-verkeerd-spoor [Accessed 12 January 2021].

Nosek, B.A., Spies, J.R. and Motyl, M. (2012) 'Scientific utopia: II. Restructuring incentives and practices to promote truth over publishability', *Perspectives on Psychological Science*, 7: 615–631.

Omroep Brabant (2012) 'Diederik Stapel leest verklaring voor over zijn fraude', 2 min. Available at: https://www.youtube.com/watch?v=VA50L-fzExI [Accessed 12 June 2020].

Popper, K.R. (2002) *The Open Society and Its Enemies* (7th edn), London: Taylor & Francis.

Redactie (2014) 'Stapel en Dautzenberg alweer weg bij Fontys Hogeschool' *Volkskrant*, 8 October. Available at: https://www.volkskrant.nl/gs-b388933b [Accessed 16 December 2019].

Schnall, S. (2014a) 'An experience with a registered replication project', *Character Context*. Available at: http://www.spspblog.org/simone-schnall-on-her-experience-with-a-registered-replication-project/ [Accessed 25 May 2014].

Schnall, S. (2014b) 'Social media and the crowd-sourcing of social psychology', *Cambridge Department of Psychology Blog*. Available at: http://www.psychol.cam.ac.uk/cece/blog [Accessed 17 February 2015].

Schnall, S., Benton, J. and Harvey, S. (2008) 'With a clean conscience: Cleanliness reduces the severity of moral judgments', *Psychol. Sci.*, 19: 1219–1222. https://doi.org/10.1111/j.1467-9280.2008.02227.x

Somers, M. (2012) 'Autobiografie Stapel is een narcistisch mea culpa', *NRC Handelsbl*, 30 November. Available at: https://www.nrc.nl/nieuws/2012/11/30/autobiografie-stapel-is-een-narcistisch-mea-culpa-a1468476 [Accessed 11 July 2019].

Stapel, D.A. (2000) 'Moving from fads and fashions to integration: Illustrations from knowledge accessibility research', *European Bulletin of Social Psychology*, 12: 4–27.

Stapel, D. (2012) *Ontsporing*, Amsterdam: Prometheus.

Stapel, D. and Dautzenberg, A. (2014) *De fictiefabriek: een bevrijdingsroman in brieven*, Amsterdam: Uitgeverij Atlas Contact.

Stroebe, W., Postmes, T. and Spears, R. (2012) 'Scientific misconduct and the myth of self-correction in science', *Perspectives on Psychological Science*, 7: 670–688. https://doi.org/10.1177/1745691612460687.

van Hintum, M. (2012) 'Levelt maakt ten onrechte gehakt van ons', *Volkskrant*, 15 December. Available at: https://www.volkskrant.nl/gs-bb1dba0f [Accessed 17 December 2019].

Vazire, S. (2014) 'Life after bem', *Sometimes Im Wrong*. Available at: https://sometimesimwrong.typepad.com/wrong/2014/03/life-after-bem.html [Accessed 13 December 2019].

Zwart, H. (2017) *Tales of Research Misconduct, Library of Ethics and Applied Philosophy*, Cham: Springer International Publishing.

4

Learning from Fakes:
A Relational Approach

Catelijne Coopmans

Each society, each generation, fakes the thing it covets most.
(Jones, 1990: 13)

Introduction

Imposters, as the contributions to this volume show, are promising figures to think with. Such thinking can take multiple forms and deliver various kinds of pay-offs for how we make sense of socio-material relations, of society. This chapter is about non-human imposters. The pay-off it aims to deliver comes from putting these 'imitations of something cared about' at the centre of the analysis, as protagonists rather than receptacles or intermediaries of the action. Doing this allows us to let them *teach us* – chiefly about two things: (1) what, in a given setting, is *valued*, and (2) how deception and its interception are distributed across ever-changing socio-material alliances.

Thinking with non-human imposters – which from now on I'll call *fakes* – in this way starts by treating them as a recognizable class of objects, or more precisely: as objects recognized in the settings in which they operate as playing the part of 'resembling the real thing but not being it'. Fake museum objects, online reviews, educational certificates, specialty coffee and archaeological artefacts are examples of such objects. Each will be featured in this chapter as the central actor in a specific story. This analytical move is warranted by how, in semiotics and its uptake in actor-network theory, an *actant* is defined as '[w]hatever acts or shifts actions, action itself being defined by a list of performances through trials' (Akrich and Latour, 1992: 259).[1] Fakes contribute to the action by bringing friction. 'Resembling the real thing but not being it' is ripe with narrative tension. Putting fakes at the centre of the action means we can learn from their adventures, from the responses to the appearance-reality puzzles they embody and the socio-material defences and countermeasures they evoke and anticipate. This is what I mean by 'a relational approach': following the fake brings various relations into relief. In and through their encounters, fakes can teach us how some corner of our world operates and is stitched together.

Such an approach might raise red flags for some, because it means solidifying and granting an identity to what many critical social scientists would keep in the realm of contested designations. 'Fake' tends to come with a question mark, so as to avoid naturalizing or granting uninterrogated authority to its opposites: the real, natural, authentic, genuine, true, and so on. Winnie Wong's book *Van Gogh on Demand* (2013) is a good example of a critique of what the contrast between 'real' Western art and 'fake' Chinese art copies does to normalize and perpetuate relations of inequality, appropriation and exploitation. Critical thinking about fakes in this sense involves destabilizing or questioning the boundary between fake and real as it is rhetorically and materially effected. Who benefits from effecting it in this way? What relations of difference and hierarchy are reproduced by holding it in place? These are important questions, without a doubt. Yet there is one thing such an approach does not allow for: keeping the spotlight on fakes, studying what they do in the world *qua* fakes.

Substantiating fakes in the way that I propose here, that is, treating them as objects recognized as playing the part of 'resembling the real

---

1   For science and technology studies (STS) scholars, semiotics' main attraction has been its study of meaning, not as the consequence of some direct correspondence between words and objects or matters of fact, but as a result of the relations between signs, whereby '[t]he value of any one sign depends on that of the others to which it relates' (Mol and Mesman, 1996: 429). STS scholars have extended semiotics from a study of 'signs' and 'linguistic meaning building' to one that examines the relations between all sorts of entities in the process of 'order building'. Instead of finding out what words make each other mean, the question becomes 'what elements, of whichever character, associated in whichever way, make each other *be*' (Mol and Mesman, 1996: 429). This general formulation has been worked out in different ways in actor-network theory/material semiotics (Law, 2009) and allied approaches, such as the textual analysis of Steve Woolgar (1991).

thing but not being it', doesn't question the boundary between fakes and their opposites in the same way, but it does harbour a form of methodological symmetry.[2] In 'Does meat come from animals?' (2015) Emily Yates-Doerr tells the ethnographic story of a Guatemalan family preparing and eating a festive dish of peppers stuffed with ground beef made from soy. She argues that understanding the substance at the heart of the meal as 'imitation or artificial meat' misses the point: the way this substance is presented and treated, both in the market and in the family's home, *renders* it meat. 'Real' meat, in other words, 'can be classified through priorities other than origin' (2015: 312). Yates-Doerr's analysis, like Wong's, destabilizes the fake–real dichotomy, but it also makes it possible to see a symmetrical line of analysis that, if there were ethnographic materials to support it, could elaborate things or substances that in their presentation and treatment were *rendered fake*.

This is where my contribution to thinking with imposters is situated. Following fakes as protagonists allows them to be rendered in their socio-material relations: as things that circulate, that slip through the net, that get stopped, whose suspected or confirmed presence evokes defensive action; as things that 'do' difference and similarity, that *teach* about difference and similarity by being, in their presentation and treatment, *not quite* what was expected, hoped for or wanted. This idea of fakes as teachers underlies the two-part framework introduced in this chapter: a framework that purports to be applicable to many different kinds of fakes and that follows them into the relations they engender and are rendered by. Relations that are specific to each case or example, yet also circumscribed by the fake's 'fakeness'; that focus on what fakes do and bring about in their role of 'resembling the real thing but not being it'.

The quote from Mark Jones at the start of this chapter provides the basic contours of what follows. Let's look at that quote a little more closely: 'Each society, each generation, fakes the thing it covets most.' In his foreword to the catalogue of the first-ever British Museum exhibition on the subject of fakes, Jones suggests that over the course of history, the waxing and waning of different kinds of fakes provides 'an ever changing portrait of human desires' (Jones, 1990: 13). In the European Middle Ages, relics of saints with miraculous powers were sought after; in the Renaissance, items from ancient Greece and Rome became hot collectibles; in the 18th century, works and things associated

---

[2]   Thanks to Else Vogel for helping me see the methodological symmetry I speak of in this paragraph.

with famous people started to become desired possessions. What was coveted was faked, and as reports of fakery spread, this in turn sparked scrutiny, warnings and protections surrounding such goods. At the time of the catalogue's publication in 1990, brand-named goods – Louis Vuitton and the like – were highlighted as the era's most coveted and faked. Today's most age-defining fakes are probably in the sphere of social and traditional media: fake news (Ball, 2017), *deepfakes* (Toews, 2020) and the simulation of online influence and popularity through 'likes', 'clicks' and 'followers' (Lindquist, Chapter 12, this volume).

Besides a way to view history, what is offered by Jones' naming of faking and coveting in one breath is also a methodological principle: the presence of fakes is indicative of coveting. Fakes can teach us what we value: by playing the role of the similar-yet-different, they force us to define and sometimes refine – through the socio-material work of scrutiny, warning and protection – what it is about the coveted thing we covet so. A second methodological principle can be derived from Jones' suggestion that fakes, like fashions, change *over time*. The landscape in which fakes operate is not a static one. To continue acting or shifting the action by 'resembling the real thing but not being it', fakes need to move and change with society. Following how they do that allows us to learn about the material effects of deception and its interception on the 'assemblies of human and nonhuman actants' (Akrich and Latour, 1992: 259) through which livelihoods are gained and lost.

The framework

The relational approach for learning from fakes articulated here is informed not only by the analytical considerations outlined earlier, but also by pedagogical commitments. From 2011 to 2017 I taught an undergraduate seminar called 'Fakes' at Tembusu College, a residential college with its own general education programme at the National University of Singapore. At the college, our mandate was to offer residents – undergraduates belonging to any of the faculties at the university – small discussion-based classes on interesting topics that weren't easily classed as the province of one academic discipline. Students were meant to gain analytical skills as well as benefit from interacting with peers from other faculties. The way I designed my Fakes class to meet that mandate was informed by my background in science and technology studies (STS) and my interest in thinking about socio-material situations that pose the question how appearances are related to realities (what we might call 'appearance-reality puzzles'). What is more, the design changed as a result of my growing experience and professional development as a teacher during the six years that I taught the class.

The framework presented here is where things ended up. It consists of the two methodological principles mentioned in the introduction:

- *Follow the thread between faking and coveting.* What are the *stakes* of fakery? By 'resembling the real thing but not being it', how does the fake teach us what is *not* valued or wanted, and thereby what *is* valued or wanted? Such questioning involves pinpointing as precisely as possible the order (the implicit or explicit assumptions and expectations) betrayed or disrupted by the fake.
- *Follow the thread between the inside and the outside.* The success and failure of any given fake, the shape it takes – these are the product of a dynamic set of relations between human and nonhuman actants. To continue to have a bearing on the action by 'resembling the real thing but not being it', how does the fake anticipate and change in response to this set of relations? Such questioning involves tracking the changing shape (inside) and fate (outside) of fakes over time.

I speak of these principles as a 'framework' in the sense that they provide a way to work with many different kinds of examples (not necessarily limited to non-human fakes, though the emphasis of the seminar was on those). Limiting the analytical focus in this way meant that other fakes-related questions, particularly those concerned with the politics of the fake–real dichotomy, were less well supported. What made up for this loss was the framework's utility in qualifying virtually all accounts of situations in which fakery was a concern or had played a part, as materials for analysis. The seminar, then, did not follow a fixed syllabus, but became – after an introductory set of lectures – a live laboratory for practising the two foci of the framework on examples students had selected.[3]

---

[3] I was influenced by Rick Glofcheski, a law professor at the University of Hong Kong who also made a career as an educational innovator and proponent of a course design approach called 'constructive alignment' (Biggs and Tang, 2011). In a 2016 seminar I attended, he shared an important change he made to the design of his introductory course on tort law (laws having to do with health and safety). From teaching tort law through iconic cases, the standard approach, he switched to giving students a basic orientation and then having them source newspaper articles or examples from a walk around the city such as unsafe scaffolding. In class, they would, with Glofcheski's help, analyse these 'unflagged materials' from a tort-law perspective. That the examples were typically mundane, marked by incomplete information, and that the professor hadn't seen them before was exactly the point.

The double focus on fakes and socio-material relations provided a practice in drawing conceptual thinking and empirical examples together – a major strength of semiotically informed STS made learnable in a general-education setting. As my students and I worked our way through examples as wide-ranging as fake wireless access points, fake social-media profiles, fake wine, fake news, phishing websites or emails, fake credentials, fake Stradivarius violins, counterfeit medicines, fake doctors, fake marriages, fake orgasms, counterfeit make-up, adulterated food, fake celebrities, Ponzi schemes, knock-off sneakers and mimicry in animals (and this is just a sample), the two parts of the framework helped to direct our attention to the details of each case and to organize what we would emphasize in our discussions.

Because working with fresh examples *is* doing live analysis, we were in effect also doing research together. By treating fakes as a recognizable class of objects yet attending, in each example, to how they matter in practice – what they provoke and encounter and what this leads us to observe about society – we engaged in the kind of sociological analysis that renews itself with each example worked through: the work being that of recovering 'the doing' of social order.[4]

In the rest of the chapter I will show how to apply this framework to specific examples. I hope that you, the reader, will find the cases and their details evocative and illustrative of a relational approach to 'learning from fakes' that you might choose to try with your own examples.

Part one: following the thread between faking and coveting

Consider the following newspaper article that appeared in the British newspaper *The Telegraph* in 2013: 'Chinese museum found with 40,000 fake exhibits forced to close' (Phillips, 2013, see Figure 4.1). I suggest that you read it in its entirety before continuing. This is a big ask, I realize: it will interrupt your flow and make reading this chapter more demanding by getting you to attend to a text within a text. Yet it is in working through actual examples that the approach's strengths and limitations, and its potential for teaching and research, become most vivid and are best communicated.

This, he argued, made students begin to think as lawyers rather than try to emulate the teacher's analysis of contrived textbook cases.

[4] Thanks to Steve Woolgar for helping me articulate this point.

Figure 4.1: Chinese museum with fake exhibits

The Telegraph UK

Home» News » World News» Asia» China

Chinese museum found with 40,000 fake exhibits forced to close

A Chinese museum has been forced to close after claims that its 40,000-strong collection of supposedly ancient relics was almost entirely composed of fakes.

By Tom Phillips, Shanghai

1:24PM BST 16 Jul 2013

The 60 million yuan (£6.4 million) Jibaozhai Museum, located in Jizhou, a city in the northern province of Hebei, opened in 2010 with its 12 exhibition halls packed with apparently unique cultural gems.

But the museum's collection, while extensive, appears ultimately to have been flawed. On Monday, the museum's ticket offices were shut amid claims that many of the exhibits were in fact knock-offs which had been bought for between 100 yuan (£10.70) and 2,000 yuan (£215).

The museum's public humiliation began earlier this month when Ma Boyong, a Chinese writer, noticed a series of inexplicable discrepancies during a visit and posted his findings online.

Among the most striking errors were artifacts engraved with writing purportedly showing that they dated back more than 4,000 years to the times of China's Yellow Emperor. However, according to a report in the Shanghai Daily the writing appeared in simplified Chinese characters, which only came into widespread use in the 20th century.

The collection also contained a "Tang Dynasty" five-colour porcelain vase despite the fact that this technique was only invented hundreds of years later, during the Ming Dynasty.

Museum staff tried to play down the scandal.

Wei Yingjun, the museum's chief consultant, conceded the museum did not have the proper provincial authorizations to operate but said he was "quite positive" that at least 80 of the museum's 40,000 objects had been confirmed as authentic.

"I'm positive that we do have authentic items in the museum. There might be fake items too but we would need [to carry out] identification and verification [to confirm that]," he told The Daily Telegraph.

Mr Wei said that objects of "dubious" origin had been "marked very clearly" so as not to mislead visitors and vowed to sue Mr Ma, the whistle-blowing writer, for blackening the museum's name.

"He [acted] like the head of a rebel group during the Cultural Revolution – leading a bunch of Red Guards and making chaos," Mr Wei claimed.

Shao Baoming, the deputy curator, said "at least half of the exhibits" were authentic while the owner, Wang Zonquan, claimed that "even the gods cannot tell whether the exhibits are fake or not," the Shanghai Daily reported.

(continued)

Figure 4.1: Chinese museum with fake exhibits (continued)

China's vibrant online community begged to differ, reacting with its customary barrage of disgust and ridicule.

One micro-blogger urged local authorities to re-open and re-brand the museum as "China's biggest fake item museum." "If you can't be the best, why not be the worst?" mused the user, "Jizhou magistrate".

China is currently in the midst of an unprecedented museum boom with nearly 400 new museums opening in 2011 alone, according to government figures.

But fake relics have proved a persistent thorn in the industry's side. In 2011, state media reported claims that 80 per cent of the fossils in Chinese museums were fake.

"Fake fossils are like poisoned milk powder that injure and insult visitors," a scientist from the Chinese Academy of Social Sciences was quoted as saying.

Source: Phillips (2013), © Tom Phillips / Telegraph Media Group Limited 2020.

You may have picked up all sorts of interesting aspects of fakery from this article. I invite you to pay attention to two aspects in particular. Firstly, the sentiment: a museum faced 'public humiliation' and a 'scandal' that, even though museum staff 'tried to play [it] down', eventually forced it to close. The museum was outed by a 'whistle-blowing writer' and subjected to a 'barrage of disgust and ridicule' from 'China's vibrant online community'. The kinds of offences it had committed to 'injure and insult visitors' were even compared to the harm done by 'poisoned milk powder'. Clearly, something went very wrong to draw such strong rebuke. Clearly there is something *at stake*.

Having noted this, we can move to the second aspect, namely the indications of the shared expectations that are being violated here. Museums are supposed to harbour 'unique cultural gems', not 'knock-offs which had been bought for between 100 yuan (£10.70) and 2,000 yuan (£215)'. Museums are also supposed 'not to mislead visitors'. Museum staff members uphold those expectations even in their partial admission of something having gone awry, by saying that not *all* items were fake and that 'dubious' items were clearly labelled. The comment by owner Wang Zonquan that 'even the gods cannot tell whether the exhibits are fake or not' sets the whole discussion on a different footing by trying to make relevant whether or not we can *tell*. Yet in the narrative of the article this remark takes on the status of a deflection of the real issue and of responsibility, the fakery having already been established as undeniable through the demonstrable 'discrepancies' brought to light.

The strength of the reaction to the items recognized as fakes, and the shared expectations that become conspicuous by having been violated, are two sides of the same coin. Observing one – locating it in what the article presents as news – leads to the other. Two more short newspaper articles now follow, with the invitation to analyse them in a similar manner, namely by focusing on (1) the reactions that indicate fakes are a problem and (2) the values and expectations that make them so.

Figure 4.2: Fake reviews

The Economist

Business, Oct 22nd 2015 edition

Reviews on Amazon

Five-star fakes

The evolving fight against sham reviews

"I WILL post awesome review on your amazon product," bess98 declared on Fiverr, a website where individuals sell freelance services for $5 or more. On October 16th Amazon charged that bess98 and more than 1,000 others were illegally hawking customer reviews. The case comes just six months after Amazon sued the operator of four sites peddling similar stuff, including the subtly named buyamazonreviews.com.

Like Amazon, other websites have fought fakes with lawsuits, carefully honed algorithms and even sting operations—Yelp, a popular review site, has had undercover staff answer ads from firms seeking glowing write-ups. Yet the problem persists.

For as long as there have been online reviews, there have been fakes. The motivation is clear: for example, one extra star on a restaurant's Yelp rating boosts revenue by 5-9%, according to Michael Luca of Harvard Business School. Mr Luca and Georgios Zervas of Boston University have shown that restaurants seeking fake acclaim are likely to be independent—online reviews matter more to them than to chains with established reputations. So some businesses ask friends to post raves, seek reviewers-for-hire and offer customers discounts in exchange for praise.

For websites that claim to be an impartial resource, such practices are troubling. "While small in number," Amazon contends in its new suit, "these reviews can significantly undermine the trust that consumers and the vast majority of sellers and manufacturers place in Amazon." The problem is particularly irksome for sites dedicated to offering reviews, such as Yelp and TripAdvisor. Amazon sells everything from books to lawnmowers; Yelp's main offerings are its 83m reviews. "If consumers can't rely on the content," says Vince Sollitto of Yelp, "then the service is of no value."

So websites have tried to fight fakes. Algorithms comb reviews for suspicious wording. Expedia allows hotel recommendations only by those who have paid for a room there. Amazon tags a review as "verified" if the writer has indeed bought the product. Presumably such reviews are more reliable, though bess98 is one of many who claim to be able to game Amazon's system.

(continued)

Figure 4.2: Fake reviews (continued)

> Yelp may have the most aggressive strategy. An algorithm removes a whopping 30% of posts from Yelp's list of "recommended" reviews, though consumers can still see the suspicious ones if they like. Businesses that try to weasel their way to a higher rating (paying off grumpy clients, for instance) have their Yelp pages branded with a red warning.
>
> Despite all this, some false acclaim and critiques inevitably slip past firms' defences. For websites, fake reviews will remain a stubborn headache. Meanwhile businesses are finding new ways to boost their reputations online. Social bots—lines of code that pose as real accounts—can build buzz on social-media sites like Twitter and Facebook. For the average consumer, it may become ever harder to distinguish real praise from puff.

Source: 'Five star fakes' (2015), © The Economist Group Limited, London (2015).

What is at stake in Figure 4.2? From the perspective of the websites, it appears to be the trust that customers put in the reviews of other customers. Companies such as Amazon and Expedia do not just offer products or services. They also offer confidence, the reassurance that what customers buy after doing research (that is, after reading the independent opinions of others) will match expectations. How do we know this is at stake? Well, companies are in a 'fight' against fake reviews that blur the distinction between customer opinion and company advertising. The measure of upset is provided by the 'lawsuits, carefully honed algorithms and even sting operations' mounted against the fakes. Yelp's 'aggressive strategy' marks out as 'suspicious' nearly a third of its reviews, presumably risking the ire of some of the consumers who wrote them, for the sake of upholding the line between opinion and advertisement on which people's confidence in the site depends.

Figure 4.3: Fake educational certificates

The Straits Times

Singapore, Apr 6, 2017, 5:38 pm SGT

Three jailed 10 weeks for using fake certificates to apply for work passes

Chew Hui Min

SINGAPORE – Four foreign workers were charged on Thursday (April 6) over submitting fake educational certificates for their work pass applications, the Ministry of Manpower (MOM) said.

The four used forged academic certificates to obtain and renew their Employment Passes (EP) and S Passes, and held jobs such as restaurant manager, assistant manager, chef and facility executive.

Figure 4.3: Fake educational certificates (continued)

They include two men from India, aged 35 and 27, a 28-year-old woman from the Philippines and one Vietnamese man, 24.

The three men pleaded guilty and were sentenced to 10 weeks' jail each.

The Filipino woman's case was adjourned to April 27.

MOM said it verified with the academic institutions that the certificates were forged.

"Using forged educational certificates to obtain work passes is a serious offence. We will prosecute the foreigners and permanently bar them from working in Singapore," said Mr Kandhavel Periyasamy, director of employment inspectorate at MOM.

MOM reminded employers to ensure that their selection and recruitment processes of foreigners are rigorous.

"They should check the authenticity and quality of their work pass applicant's academic qualifications," the ministry said in its statement.

As an additional safeguard, MOM conducts additional checks and verifications on applications.

Anyone convicted of submitting forged academic certificates may be fined up to $20,000, and/or jailed for up to two years.

In the last two years, 73 foreigners were convicted and permanently barred from working in Singapore.

Source: Chew (2017), The Straits Times © Singapore Press Holdings Limited. Reprinted with permission.

In Figure 4.3 again a lawsuit tells us something is amiss. A government is trying to safeguard its mechanism for determining which foreign workers are allowed into the country. The word 'safeguard' is literally used. What is valued are academic qualifications, the implicit assumption being that these provide a useful proxy for the quality of a foreign work pass applicant. Upholding this system is very important, seen from the way employers are petitioned 'to ensure that their selection and recruitment processes of foreigners are rigorous', the Ministry's 'additional checks and verifications', and the fines and punishments meted out to those submitting fake certificates. Forged academic certificates are considered a threat for undermining a system through which the government makes its policy regarding foreign workers accountable to citizens.

In each of the cases presented in Figures 4.1, 4.2 and 4.3, we have paid attention to the strength of the reaction and the stakes elicited by fakes. The job of analysis has been one of recovery more than discovery. More than telling us something new about the audacity of, and public outrage against, fake museum objects in China, about fake reviews undermining trust in websites, or about the Singapore government's valuing of academic credentials, learning from fakes takes the form of witnessing social organization. It is remarkable that even the briefest of articles allows us to establish in quite a precise way both *that* and *how* fakes are a problem within the relationships portrayed – thereby providing a window into the social organization of important aspects of life.

Take the article on fake certificates: even if you come from a culture in which formal credentials do not matter quite so much, or if you don't personally care or think it is right that so much emphasis is placed on them, the article shows, through the strength of the sentiment provoked by these fakes, that Singapore's society is organized to value them a great deal.[5] Fakes, by presenting a crisis of sorts (Akrich and Latour, 1992: 260), provide a narrative newspapers can write about, one that charts the stakes and brings into relief the relationships out of which social order is constituted. This manner in which we can learn from fakes resembles Harold Garfinkel's (1967) 'breaching experiments' (also discussed in Chapter 1) – examples of which include issuing a greeting at the 'wrong moment', standing very close to someone while maintaining otherwise normal conversation, acting as a lodger in one's own family's home, and so on. Such actions, by the reactions they evoke, bring attention to the unspoken norms, rules and methods that produce stability in social interaction. Even though those norms, rules and methods aren't 'new' to us, their disruption brings a vivid, even visceral experience of them, which reminds us that social order is not static but alive in its upholding. The norms, rules and methods recovered by doing breaching experiments are familiar ones, but their familiarity becomes new, and can be engaged anew, in this practical, empirical recovery. In our case, the disruption introduced by the fake, through the reactions elicited, shines a light on the unspoken expectations and assumptions on the basis of which everyday transactions are created and maintained.

---

[5]   We could go into another layer of analysis by arguing that the very existence of the report as 'news' is a way for the government to talk to its citizens, *The Straits Times* being a government-controlled newspaper. I will not deal with this aspect so as not to distract from the focus on what the narration makes available in terms of the relations and contrast points brought into relief by the fakery. Another layer of sophistication could come from treating the text as organizing the relations described in it, thus not merely *describing* how fakes cause disruption but *organizing* how fakes are to be read as disruptive. I thank Ivanche Dimitrievski for pressing me on this (Dorothy Smith-inspired) point, which, again, I do not take up in this chapter.

'What is its true value to me?' Researching civet coffee

In the newspaper stories presented in the previous section, the fact of the fakery was treated as firmly established. Their newsworthiness lay chiefly in what the recognized presence of the fakes had other parties to the action do *next*: close the museum, deploy algorithms and verifications, or mete out prison sentences. The following case will illustrate how our inquiry can be adapted for cases in which it is unclear if we are even dealing with a fake.

In 2015, Wilson Tan, an economics student in my class on Fakes, shared the following account:[6]

> Kopi Luwak, also known as civet coffee, is one of the most expensive (processed) coffees in the world. On multiple occasions I have seen in documentaries different types of Kopi Luwak, farmed or procured from the wild, and heard the unanimous exclamations regarding its sweetness and fragrance. Just late last year I even watched one documentary that featured elephant dung coffee! While some coffee connoisseurs have labelled Kopi Luwak as simply a novelty coffee with no significant improvement in taste and quality (before and after its processing by civets), there are many in Vietnam who feel that the only way for Vietnamese coffee to establish its quality in the world market is to proliferate cà phê Chồn (loosely translated: cat coffee, or weasel coffee).
>
> This is probably the reason why my parents and I were somewhat excited when we arrived at the Ben Thanh Market in Ho Chi Minh City, with endless rows of stalls selling the prized Kopi Luwak. We were wary, but only for a short while because the onslaught began. After being 'assaulted' by many shopkeepers who tried to convince us in barely comprehensible English that their Kopi Luwak was genuine, we finally chanced upon a stall that was run by a Chinese-Vietnamese shopkeeper. A brief exchange of Chinese words assured us that we would not be victims of a scam at her stall because we are all Chinese, and after a few quick sniffs we walked away with 500 grams worth

---

[6] This was in response to an assignment that asked students to write a few paragraphs about how fakes, fakery or concerns about fakes had touched their lives. I am grateful to Wilson Tan for allowing me to share his work.

of genuine (!!) civet coffee beans that had also required us spending a hefty sum of US dollars. The price was naturally discounted specially for us thanks to our good fortune of being Chinese.

That is the end of my story for now, because I have yet to try the coffee that we purchased. However, after doing some research I do know for a fact that it is highly difficult to distinguish between civet coffee that is real and organic, real and farmed, and fake. Therefore, there is a good chance I will never know the true nature of the Kopi Luwak that I have at home, and it will be interesting to wonder: What is its true value to me?

In this account, neither we nor its teller, Wilson, know the 'true nature' of the *Kopi Luwak* in question. The situation of excited-yet-wary tourists at a market where the coveted local product is sold, is vividly sketched: the family are interested in buying this exclusive product here and now; they are also concerned about getting scammed. The shopkeepers acknowledge, and try to alleviate, this concern by emphasizing the genuineness of their merchandise. The Chinese-Vietnamese shopkeeper, having built trust based on their shared Chinese ethnicity, wins the business of Wilson's Singaporean-Chinese family. In hindsight, Wilson isn't sure what to make of the transaction: Is the coffee she sold them indeed genuine? Was it really a discount they got? The very possibility of the coffee being fake, and the prospect of not being able ever to know for sure, also provides an impetus for asking why the family had wanted to buy it in the first place. What were they looking for? What kind of 'achievement' did they want this coffee to represent?[7]

Our entry into analysing this case is to follow three different scenarios for how this coffee might become recognizable as a fake to its buyers. This is like following trails outwards from Wilson's story into possibilities of testing and knowing, envisaging *trials of strength* (Akrich and Latour, 1992: 259) that, even if they stay hypothetical, recover the expectations and assumptions invested in the transaction.

The first is the exotic production process of *Kopi Luwak*. When civet cats eat coffee cherries and excrete the hard beans inside they can't digest, something apparently gets imparted to the bean that lends coffee made from it a sweet and fragrant taste. Wilson's family

---

[7]   I am drawing here on Denis Dutton's notion that fakery entails a misrepresentation of achievement. He developed this for the case of fake art (in Dutton, 2008 [1979]).

wanted to experience this delicacy, and it was important to them that the origin story checked out: if the coffee beans they bought had not, in fact, passed through a civet's gut then this would be a clear–cut misrepresentation. A test to determine this had also been of interest to others, given the exclusive and expensive nature of the product. In 2013, the *Telegraph* newspaper in the UK reported that researchers at Osaka University in Japan had 'identified a unique chemical fingerprint that exists in coffee that has been excreted by a palm civet' (Gray, 2013). They translated the 'black box' process of animal digestion into traceable 'physical and enzymatic consequences to the coffee bean' (Jumhawan et al, 2013: 7994). With the aid of GC–MS-based multimarker profiling, the real thing could be set apart from coffees designed to mimic the flavour ('processed to approximate the sensory profile of Kopi Luwak') such as the 'fake coffee' in Figure 4.4 (from

Figure 4.4: Principal Component Analysis (PCA) of four different types of coffee. The dotted circle in the lower right quadrant designates the non-commercial subset of authentic Kopi Luwak coffee

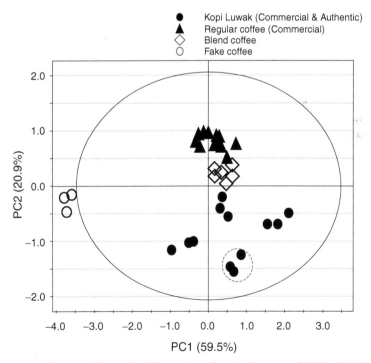

Source: Reprinted (adapted) with permission from Jumhawan et al (2013: 7998, fig 3). Copyright 2013 American Chemical Society.

Jumhawan et al, 2013: 7998). The article speculated on differences in the amounts of 'citric acid, malic acid, inositol and pyroglutamic acid' being responsible for the famed taste of *Kopi Luwak*.

But the passage through animal guts is only one stage of the provenance of civet cat coffee beans. Civet coffee gained its reputation based on beans harvested from the excrement of civets roaming in the wild. However, investigative journalists have exposed industrial practices as more typical: a 2016 *National Geographic* magazine article revealed that *Kopi Luwak* production on Balinese coffee plantations involved civets kept in cages in conditions detrimental to animal welfare. Besides its impact on the animals, this also has a negative effect on the coffee: 'Part of what makes kopi luwak so special, experts say, is that wild civets pick and choose the choicest coffee cherries to eat. Keeping civets in cages and feeding them any old cherries leads to an inferior product' (Bale, 2016). Consumers were also warned of labelling fraud, whereby coffee from caged civets is passed off as wild civet coffee. Exposés like these don't provide a direct way of knowing what Wilson and his family bought in that market, but they do provide another distinction with which to assess what they wanted from that purchase in the first place. Even if those beans *did* pass through a civet's guts, a product from civets kept in caged conditions and fed 'any old cherries' may be considered short of the mark: not quite the delicacy that, in its rarity and cultural authenticity, the family had wanted to sample.

So what about the taste? Wilson and his family were no coffee connoisseurs, but would an expert be able to tell whether they bought the real deal from the smell and taste of the coffee alone? We asked Cedric Chin, a coffee connoisseur and Tembusu College alumnus living in Vietnam at the time, to provide some input. He pointed to another stage in the coffee production process that should be taken into account: roasting. The coffee that connoisseurs like is medium roasted 'to preserve some of the acids and more delicate flower-like flavours'.[8] For commercial sales of coffee, this poses a problem because, as medium-roasted coffee is highly perishable, it needs to be consumed within fourteen days. Like most coffees produced and sold on a large scale, commercial civet coffee is 'roasted to hell' (at a high heat) to make the chemical composition of the beans more stable and therefore

---

[8] Cedric Chin's quotes are from the online chat he had with Arjun Saha, a senior student in the Fakes class who became my informal teaching assistant that semester. I thank them both.

preserve the coffee for longer. However, this also negates whatever flavour enhancements the civet's digestive processes impart to the bean. Civet coffee, when roasted in this way, is as uninteresting as other commercial coffees to the connoisseur. Rather than a genuine taste experience, it is, in Cedric's words, 'a total gimmick'. To the news that Wilson's family had bought a pack he replied: 'what a waste'. In this account, *all* commercially sold civet coffee is fake.

These various trials – principal components analysis, investigative journalism about civets on coffee plantations, and a connoisseur's perspective on flavour and roasting – complicate the *Kopi Luwak* by opening up more and more distinctions that could be brought to bear on the question of both its fakeness and its value. At each turn, a new possible way in which the product at hand might be a misrepresentation of 'the real thing' crops up, and at each turn the expectations and assumptions of the family come up for clarification. This kind of investigation – identifying different tests and methods to resolve the coffee's status – thus brings in its wake a detailed understanding of how and why it might matter that this coffee is what it claims to be.

Towards the end of that semester, I received an email from Wilson reporting the family's disappointment with the taste of the coffee they had bought. Even his parents, who had not been following our investigation, considered it 'not only not on par, but inferior' to good-quality regular coffee. He concluded: 'One thing for sure is that it is unlikely that we will be trying to purchase civet coffee in the future, if only for the reason that it is exceedingly difficult to procure a legitimate sample.' In whichever way the question of value was invested (in taste, in the coffee's origin story, in a combination of both), expectations were disappointed to the point that for the family, civet coffee lost its status as a coveted delicacy.

In summary

Working back-and-forth between the sentiments provoked by fakes and the question of what constitutes fakery (what assumptions and expectations are violated, or at risk of being violated) has given us a way to analyse newspaper reports and encounters with possible fakes such as Wilson's civet coffee. Following the thread between faking and coveting has helped us cultivate an eye for the stakes of fakery. It has led us to describe what the fakes in each story set in motion, most particularly what responses they effect that underwrite the importance of certain distinctions: the distinction between genuinely old artefacts and imitations, between independent opinions and advertising, between earned educational qualifications and just paper. Whenever fakes are recognized as a class of objects against which measures (such as closure, new technology, jail sentences) are to be taken, these distinctions are affirmed in practice. And when the status of a particular specimen is unclear, as in the case of the coffee, the exploration of hypothetical trials of strength generates the distinctions that, in turn, recover the expectations and assumptions invested in the transaction. With each new distinction brought to bear on civet coffee – gut/no gut, wild/caged, medium roasted/high-heat roasted – the double question of what would count as fakery and what was coveted in making the purchase, is made relevant again.

Part two: following the thread between the inside and the outside

Let's go back once more to Mark Jones' statement: 'Each society, each generation, fakes the thing it covets most.' Besides the relationship between fakery and coveting, another message is implied in this short phrase: fakes are dynamic, they are the product of circumstances that, in turn, are subject to change. Jones doesn't specify this, but the temporal dimension complements the relation between fakery and coveting not only in allowing us to track changing trends over time. It also provides for studying the fake's attuning to its environment, and how its fate – success or failure – is a function of ever-changing socio-material alliances.

We can approach this as following a thread between the inside and the outside in that fakes both *contain* – in their design and material composition – and *change* the relations of trust and suspicion that make up a setting. Cuckoo eggs, as discussed in the work of evolutionary biologist Martin Stevens (2016) (see also Abbott and Large in this volume), illustrate the idea well. Some cuckoo species lay eggs that look very similar to those of their hosts, but others do not – cuckoo eggs come in a wide variety of appearances. In each case, the egg's characteristics seem to be a function of the countermeasures (against the cuckoo's deception) that a particular species has encountered in its environment over time. The more actively its chosen host has fought back against the threat of raising cuckoo young in its nest, the more refined the mimicry tends to be. 'This type of interaction,' writes Stevens, 'is referred to as co-evolution because it involves reciprocal changes in each party (cuckoo and host) brought about by changes in the other species' (Stevens, 2016: 195). The balance of power is constantly subject to being redressed by the way better fakery triggers the need for better defences, and vice versa. This is not just a two-way relation but a matter of the resources and constraints of both hosts and cuckoos in the ecosystem that sustains them and in which they vie for survival. It is in this sense that the internal make-up of a fake encodes the countermeasures it has encountered and expects to encounter out in the world.

This brings to mind how, in actor-network theory, Madeleine Akrich (1992) speaks of 'inscription' and 'de-scription'. In the technical content of a new object we find *inscribed* a forecast of, and attempt at management of, its reception in the world. This 'scripting' of relations makes room in an existing order for the new object, enabling it to 'do its thing'. Once it starts to circulate, the object's success or failure is a function of whether or not its actual users – and the entire assembly of human and non-human actants by which it aims to be received – behave according to this script (this is 'de-scription'). If they don't, this may in turn prompt adjustments to the design. Latour's (1988) language of 'alliances', too, gets at the temporal dynamics that tie content and context together in the achievement of socio-material change.

Getting back to our fakes, we can think of a successful fake as one designed and operating in such a way that it is carried forward and propped up by relevant entities in its environment (whether they do so wittingly or unwittingly). Conversely, an unsuccessful fake is one that meets effective resistance from the entities in its environment that have a bearing on its fate. Just as a bird host may at one point tolerate a cuckoo's eggs and at another point throw them out of its nest, there is a dynamism at work, whereby relationships can change from supportive 'ally' to resistive 'adversary', and vice versa. So the successful fake then is the one that, in its design and the way it operates, enjoys an alliance in its favour, one that is stronger than the counter-alliance that also must be reckoned with.

In the case of the Chinese museum discussed in the previous section, rather than focusing on the sentiment of protest and its indication of breached expectations, this second part of the framework has us focus on *what changed*. What made it so that an alliance that once worked in favour of the fake – 40,000 fake relics were made, bought, given a roof over their heads and beheld as the real thing by scores of museum-goers – fell apart? Drawing on details of the article, we can trace the formation of a counter-alliance: something in the physical form of the fake (namely 20th-century writing on supposedly ancient artefacts) became a crucial weakness when a writer with access to social media mobilized enough of China's online community to pressure the authorities overseeing the museum to close it. The pressure came about through an alliance of adversaries that included a whistleblowing writer, digital social media and vocal netizens. This museum closed ... but that may not be the end of the story: another one may open with 'better' fakes. Such dynamic adaptations are further illustrated in our final example.

From looting to faking: tracing a shift in the black market for antiquities

In 2009, the archaeologist Charles Stanish wrote an informal piece for the journal *Archaeology* on the impact of the auction-site eBay on the global trade in antiquities and the black market associated with that trade. The article, 'Forging ahead: Or, how I learned to stop worrying and love eBay', started from a worry: 'We [archaeologists] feared that an unorganized but massive looting campaign was about to begin, with everything from potsherds to pieces of the Great Wall on the auction block for a few dollars' (Stanish, 2009: np). This, however, was not what happened. About five years after eBay's inception, around the year 2000, Stanish, who specializes in Andean archaeology, began to observe a massive rise, not in the traffic of looted artefacts, but in that of fakes of 'every grade and kind of antiquity' (Stanish, 2009: np).

> In an antiquities store in La Paz, I recently saw about four shelves of supposed Tiwanaku (ca. A.D. 400–1000) pottery. I told the owner that most were fakes and she became irritated and called me a liar. So I simply touched one at a time, saying 'fake,' 'real,' 'real from Tiwanaku,' 'fake,' 'fake made by Eugenio in Fuerabamba,' and so forth. She paused for a moment, pulled one down that I said was real, and told me that it was also a fake. I congratulated her on the fact that her fakes were getting better and she just smiled.

Stanish tells a multifaceted story, culminating in the scene just described in which the fakes are winning. He points to eBay as crucial in changing the landscape. Before the advent of the online auction site, where did local people sell cheap reproductions of Andean antiquities? They did so outside tourist attractions, as 'tacky tourist art'. And where did they sell any looted items they could get their hands on? Through a very different channel: an elaborate system of middlemen adept at smuggling such artefacts across borders and, for top quality pieces, through 'backdoor channels' to high-end dealers in Europe or North America who would place the items with wealthy collectors or museums. There was more profit in the second approach, though also a lot more risk, because genuine antiquities are scarce and agents of the law try to stop them from leaving Bolivia or Peru.

The uptake of eBay around the year 2000 disrupted the alliance that was holding up the trade in looted artefacts. Local people, both craftsmen and people with internet skills and eBay accounts, realized they could 'make more money cranking out cheap fakes than they can by spending days or weeks digging around looking for the real thing' (Stanish, 2009: np). These fakes could be shipped to customers without the involvement of middlemen, and even if the shipment were stopped at customs, the risk of prosecution for selling fake artefacts would be low. When this was put into practice, the supply of 'antiquities' on eBay increased rapidly. Then the gap between tourist souvenir-buying and high-end antiquities collecting, previously so sharply demarcated, began to close:

> Surely the sophisticated high-end buyers would not be affected by the rubes who pay $223 (plus $30 shipping from Lima) for a 'genuine pre-Columbian Moche III Fineline' piece (which, by the way, can also be bought for $15 from the woman selling pottery outside the tourist buses in the Peruvian city of Trujillo). But the high-enders are indeed affected. It was only a matter of time before a few workshops producing the cheap fakes started turning out reproductions that can fool even supposed experts like me. A number of these workshops have swamped the higher-end market with beautiful pieces that require intensive study by specialists and high-cost tests to authenticate. (Stanish, 2009: np)

How is this possible? In traditional workshops in places with a rich archaeological heritage, people have access to local cultural knowledge and know-how of ancient techniques for firing and painting pottery. They also have access to ancient source materials, and to information about what experts would expect of a genuine piece:

> I know, for instance, of one fellow who makes grass-tempered reproductions of a 2,000-year-old pottery style. Having worked on archaeological projects for years, he learned to get the grass for his fakes from ancient middens near his house. If fired properly, and if the organic residue in one of his pots were carbon dated, it would appear to be a very old piece indeed. Looters on the north coast of Peru have discovered not only the famous 12th-15th-century

A.D. Chancay anthropomorphic vessels, but also the original molds used to make the vessels. Thanks to publicly available archaeological reports, they also now use the original clay sources and minerals to make and paint the pottery. They can create virtually perfect reproductions. (Stanish, 2009: np)

In their material design, the fakes created in these workshops anticipate the force-field of relations on which their distribution and success depends. Techniques such as carbon dating (for age) and thermoluminescence microscopy (to find out how long ago a clay object was fired) are subverted by mixing grass from ancient middens or fragments of ancient pottery in the forgeries. Publicly available archaeological reports are enrolled to pinpoint the local materials that experts would expect to find in genuinely old specimens. Because eBay provides a direct channel to sell fakes 'of every grade and kind of antiquity', cheap ones that can be cranked out by the hundreds provide the capital to invest in continuous experimentation and improvement at the top end of the market. By 'swamp[ing] the higher-end market with beautiful pieces that require intensive study by specialists and high-cost tests', fakes intervene in the relations between scholars and the objects they study, recalibrating dynamics of trust and suspicion. Stanish even cites research that shows that 'the experts who study the objects are sometimes [unwittingly] being trained on fakes', and then go on to perform authentications based on false reference models. Even the 'backdoor channels' that have provided high-end dealers and customers with a way to source looted works through the black market are being tapped as distribution channels for now 'virtually perfect reproductions' (Stanish, 2009: np).

What Stanish describes is the reconfiguration of an ecosystem that propped up looting into one that props up faking. He 'loves' eBay for this, considering fakery the lesser evil, though the material effects of the new market for illegal antiquities are wide-ranging. For our purposes, if we position the fakes as protagonists of the story, they have 'won' by amassing a powerful alliance that includes local people in the Andes, local materials (natural and man-made), cultural knowledge and know-how of ancient techniques, the law, experts and their research, buyers worldwide, and, of course, eBay, which in this account, provides the missing link to give fakes their advantage. Following the thread between the inside and the outside, we see that the lowly tourist trinket, bought on eBay as a genuine antiquity, starts to proliferate and generate spin-offs that further extend the fakes' reach. In their different qualities, and abundant quantity, the fakes both capitalize on the new assembly of human and non-human actants that has come into play, and further it. One main way in which they do so is by building into the design and composition of the high-end fakes features that allow them to withstand tests and probes, getting even experts on their side. The quantity of fakes, their quality, their material composition – all these both *reflect* (inside) and *reconfigure* (outside) the black market for antiquities as a dynamic ecosystem.

In summary

The examples presented in this section have shown how we can learn from fakes by tracking their fate over time. An alliance that holds for some time, such as the one propping up the museum with 40,000 fake relics in China, gets dismantled when the strength of the socio-material relations that oppose it makes its continued existence untenable. In Stanish's account, the fakes' reversal of fortune went the other way when eBay helped forge a strong alliance in their favour, and fake antiquities from the Andes of all grades and qualities began to proliferate. As an analytical redescription informed by actor-network theory, this is a way of thinking together the material make-up and design of fakes with the settings in which they circulate and evolve.

Conclusion

This chapter has introduced a way of thinking with imposters that puts *non-human imposters*, or fakes, in the spotlight. Beginning from the broad question of what they can teach us, what we can learn from them, the approach treats fakes as a recognizable class of objects and as protagonists, rather than receptacles or intermediaries, of the action in a story. The relational framework introduced here provides two ways of thinking with fakes that foreground the fake's 'fakeness' while remaining attuned to the empirical detail of what it does and brings into relief. The *first* threads together fakery and and valuing – what we gain from it is a recovery of unspoken expectations and assumptions on the basis of which everyday transactions (as well as distinctions that matter) are created and maintained. The *second* threads together the design of a fake and its environment – what we gain from it is an understanding of the status quo (at any point in time) as a precarious balance of antagonistic relations that shape and are shaped by the fake.

Learning from fakes in these ways involves letting them lead us from one thing to the other – from what's been imitated to what's coveted, and from the design and material properties of an artefact to the setting in which it seeks to succeed – and back again. This is how they help bring everyday interactions and configurations of discernment and of valuing, of trust and suspicion, into view. Some examples lend themselves more to an analysis of the values and distinctions brought into relief by fakes. Others lend themselves more to the tracing of the fake's fate over time – and sometimes we can do either, or both. Because of their specific entanglements as objects playing the part of 'resembling the real thing but not being it', fakes are not going to run out of steam. Because they track what we covet and move with the times, because of the way they shadow and insert themselves in human affairs, and because they make for a good story, there will always be materials aplenty for the analyst of fakes, wherever we look. This is perhaps the most exciting aspect of the framework introduced here: the opportunity it provides to turn ever-renewing fakery stories into ever-renewing understandings of the relational mechanisms of ordering and valuing.

Acknowledgements

I developed my thinking on Fakes by teaching an undergraduate course on the topic for six years. Thanks to all the students who came on the journey: from our interactions emerged this chapter. Thanks to Sara Bea, Kristina Grünenberg, Ting Jun Heng and the editors for helpful comments on earlier drafts.

References

Akrich, M. (1992) 'The description of technical objects', in W.E. Bijker and J. Law (eds) *Shaping Technology/Building Society: Studies in Sociotechnical Change*, Cambridge: MIT Press, pp 205–224.

Akrich, M. and Latour, B. (1992) 'A summary of a convenient vocabulary for the semiotics of human and nonhuman assemblies', in W.E. Bijker and J. Law (eds) *Shaping Technology/Building Society: Studies in Sociotechnical Change*, Cambridge: MIT Press, pp 259–264.

Bale, R. (2016) 'The disturbing secret behind the world's most expensive coffee', *National Geographic Magazine*, 29 April. Available at: https://www.nationalgeographic.com/news/2016/04/160429-kopi-luwak-captive-civet-coffee-Indonesia/ [Accessed 30 June 2020].

Ball, J. (2017) *Post-Truth: How Bullshit Conquered the World*, London: Biteback Publishing.

Biggs, J. and Tang, C. (2011) *Teaching for Quality Learning at University: What the Student Does* (4th edn), Maidenhead/New York: McGraw-Hill/Society for Research into Higher Education/Open University Press.

Chew, H.M. (2017) 'Three jailed 10 weeks for using fake certificates to apply for work passes', *The Straits Times*, 6 April. Available at: http://www.straitstimes.com/singapore/three-jailed-10-weeks-for-using-fake-certificates-to-apply-for-work-passes [Accessed 30 June 2020].

Dutton, D. (2008 [1979]) 'Artistic crimes', in A. Neill and A. Ridley (eds) *Arguing about Art: Contemporary Philosophical Debates* (3rd edn), London/New York: Routledge, pp 102–114.

'Five-star fakes' (2015) *The Economist*, 22 October. Available at: https://www.economist.com/business/2015/10/22/five-star-fakes [Accessed 30 June 2020].

Garfinkel, H. (1967) *Studies in Ethnomethodology*, Englewood Cliffs: Prentice Hall.

Glofcheski, R. (2016) 'Authentic learning and task-based self-directed learning', workshop organized by the National University of Singapore Teaching Academy, 20 May.

Gray, R. (2013) 'Test to prove the world's most expensive coffee really has come from civet poo', *The Telegraph*, 22 August. Available at: www.telegraph.co.uk/news/newstopics/howaboutthat/10259984/Test-to-prove-the-worlds-most-expensive-coffee-really-has-come-from-civet-poo.html [Accessed 30 June 2020].

Jones, M. (1990) 'Why fakes?', in M. Jones (ed) *Fake? The Art of Deception* (exhibition catalogue), Berkeley/Los Angeles: University of California Press, pp 11–16.

Jumhawan, U., Putri, S.P., Yusianto, Marwani, E., Bamba, T. and Fukusaki, E. (2013) 'Selection of discriminant markets for authentication of Asian palm civet coffee (kopi luwak): A metabolomics approach', *Journal of Agricultural and Food Chemistry*, 61: 7994–8001.

Latour, B. (1988) *The Pasteurization of France*, translated by Alan Sheridan and John Law, Cambridge: Harvard University Press.

Law, J. (2009) 'Actor network theory and material semiotics', in B.S. Turner (ed) *The New Blackwell Companion to Social Theory*, Chichester: Wiley-Blackwell, pp 141–158.

Mol, A. and Mesman, J. (1996) 'Neonatal food and the politics of theory: Some questions of method', *Social Studies of Science*, 26: 419–444.

Phillips, T. (2013) 'Chinese museum found with 40,000 fake exhibits forced to close', *The Telegraph*, 16 July. Available at: http://www. telegraph.co.uk/news/worldnews/asia/china/10182088/Chinese-museum-found-with-40000-fake-exhibits-forced-to-close.html [Accessed 30 June 2020].

Stanish, C. (2009) 'Forging ahead: Or, how I learned to stop worrying and love eBay', *Archaeology*, 62(3). Available at: http://archive. archaeology.org/0905/etc/insider.html [Accessed 30 June 2020].

Stevens, M. (2016) *Cheats and Deceits: How Animals and Plants Exploit and Mislead*, New York: Oxford University Press.

Toews, R. (2020) 'Deepfakes are going to wreak havoc on society: We are not prepared', *Forbes*, 25 May. Available at: https://www.forbes. com/sites/robtoews/2020/05/25/deepfakes-are-going-to-wreak-havoc-on-society-we-are-not-prepared/#9db0fee74940 [Accessed 30 June 2020].

Wong, W. (2013) *Van Gogh on Demand: China and the Readymade*, Chicago/London: University of Chicago Press.

Woolgar, S. (1991) 'Configuring the user: The case of usability trials', in J. Law (ed) *A Sociology of Monsters: Essays on Power, Technology and Domination*, London: Routledge, pp 58–97.

Yates-Doerr, E. (2015) 'Does meat come from animals? A multispecies approach to classification and belonging in Highland Guatemala', *American Ethnologist*, 42(2): 309–323.

5

Imitations of Celebrity

Mandy Merck

There are many current debates over the influence of the famous, the agency of fans, and the consequences of the emulation of prominent people. One of the most pervasive anxieties these address is that those who admire the famous, particularly young admirers, will imitate their 'idols'. While Western commentators investigate the connection between 'celebrity worship' and cosmetic surgery (Maltby and Day, 2011), their Asian counterparts explore the relation of the 'imitation of celebrity models and materialism' among Chinese youth (Chan and Prendergast, 2008). Psychologists have developed an entire literature on the mimetic impulse in fandom, charting the descent from admiration to empathy to over-identification to obsession (Giles, 2000; McCutcheon et al, 2002, 2003). Their pathologization of this progress is echoed by the Christian motivational writer Kimberly Davidson on what she calls the 'Celebrity Imitation Complex':

> Young people mirror what they see through the media and the Internet. A celebrity fits with their human desire to be approved, applauded and considered special. Many teenagers truly believe emulating the lifestyle of their favorite celebrities is the only way to form an identity. ... Unfortunately, and unbeknownst to the stargazer, there are far too many awful celebrity role models being emulated with disastrous consequences. (Davidson, 2011: 8–10)

But there is, Davidson goes on to claim, an alternative – not the abandonment of imitation, but its purposeful practice. Citing the

15th-century Latin devotional text *Imitatio Christi*,[1] a collection of biblical and early Christian teachings by the Augustinian monk Thomas à Kempis, she maintains:

> God did not create us to impersonate or obsess after other flawed human beings. The purpose of life is to live a life of purpose, which is to model to the world a truthful reflection of who Jesus Chris is and what he is like. (Davidson, 2011: 11)

Paradoxically, in order for 'Teens to Live Authentically in a Celebrity-Obsessed World' (the subtitle of Davidson's polemic), they must imitate a very famous figure indeed.

Despite Davidson's self-contradiction, her concerns about the imitation of prominent people reflect pervasive worries about people 'passing' into social spheres or occupations barred to them, or presenting themselves as materially or morally superior to their actual condition, convincing others and even themselves that they are different than they are. When such an imitation extends to passing oneself off as a specific individual, it may be called an imposture, the pretence to be someone else in order to deceive others, or even to deceive oneself. Imitation is always in danger of becoming imposture, while imposture necessarily involves the imitation of another's likeness, credentials and characteristics. This chapter explores the relationship between imitation and imposters through the genre of the melodrama, particularly the literary and film versions of *Imitation of Life*.

The discussion of imitation is ancient in origin, encompassing the spheres of education, politics and aesthetics, with the two most influential figures, the 5th century BCE Plato and the 4th century BCE Aristotle, disagreeing radically. A key term, *mimesis*, remains impervious to consensus and almost impossible to gloss. 'Mimesis and its Greek cognates defy translation', observes Paul Woodruff (1992: 73), offering 'imitation', 'image-making', 'representation', 'reproduction', 'expression', 'fiction', 'emulation' and 'make-believe' as popular English renderings. In the works of both philosophers dramatic performance involving the impersonation of a character is a pervasive example of mimesis, perhaps unsurprisingly since the word is derived from the theatrical 'mime' (Else, 1986: 27). But the

---

[1]   See Thomas à Kempis ([1418–1427] 1999). The first chapter of this text admonishes the reader that 'It is vanity, too, to covet honours, and to lift ourselves on high'.

two differ on its moral implications, Plato warning (in a work on just government, *The Republic*) that such a performance can challenge the ability to discriminate between truth and illusion (Lee, 1974), with Aristotle responding (in a work on drama, *Poetics*) that imitation facilitates learning, enabling people 'to understand and work out what each thing is'. Thus, he maintains, human beings learn 'their earliest lessons through imitation' and take 'universal pleasure' in it (1996: 6–7). This ancient attention to dramatic imitation anticipates its modern adoption in the sociology of Erving Goffman, whose analysis of social interaction in *The Presentation of Self in Everyday Life* (1959) compares it to a theatrical performance, in which individuals attempt to present themselves in ways they wish to be viewed (on Goffman and performance, see also Chapter 1 in this volume).

Aristotle's claim that the child forms its individual subjectivity through imitation also survives as a tenet of psychology, although not without complexity. In his influential study *Play, Dreams and Imitation in Childhood*, Jean Piaget argues that the acquisition of intelligence requires both the infant's acting upon the external world of objects, which he identifies with play, and its accommodation to that world, which he terms imitation. 'Since representation involves the image of an object', he claims, it can itself 'be seen to be a kind of interiorized imitation' (Piaget, 1991: 5). Where the psychologist moves from the child to representation, the playwright Bertolt Brecht (2001: 152) reasons in reverse, observing that the child learns how to behave 'in a quite theatrical manner', as though it is a member of an audience. 'It joins in when there is laughter, without knowing why; if asked why it is laughing it is wholly confused … The human being copies gestures, miming, tones of voice. And weeping arises from sorrow, but sorrow also arises from weeping.'

The stimulation of emotion is a noted feature of the melodrama, the characteristic genre of the imitation narratives to be considered here. If, in Brecht's dramaturgical characterization, emotions can be experienced through imitating their expression, it is not surprising that the spectator responds similarly to its performers. Reflecting on the recollections of female fans whose film-going was conducted in the era of the 1940s and 1950s melodrama termed 'the woman's weepie', Jackie Stacey (1994: 161–170) considers their memories of *resembling* its stars ('I could be Bette Davis' double'), *imitating* their speech and gestures ('I would tigerishly pace about like Joan Crawford') and *copying* their hairstyles, make-up and clothes. (The frequency with which Stacey's correspondents mention Davis and Crawford tells its own story about the genre's mimetic lure.) Stacey challenges the condemnation of identification in the film theory of Brecht's descendants, denying that such forms of engagement necessarily 'fix identities, destroy differences and confirm sameness' (1994: 172). Invoking Jessica Benjamin (1990: 106) on how the love of the child for the same-sex parent can be both narcissistic and erotic, she stresses the transformative interplay of desire and identification in subject formation. As I hope to show, the complex ways that imitating others may further differentiation and conformity, truth and falsity, self-affirmation and subordination, come into view in melodramas that take imitation as their titular theme and celebrity as a recurring narrative. Both the ontological and ethical status of 'imitation' vary within and between these texts, putting into question any simple condemnation of the 'imitation of life'.

Imitation of life: Fannie Hurst (1933)

The sequence of texts discussed here – itself one of an original and its mimetic adaptations, remakes and reinterpretations – begins with a 1933 novel by the American author Fannie Hurst. Originally serialized under the title *Sugar House*, it chronicles the fortunes of a young widow who supports her infant daughter Jessie and her invalid father by taking over her husband's maple syrup concession in Atlantic City, New Jersey. To secure access to the male-dominated marketplace of early 20th-century America, the late Benjamin Pullman's business cards, engraved 'B. Pullman', are appropriated by his wife Beatrice. Soon the desperately busy Bea (pun intended – she will later be called 'Honey Bea') needs domestic assistance. She hires Delilah Johnson, a widow 'with a round black moon face that shone above an Alps of bosom' (Hurst, [1933] 2004: 75), as a live-in maid. With Delilah comes her own baby Peola, whose 'pale and black' appearance will strain Hurst's descriptive powers. The 'white' Bea and her child acquire 'black' doubles early in a story in which various forms of simulation, replication and 'copy' become strategies for economic and social advancement as well as the central structuring device of the narrative. In time Bea enlarges her syrup business to include a flourishing Atlantic City coffee shop, done up in imitation of the railway dining cars traditionally staffed by Black attendants (as were 'Pullman' luxury sleeping compartments), selling candies and waffles cooked by Delilah. Not only does Delilah provide the recipes for these products, her photograph – 'in her great fluted white cap, and her great fluted white smile' – is adopted as their trademark. The company becomes an empire of identical diners named B. Pullman but signified by countless images of 'the chocolate-and-cream effulgence that was Delilah' (86–7). The humble Mammy is replicated in its ubiquitous logo, 'Delilah's face, Delilah's name, Delilah's smile ... reached from coast to coast' (144), and the sustenance it offers to a generation wounded by war renders the brand a national 'institution' (268) and Delilah a rare Black celebrity.

Written in the midst of the cultural change that legitimized synthetic materials – patent leather, Bakelite plastics and, most significantly for a later version of this story, costume jewellery – Hurst's novel can be seen as a rumination on what the Marshall Field department store in Chicago proclaimed to its customers in the 1920s: 'The imitation is no longer a disgrace' (Schwartz, 1996: 195). The Pullman company mass-produces sweet foods and the replica railway cars in which they are consumed, served by buxom Black women chosen for their resemblance to Delilah and schooled in the role of the Mammy. (Their need to learn this role has itself been adduced as evidence of the infinite regress of imitation in this narrative, as well as its recognition that servility is an acquired disposition.) The hive of the Honey Bea is enlarged by identical new cells, advertised across the country by full-page 'copy' personally written by the proprietor, who takes great pleasure in composing her 'hokum' (145), early 20th-century slang for faked feeling, hackneyed sentiment. Meanwhile this Queen Bea acquires cultural as well as financial capital by further imitation, marked by her changing taste in fashion, interior decoration and architecture.

B. Pullman's headquarters are established in New York's Flat Iron Building (a landmark of commercial success in the period) and her household decamps to an apartment overlooking Central Park. But despite wealth and fame, life does not go smoothly in this prototype of the Women Can't Have It All novel. The sugar house is a fragile edifice. In imitation of the children she overhears in the park, Jessie tries out the word 'n★★★★★'[2] in a row with Peola. The 'whitish' Peola's anguished response is to blister her scalp ironing her hair. Steeped in Christian submission, Delilah counsels her daughter that God 'had some good reason for makin' us black and white' in terms that Kimberly Davidson would endorse; but the enraged child is all too aware of the metonymic effect of her mother's appearance. Acutely aware of their dissimilarity, she protests, 'You're so black! That's what makes me n★★★★★' (148) before fainting into 'a pallor that made her whiter than chalk' (150). The impossibly hard-working Delilah, still caring for Bea's family while opening diners all over the country, cherishes her own

---

[2] It is Bristol University Press policy to asterisk words which might be deemed to cause offence.

White aspirations, imagining the splendour of her heavenly reward. Spurning the discomforts of their new living room furniture, 'a replica of a first-class department store window display' (137), she saves her money for her final rest: 'Ridin' up to heaven in a snow-white hearse wid de Lawd leanin' out when He hears de trumpets blowin' to see if I's comin' in a white satin casket pulled by six white hosses' (143).

Bea discovers her own dissatisfaction when she is befriended by an even more successful counterpart – Virginia Eden, a celebrity cosmetics mogul modelled on Hurst's real-life friend Elizabeth Arden. Enriched by selling calories labelled with the image of an 'immense woman' (79), Bea has begun to worry about her waistline and that of her (White) female employees, installing a staff gym complete with an extensive array of beauty aids courtesy of their publicity-conscious manufacturer. When the two entrepreneurs first meet for lunch at the Waldorf Astoria, Eden prefaces a real estate proposal with a declaration of their similarities:

> Want what you want when you want it. That's you. That's me. I want love. I want money. I want success … You and me ought to work together, Pullman. You make women fat and comfortable. My job is to undo all that and make them beautiful. You're grist to my mill. I want to be grist to yours. (159–160)

Even more than B. Pullman, Virginia Eden is revealed to be a pseudonym, for Sadie Kress of Jersey City. An imitation in ethnic, class and – in her snappy salesman's patter – gender terms, she seeks Bea's backing to transform an East River tenement block into a faux-Georgian fabrication, 'the London Embankment all over again'. Later her staff will 'do over' Bea herself, applying 'their accouterments of dress and good grooming' (154) to make her appear to the admiring Jessie 'so young' (274). Famed for its theme of racial 'passing', the novel identifies many objects of simulation, including the youth promised by the elixirs of the aptly named Eden, whose own face has 'the look of fine-grained writing paper with careful erasures' (158). Ironically, it is when this mistress of the masquerade offers to help Bea 'manufacture some of your own destiny', that she suddenly perceives the falsity of her existence: 'Love and happiness, as she said them, made what had been going on through years of a petty and mundane routine seem imitation of life' (161).

What is the 'life' that Bea believes she lacks? In parallel with the increasing prosperity and social license sweeping the US in the 1920s, her desires, and the governing metaphors of the novel, move from basic sustenance to romance. Hurst portrays her single heroine as an industrious and generous woman, but she is also shown to maintain the conventional colour line – employing Black women in her restaurants and Whites in her headquarters, assigning Delilah and Peola to the smaller bedroom in their Manhattan apartment, relying upon her business partner's devoted ministrations and maternal care of Jessie, and eventually sanctioning the segregation of their children in separate and very unequal boarding schools. Jessie, awed by her mother's eminence, becomes increasingly distant. Peola escapes as soon as she can to Seattle, where she 'passes', becomes a librarian, falls in love with a Bolivia-bound engineer and has herself sterilized to prevent any telltale offspring. Returning to New York, she bids Delilah and Bea goodbye forever with the novel's litany of desire: 'I want happiness. I want my man. I want my life' (247).

If the reader hasn't guessed from the mounting press requests for Bea's views on the price of a woman's career, Peola's forfeit anticipates the symmetry to come. In a formal equivalent of its mimetic theme, Bea's fate imitates hers, but in reverse: where Peola loses the prospect of children to romance, Bea loses romance to her child. After many pages of prevarication, she approaches her 40s resolved to ignore the eight years between herself and her handsome young business manager – 'a woman with a sublime need to keep young, can' (233) – only to discover that he and Jessie have fallen in love.

Commentators have described Hurst's novel as 'melodrama', and in this ironic reversal, its focus on the clash between public responsibilities and private desire, and its particular attention to relations between mothers and daughters, it certainly fits the genre. And yet, as Molly Hiro (2010: 95) points out, in important ways Hurst creates 'a distance between that form' and its narrative. Unlike its two famous film adaptations by John Stahl in 1934 and Douglas Sirk in 1959, her novel does not succumb to the traditional pathos of the 'tragic mulatta', the beautiful woman of mixed race whose appearance creates an ethical crisis. In linking Bea's financial aspirations to Peola's desire to manufacture her own destiny by escaping racial subordination and her oppressively self-sacrificing mother, it is markedly less moralistic. Not only does Peola plead with Delilah for her 'freedom' in the language of the desperate slave – 'Let me go. Let me pass' (248) – she coolly rebuts her moral condemnation: 'there's nothing wrong in passing. The wrong is the world that makes it necessary' (244).

Imitation of life: John Stahl (1934)

When Hurst's novel followed Bea's fictional restaurant to Hollywood, the prospect of its motion picture adaptation presented Universal Pictures with both a problem and an incentive. From 1930 the Hollywood studios were governed by guidelines for the self-censorship of film narratives, the Motion Picture Production Code, rigidly enforced from 1934. Until 1956 the Code's 'miscegenation' clause prohibited any representation of a 'sex relationship between the white and black races' (Doherty, 1999: 363). Although no such union occurs in the film, Peola's ambiguous appearance was interpreted by the Code's administrators as evidence of a '*suggested* intermingling of blacks and whites' (Courtney, 2005: 145, emphasis in original). Moreover, the ability of this 'Black' character to pass plausibly as 'White' raised the issue of her casting. Too close a simulation in this regard might itself suggest an unmentionable form of social intercourse in the ancestry of the designated actress. In both cases the Code's imperative to draw a clear colour line was contradicted by the narrative of the novel, as indeed the term 'passing' contradicts itself – since in asserting her White ancestry its mixed race subject is not passing herself off as anything other than what she is.[3] Yet it was just this cluster of contradictions that had established the 'tragic mulatta' as a compelling character of American literature and theatre. Introduced by the abolitionist Lydia Maria Child in her 1842 short story 'The Quadroons', this usually female figure was portrayed as the virtuous victim of sentimental literature. In Child's story she is the fair daughter of a Georgia gentleman who is schooled in the accomplishments of plantation society and courted by her White music teacher. Only after her father's death does she discover that her ancestors were never freed. Sold to cover his debts, she dies a slave.

Combining oppressed innocence, social dislocation and epistemological undecidability, the tragic mulatta continued to fascinate both sides of the racial divide well into the 20th century. Commentators stress Hurst's debt to four novels written by members of the New York Black arts movement, the Harlem Renaissance, which engage with passing – Walter White's 1926 *Flight*, Jessie Fauset's *Plum Bun* and Nella Larsen's *Passing* (both 1929) and George Schuyler's *Black No More* (1931). A civil rights activist and cultural patron who

---

[3]   As Kaufman and Mora-Gámez note in this collection (Chapters 8 and 13), the effort to prevent imposture effectively incites it, encouraging the ambiguous individual (character and/or actor) to assume a more acceptable identity.

awarded the young Zora Neale Hurston a literary prize in 1926 and then befriended and employed her to fund her college studies, Hurst had an engaged, if arguably ambivalent, relation to this movement and its members to her. Hurston waspishly observed that their contrast in colouring flattered her mentor, making her seem paler and perhaps less Jewish. Yet she also told Hurst that when she was accused of 'having furnished you with the material of "IMITATION"', she accepted the accusation as a tribute to 'the truth of the work' (Harrison-Kahan, 2011: 119, emphasis in original). In 1938 the poet and playwright Langston Hughes would stage 'Limitations of Life', the first of many parodic imitations of the story, this one reversing the central roles to have a White maid lovingly massage the feet of her big Black mistress; but three years earlier Hughes' own drama of racial 'mixtries' (Hughes, [1935] 1963: 31) in rural Georgia, *Mulatto*, gave voice to its doomed hero's insistence on using the front entrance to his father's mansion, literalizing the desire to 'pass': 'Don't I look like my father? Ain't I as light as he is? Ain't my eyes gray like his eyes are? Ain't this my own house?' (Hughes, [1935] 1963: 19).

In transforming Hurst's novel into the maternal melodrama popular in the 1930s, John Stahl's film takes up the saddened Black mother's viewpoint after her disconsolate daughter looks into the mirror and asks a similar question – 'Am I not white? Isn't that a white girl there?' The Production Code's miscegenation ban ruled out Peola's escape into a White marriage, as well as her final refusal of her mother's moral judgement. Instead, the film's Peola (Fredi Washington) returns to New York for Delilah's (Louise Beavers) funeral, overcome with grief and remorse. In the same normative revision, Stahl's heroine is not Hurst's 'social chameleon' (Itzkovitz, 2004: xxiv). Bea's acquired wealth and status is signalled by a formal party at her Manhattan home, but in Claudette Colbert's portrayal she remains the unpretentious woman next door even in glamorous evening dress. This enables a 'cute meet' in which the now eminent business woman is derided as 'the pancake queen' by her guest Stephen Archer (Warren William), whose failure to recognize his hostess does not impede their romance. Jessie will also fall for Stephen, and to assuage her feelings, Bea will ultimately part from him in her own doubling of Delilah's more profound sacrifice. Although a possible mother–daughter rivalry is set up, the 37-year-old Stephen is suitably mature (Colbert was then in her early 30s) and remains devoted to Bea. Thus the symbolic cosmetics mogul can be eliminated, along with the novel's theme of conscious self-fashioning. In this radical departure from the moral ambiguity of the novel, there is no successful passing, no imposture in the film. But if these changes purge the imitation from Bea's characterization, it is conversely intensified in the film's rendering of Delilah as an object of commercial replication.

American readers and spectators of the 1930s could not fail to understand that Delilah is herself a copy – of the trademark mammy 'Aunt Jemima', a character borrowed from a minstrel show in 1889 to represent an actual ready-made pancake mix. Like the fictitious Black cooks fronting the contemporaneous convenience foods Uncle Ben's Rice and Cream of Wheat, this figure was devised to personify the smiling service of instant sustenance. To her credit, Hurst is at pains to show that this image does not come naturally to Delilah any more than to her avatars. When Bea makes her pose for their first candy box in her large cook's cap, she becomes as 'irate as a duchess' and pleads to be photographed in her wedding hat as a remembrance for her daughter. But the employer prevails, and the resulting portrait is said to 'illuminate and reveal' (87). What exactly is revealed varies in Hurst's and Stahl's versions of the story.

In the novel the Delilah captured by 'the clicking of the camera' is a comforting grotesque, her 'shellacked eyes' and 'huge upholstery of lips' (87–88) confirming her function as the furniture of Bea's new enterprise and thus her suitability to become, in Stahl's adaptation, a literal sign. Hiro (105) argues that Hurst's objectifying descriptions of her Black characters, who even when weeping can seem 'little more than faces, masks, or surfaces', are directly opposed to the moral physiognomy of melodrama, in which ethical status is revealed by appearance, particularly facial features and expressions. Where Hurst's language denies this visual epistemology, Stahl's filming of the large-eyed Louise Beavers in tight close-ups is clearly designed to convey Delilah's virtuous suffering, and thus make the spectator feel the pathos of her daughter's rejection. In Brecht's apt description, the spectator's sorrow is stimulated by the weeping of the film's characters – ensuring cross-ethnic identification by amplifying Delilah's tears with Bea's at her deathbed. But Stahl also reverses melodrama's poetics of disguise and disclosure to demonstrate how the authentic individual can be made to assume a mask. So doing, he engineers a far more reflexive meditation on the racial representation of his own medium than Hurst ever attempts, notwithstanding the modernist distancing of her descriptions.

Both versions of the story acknowledge the mistress's determination of the maid's image, but only the film maker shows, in reverse shots, reframings and temporal duration, how Bea 'directs' (Courtney, 2005: 159) Delilah to assume a broad, blackface smile for the sign she wants painted and hold it. And hold it. Delilah's freeze into this wide-eyed rictus – for 30 seconds as Bea instructs her decorator – introduces

Figure 5.1: Delilah Johnson (Louise Beavers) poses for the shop sign in *Imitation of Life* (John Stahl, 1934)

Source: Courtesy of Universal Studios Licensing LLC.

an uncharacteristic stillness into the motion picture, anticipating her ultimate loss of life as well as the loss of personhood perpetrated by the painted, printed and neon-lit caricatures that follow (see Figure 5.1). When Bea tells the decorator that she wants the sign to read 'Aunt Delilah's Homemade Pancakes', the film acknowledges the emblematic figure that Delilah is instructed to copy (see Figure 5.2). 'I'm going to make you famous yet', the laughing Bea promises, in a phrase that suggests both the film director and the film industry's own contribution to these stereotypes, displaced on to the labels of the pancake mix soon to be speeded along a production line. In an expansion of the novel's critique, its definition of imitation as the mass production of commoditized semblance is enlarged to indict Hollywood's manufacture of racist stereotypes.

Stahl's film provoked a wide variety of responses, pleasing many Black spectators with its sympathetic treatment of the victimized Delilah, while infuriating Black critics like Sterling Brown. In a 1935 slating of the film, Brown seized on a line from Bea's business manager (Ned Sparks) when Delilah refuses the offer of a 20 per cent share in the profits of their rapidly expanding company and the personal independence it would bring: 'My own house? You gonna send me away Miss Bea? … How

Figure 5.2: The completed sign for 'Aunt Delilah's Pancake Shop' in *Imitation of Life* (John Stahl, 1934)

Source: Courtesy of Universal Studios Licensing LLC.

am I gonna take care of you and Miss Jessie if I ain't here? ... I's yo' cook and I want to stay your cook ... You kin have it: I make you a present of it'. 'Once a pancake, always a pancake', the exasperated manager replies. By the 1930s, 'pancake' had become, courtesy of Aunt Jemima, a Black epithet for what Hurston's own 'Glossary of Harlem Slang' calls a 'humble type of negro' (Hurston, nd). In Brown's excoriating review, Delilah's self-abnegating goodness is traced back to 'the old stereotype of the contented Mammy'. Notwithstanding the film's welcome move from the ridiculed laziness of the comic characters of Stepin Fetchit to the serious portrayal of Black labour, Brown perceives no departure from the traditional mammy's willing sacrifice and accuses both Hurst and Hollywood of replicating racist Southern fiction (Brown, [1935] 1996: 288–289). But the care taken in the film to remind spectators of the real-life breakfast food in question (unlike Hurst's waffles) and its appropriation as a term of abuse warrants further consideration.

If Delilah – flattened by humility, exhaustion and her place at the bottom of the social hierarchy – is, as the term suggests, Black on the outside but White in her fundamental loyalties, what do we make of Peola? Her dilemma, as she points out to Bea, is the opposite: 'to be black and look white'. Courtesy of Stahl's film, 'Peola' would also become Black slang, for an African American who in Brown's description 'wants to be white in the worst way' (Brown, [1935] 1996: 287), but the irony of this 'whitish' woman is that there is no contrast between internal desire and external appearance, no attempt at imitation. Peola looks White because, in her ancestry, she is. So too the pale brunette Fredi Washington, whose ambiguous appearance was both sought in her casting and darkened by low lighting and the capacious embrace of Louise Beavers' Delilah in what may have been an attempt to obscure the racial uncertainty feared by the Production Code administrators (Courtney, 2005: 166–168). Ironically accused of racism because of her performance of Peola, the young actress refused the studios' inducements to pass and affirmed her African American identity, founding in 1937 the Negro Actors of America to campaign against prejudice in the entertainment industry. Unable to play the romantic roles proscribed by the Code and too pale to play maids, she worked in theatre, journalism and as a civil rights campaigner, declaring in 1945 her opposition to hiding 'the fact that I am a Negro, for economic or any other reasons' (quoted in Conrad, 1945).

Imitation of life: Douglas Sirk (1959)

Unlike Hurst who cites Hollywood stardom as part of the larger celebrity culture that prompts consumer emulation, and Stahl who identifies Hollywood's racist visual coding with the Black branding of quick-service convenience foods, Douglas Sirk makes the White mistress the commodified celebrity. Here the dramatic focus of public attention is literalized with the transformation of the business woman to an aspirant actress played by Lana Turner, who had taken similar roles several times in her film career, for example *Dramatic School* (Robert B. Sinclair, 1938), *Two Girls on Broadway* (S. Sylvan Simon, 1940), *Ziegfield Girl* (Robert Z. Leonard, Busby Berkeley, 1941) and *The Bad and the Beautiful* (Vincente Minnelli, 1952). In drawing a parallel between commodification and acting, Sirk invokes the historical condemnation of the actor, whose 'art', Jean-Jacques Rousseau memorably declared, is 'of counterfeiting himself … saying what he does not think … and put[ting] his person publicly on sale' (Rousseau, 2004: 309). In this indictment Rousseau effectively equates speaking fictional dialogue with lying for monetary gain, the deliberate deceit also attributed to imposture.

Rousseau is even more strident in his condemnation of the female actor, equating her self-sale with prostitution. Sirk amplifies this accusation by casting Turner, with her scandalous personal life and corresponding star persona, in contrast to Stahl's 'class product' (Everson, 1976: 19) Claudette

Colbert. A 'glamour girl' with a brassy tinge, Turner typically conveys erotic impact with a price attached. In her film noir roles, she is the femme fatale, using her sexual wiles in performances whose duplicity is often signaled by a stylized exaggeration. In her later melodramas she exhibits an uneasy sense of sexual repression, a corseted libido that typically erupts into tears at the film's climax. All of this depends upon what Richard Dyer calls our 'fascination … with the [star's] manufacture itself' (2004: 410), her soft features and curvaceous body increasingly defined by dieting, carefully structured costumes and make-up as she aged, as well as her aspirational assumption of an erect posture, sweeping gestures and formal facial expressions to punctuate scenes. If the moral authority of melodrama to reveal the truth of character was imperiled by the genre's waning in the late 1950s, Turner's reliance on its artifices could aptly convey the authenticity of imitation as a way of life.

Critics have long observed that there is no reliable way to distinguish the real Turner from her roles, a result of an intermingling of biography and character exceptional even for classical Hollywood. The actress married eight times and took many lovers, reportedly propositioning handsome stagehands with a masculine insouciance. In her autobiography she pleads her father's abandonment of his family to explain her penchant for brief relationships with abusive men, but this has simply encouraged speculation about Turner's manipulation of her publicity in order to pursue pleasure without censure. Unusually, instead of suppressing the scandal, the studios recognized its box office appeal. The resulting confabulation of art and life intensified with her role as the lead in Sirk's 1959 *Imitation*, renamed Lora (itself a combination of Lana and Cora, her signature temptress in Tay Garnett's 1946 *The Postman Always Rings Twice*).

Two years earlier, Turner had won a Best Supporting Actress Oscar for her performance in *Peyton Place* (Mark Robson, 1957) of a puritanical single mother zealously guarding the virtue of her teenage daughter while fending off the attentions of the attractive high school principal. When her daughter learns that her mother is not the widow she purports to be, but the former mistress of a married man, she angrily leaves town. The two are reunited when they are called to testify on behalf of young woman accused of murdering her abusive stepfather. Explaining on the witness stand how a mother may fail to protect her daughter, Turner's character collapses in tears. Incredibly, this courtroom drama was then repeated for real, when in 1958 Turner's own daughter Cheryl was arrested for the murder of her mother's violent lover, a minor mobster called Johnny Stompanato. The pretty 14-year-old claimed that screams from her mother's bedroom led her to grab a kitchen knife and run to her door. When Stompanato suddenly emerged with his arm raised above Turner, the girl had plunged the knife in, only to discover that what he held aloft was clothing removed from her closet. A more plausible explanation for the demise of the reckless lothario circulated in Hollywood, but Turner was never accused by the police. Since Cheryl's trial coincided with the filming of *Imitation of Life*, her mother initially declined the role. But Universal was unsurprisingly determined to sign her, and Sirk obligingly scheduled the scenes of Lora's tearful reaction to the death of her maid around Turner's testimony in court. After what was widely described as 'the performance of her life', her daughter was acquitted, just like Cheryl's predecessor in *Peyton Place*.

Sirk's resort to an extreme reflexivity to exploit this mise en abyme of imitation was not new to the filmic treatment of Hurst's narrative. As we've seen, Stahl had already called attention to the movies' complicity in the stereotype of the minstrel mammy when Bea directs Delilah's pose for her sign. Sirk repeatedly frames Lora in a screen-like mirror to reveal her professional dishonesty (ringing her daughter after her debut to promise that she'll soon be home), ambition (memorizing her lines looking into the glass while her maid tells the children about the star of Bethlehem) and regret (confessing her disappointment with success in her dressing room). In all three scenes the reflection suggests a self-division and duplicity inherent in performing a role. Vincente Minnelli had anticipated this use of mirrors in his own portrait of Turner as an aspirant actress in *The Bad and the Beautiful*, prompting Dyer (2004: 419) to marvel at a film 'giddy with reflection images … How can we, as we watch, pick out the levels of illusion here?'

In a rare excursion into film criticism, Judith Butler directs that question at Sirk's imitation of these imitations at a time when melodrama's mimetic credibility is arguably breaking down. She opens with a virtuoso reading of the film's title sequence, in which scores of diamonds fall into a darkened container as crooner Earl Grant (in a notable imitation of Nat King Cole) sings 'What is life without the living? What is love without the giving?' and answers, 'A false creation, an imitation of life'. In 1970 Sirk offered a commentary on this scene, as well as his film's repeated resort to mirrors and windows, with a citation of St Paul's neoplatonic observation in the Epistle to the Corinthians: 'There is a wonderful expression: seeing through a glass darkly. Everything, even life, is inevitably removed from you. You can't reach, or touch, the real. You just see reflections. If you try to grasp happiness itself your fingers only meet glass. It's hopeless' (Halliday, 1997: 51).

Butler's account, published in the same year as her highly influential *Gender Trouble*, draws a parallel between the slow motion heaping of the gems and the action of the camera in a film whose use of mirrors and frequently obstructed views refuses to 'give' back an undistorted image and instead 'accumulates the gaze'. Lora's tropic attraction to that gaze, her constantly turning toward it, is ascribed to a compulsion to perform the gestures of a fantasmatic femininity. Her repetition of this performance is also hopeless, 'compelled to fail' (Butler, 1990: 2), because no such ideal can be achieved. While applauding what one might call this post-structuralist Turner, and Butler's attention to the film's (and the original novel's) awareness of the performative constitution of gender as well as race, I want to consider Lora as the projection of a related but different fantasy, the spectator's imagination of stardom.

The title sequence shows us the descent of countless jewels, their brilliant facets set off by the black behind them, conveying many meanings: glamour as an exceptional beauty manifest in light, fakery if these jewels are glass, racial dominance in the hierarchy of light foreground and dark background, and sorrow, for many of the gems are tear-shaped. This crystalline weeping heralds the narrative consequences of a false life and this melodrama's pathos. But copious as they are, the tears don't begin to fill the space they fall into. Writing about the momentary incandescence of a falling star – which Lora's neglected daughter Susie (Sandra Dee) will wish on, or perhaps wish for – the poet A.E. Housman observes 'It rains into the sea/ And still the sea is salt.'[4] This sense of insignificance is enhanced in the title sequence by the absence of any human figure to provide a representation of individual agency or scale, an effect that is furthered by the edit that ends it. The diamonds dissolve into a seaside scene, beginning with four brief shots of a huge crowd on the beach at Coney Island, a public beach near New York City that in the 1940s sometimes accommodated a million people at time. (The vast assembly is made even more anomalous by its natural location, and the American idiom still retains the comparison 'like Coney Island' for packed outdoor gatherings. If this is Sirk's version of Hurst's Atlantic City, it vividly departs from the small town feel of its boardwalk.) Like the gems, the bathers are indistinguishable at this distance (see Figure 5.3). Immortalized in 1940 by the New York photographer Arthur 'Weegee' Fellig, who incidentally styled himself 'Weegee the Famous', this is a

---

4   'Stars, I have seen them fall,/ But when they drop and die/ No star is lost at all/ From all the star-sown sky./ The toil of all that be/ Helps not the primal fault;/ It rains into the sea/ And still the sea is salt' (A.E. Housman, 'A Shropshire Lad' [1896] in *Collected Poems* (1956: 167)).

Figure 5.3: Summertime crowds at Coney Island in *Imitation of Life* (Douglas Sirk, 1959)

Source: Courtesy of Universal Studios Licensing LLC.

crowd you could get lost in, one that immediately invokes metaphoric multitudes, the grains of sand on a beach.

A cut to a closer shot reveals another image of the faceless crowd, pedestrians filmed from below waist height walking on the boardwalk above the sand, frame left to right. When a shapely pair of bare legs enters the frame from the opposite side they are conspicuous for their movement against the flow of human traffic. Thus Lora emerges out of anonymity to take the first of many bows. As she bends towards the crowd over the boardwalk's railings, the camera pulls back to reveal her face and bosom, as well as a large banner announcing the year – 1947 – and the place, captioning her image, as it were, but without her name (see Figure 5.4). The crane shot follows her down the steps from the boardwalk to the beach, where she passes a man taking photographs of the bathers in imitation of Weegee, whose photographs Sirk is imitating. She then asks another man if he's seen a little girl in a blue sun suit. Without a friendly word the stranger scowls his denial and turns away. This evocation of the uncaring crowd echoes complaints about urban isolation made since the 19th century, notably Wordsworth's ([1850] 1954: 330) description in *The Prelude* (Book 7, 665–668) of 'the feeble salutation from the voice/ Of some unhappy woman, now and then/ heard as we pass, when no one looks about,/ Nothing is listened to'.[5]

---

[5]   For a discussion of this passage see Ferry (2002: 200–203). In *Anatomy of Myself* (226), Fannie Hurst similarly observes the 'bluish dead-faced murals of people with the unseeing stares, sitting in rows' in the New York subways.

Figure 5.4: Lora Meredith (Lana Turner) searches for her lost child in *Imitation of Life* (Douglas Sirk, 1959)

Source: Courtesy of Universal Studios Licensing LLC.

But someone is paying attention to Lora: Steve (John Gavin), the photographer who has already snapped a picture of her, directs her to a helpful policeman and later becomes her friend and eventual fiancé. The image maker is her deliverer from a frightening moment of social disregard, as becoming an image, an object of the gaze, will deliver Lora from the humiliating treatment often meted out to the unseen.[6]

One such image becomes the vehicle for Lora's first stage role, as she hopes when she agrees to appear in a magazine advertisement for flea powder: 'If I can get a job modeling at least I'll be seen.' And seen she is, by the playwright David Edwards (Dan O'Herlihy), who invites her to read for a supporting role in his new Broadway production. Although Lora is repeatedly accused of acting by those close to her, the audition is the only time we see her act professionally and she performs badly. But instead of losing the part, she criticizes the scene, persuades

---

[6] In *Death 24x a Second* (2006: 156–158) Laura Mulvey has demonstrated how digital imaging makes it possible to stop this scene and reveal a Black woman hidden by its movement and the multitude of White bathers. At normal speed the crowd is presented at first as a monochrome immensity, one that frightens and fascinates Lora. Here again Book 7, lines 722–728 of Wordsworth's *The Prelude* (331–332) anticipates this in its description of 'blank confusion! true epitome/ Of what the mighty City is herself/ To thousands upon thousands of her sons/ Living amid the same perpetual whirl/ Of trivial objects, melted and reduced/ To one identity, by differences/ That have no law, no meaning, and no end –'.

Edwards to omit it from the play, and is awarded an even larger role by the smitten writer. This instant success and its accomplishment by looks and personality rather than dramatic skill is part of the mythology of the movie star's discovery, to which Turner made a legendary contribution. Cutting her typing class at Hollywood High to buy a Coke at a malt shop on Sunset Boulevard, the 16-year-old was seen by the publisher of *The Hollywood Reporter* and recommended to the comedian and agent Zeppo Marx. Marx introduced her to the director Mervyn LeRoy and she was immediately cast in *They Won't Forget* (1937), where her tight-fitting knitwear won her the lifelong sobriquet of 'the sweater girl'. As ever in this film, Turner's biography guarantees the veracity of its fiction, complicating our sense of just what is and isn't an imitation.

What we can say is that Lora's rise to fame corresponds very closely to the fantasy identified by the psychologists who worry about fans imitating the famous, a fantasy in which the ordinary individual is propelled by an arresting personal style rather than painfully developed skills to the pleasures and powers that flow from public attention. Most recently this concern has been articulated about the kind of celebrity made available through social media and reality television, but it is worth pointing out that Sirk's film opened only two years before the publication of the most famous book on fame in the 20th century, Daniel Boorstin's *The Image: A Guide to Pseudo-Events in America*, with its dismissal of the modern celebrity as 'a person who is well known for his well-knownness' ([1961] 2012: 57), the phrase itself resounding like the echo chamber of public opinion. This is how Lora is known, not as an actress but as an accumulator of the gaze, a cynosure, a name in lights, bowing to applauding audiences, accepting their bouquets, and sweeping into rooms in yet another of the 34 different costumes she wears in the film. Their absurd glamour, together with the film's sets, furnishings and cars, are typical of the Universal films produced by Ross Hunter, who declared his intention to gratify his audience's desire to experience vicariously the life of 'beautiful women, jewels, gorgeous clothes, melodrama' (Gussow, 1996). This steeply rising extravagance seems to suggest that Lora herself is imitating her social superiors, in the Middle Ages a violation of the sumptuary laws which forbade the adoption of aristocratic luxuries by those of lower rank, in contemporary capitalism a consumer imperative.

The entrance is another of Lora's signature characteristics: most of the film's scenes begin with her walking into a room. It is as though the action of the narrative, 'life', cannot start without her presence, itself a narcissistic fantasy underlying the desire to be seen. But not to see. In a gesture typical of Lana (Dyer, 2004: 424), Lora repeatedly turns away from her interlocutor, facing the camera, and the spectator, in formal poses that solicit our attention and identification. Most strikingly, after her remarkably successful debut in his play, she declares her love for Edwards while looking away from him – lips parted, eyebrows raised, head up – towards the darkened buildings of the New York theatre district. Only at the film's end, when calamity arrives with the loss of those closest to her, does Lora become a spectator, part of the funeral congregation looking up at a real life star, the gospel singer Mahalia Jackson, in the choir loft above. Lora's subordination in this scene concludes the fan's fantasy of stardom, a mysterious rise that ends in a fall, gratifying the sadism that often follows celebrity idealization. So patent is this narrative and so pervasive, that we could say that Lora's fame is not her fantasy, or Lana's, but that of the spectators who behold her. She enacts our wishes and our punishment for their realization.[7]

Coda: *Imitation of Death* – Cheryl Crane (2012)

In a perfect conclusion to this series of imitations, Cheryl Crane has offered us a final meditation on Hollywood, celebrity, racism and mother–daughter relations. Born in 1943, Turner's daughter survived childhood abuse by her mother's fourth husband, Tarzan actor Lex Barker, her trial for Stompanato's murder, three years in reform school, a drug bust and an openly lesbian adulthood to become a Palm Springs estate agent and part-time mystery novelist. Her 1988 autobiography *Detour* accused her mother of Lora Meredith-style neglect. More recently, in what seems an act of equivocal reparation to the long deceased Turner and a blatant exploitation of her remaining celebrity capital, Crane has published a series of novels about a fictional detective whose mother is a retired but still beloved star of screen melodrama, Victoria Bordeaux, with a 'million dollar smile … that had taken her from an ordinary teenager on a stool in a soda shop on Sunset, to an Oscar nominated actress living in Beverly Hills' (56). Victoria's daughter, part-time sleuth and 'real estate agent to the stars' Nikki Harper, is the heroine of these mysteries, including the 2012 *Imitation of Death*.

Like Hurst's heroine, whose rise from pancake queen to entrepreneur is capped by building a block of repro-Georgian townhouses in New York, Nikki purveys the mimetic fantasies of architecture, Hollywood-style: mansions in Neoclassical, French Regency and Mediterranean

---

[7] See Caroline Rosenthal's observations in this collection on the punitive attitude of the audience to a public figure who enacts their desire but is then proved to be in some way false.

pastiche, a terrace 'which resembled a tropical island', *Eine Kleine Nachtmusik* on the doorbell. She sells simulations in a town in which fiction is the normative product. Her best friend is a closeted gay action movie hero. Her half-brother is an Elvis impersonator. In *Imitation of Death* she nearly blows the cover of a cop disguised as a neo-Nazi. She meets a TV chef who specializes in making sweets look like savouries and avoids a young woman who resembles 'every other young woman in L.A. – bleached blond, flat-ironed hair, and double-D enhancement' (56).

If there's a standard of value in this Hollywood hall of mirrors, it is celebrity. Crane is no Proust, but she evokes an elaborate order of power and patronage in which Victoria still has the pull to summon contemporary stars to her movie evenings, while gently condescending to the Aaron Spellingesque TV producer next door. Beneath them is a service sector of publicists, personal assistants, trainers, pool boys, drug dealers and ministers of the Scientology-like Church of the Earth and Beyond. Down their mean streets Nikki drives her Prius, in pursuit of the real killer of the TV producer's ne'er do well son so that she can exonerate her childhood friend, Jorge Delgado, the offspring of her mother's Latina housekeeper. (In homage to Sirk's 1956 *All That Heaven Allows*, Jorge has grown up to become, like so many other Latino Californians, a gardener.)

In an attempt to modernize the Sirk film by having the daughter of the White actress rescue the son of her servant, *Imitation of Death* mounts a polemic against the racist antagonism to Hispanic Americans:

> 'People are talking about how this is the case,' Ashley continued, 'that will finally force legislation controlling illegal aliens entering the country. Meaning Mexicans,' she intoned.
>
> 'But Jorge was born here!' Nikki protested. 'He's an American citizen, the same as you or me.' (106–107)

Nikki speaks some Spanish, befriends the Black TV chef who looks like Naomi Campbell and finds Will and Jada Pinkett Smith refreshingly natural when they attend one of Victoria's soirees. White people can oppose racism and Black people rise to stardom in today's Hollywood. But some things never change. Although Jorge declares no showbiz ambitions, neither Crane nor her heroine is immune to the celebrity imitation complex. Not only does Nikki revel in the attention bestowed on her mother and the overspill that she receives, she relies upon it to do her job: 'Celebrity clients liked working with Nikki because their celebrity didn't faze her; she'd grown up in the limelight of a celebrity among celebrities' (93). And when she discovers (in a blatant projection of the author's maternal ambivalence) that the killer is the victim's mother, it is Nikki who makes the cover of *People*, her image dwarfing the photo of Victoria inset beside it. In the final gratification of the aspiration and antagonism dramatized in all these imitations, the celebrity's daughter becomes the celebrity herself, appropriating not only her mother's fame but her famous features:

> She looked at the photos again. She was told all the time she resembled her mother, but she never saw it. Victoria was gorgeous, and Nikki ... while she may not have been the ugly duckling, she never thought of herself as beautiful. But looking at the two photos, she was shocked to see her mother's beauty in her own face. (295)

Acknowledgements
Work on this chapter was enabled by a 2011 Leverhulme Fellowship, for which I am very grateful. A longer version is included in Mandy Merck, *Cinema's Melodramatic Celebrity: Film, Fame and Personal Worth* (British Film Institute/Bloomsbury, 2020).

References
Aristotle (1996) *Poetics*, translated by M. Heath, London: Penguin.

Benjamin, J. (1990) *The Bonds of Love: Psychoanalysis, Feminism and the Problem of Domination*, London: Virago.

Boorstin, D.J. ([1961] 2012) *The Image: A Guide to Pseudo-Events in America*, New York: Vintage.

Brecht, B. (2001) 'Two essays on unprofessional acting', in J. Willett (ed and trans.) *Brecht on Theatre*, London: Methuen, pp 148–153.

Brown, S.A. ([1935] 1996) 'Imitation of life: Once a pancake', in M.A. Sanders (ed) *A Son's Return: Selected Essays of Sterling A. Brown*, Boston: Northeastern Press, pp 287–290.

Butler, J. (1990) 'Lana's "imitation": Melodramatic repetition and the gender performative', *Genders*, 9: 1–18.

Chan, K. and Prendergast, G. (2008) 'Social comparison, imitation of celebrity models and materialism among Chinese youth', *International Journal of Advertising*, 27(5): 799–826.

Conrad, E. (1945) 'To pass or not to pass', *Chicago Defender*, 16 June, cited in *Harlem World Magazine*, 2017. Available at: www.harlemworldmagazine.com/fredericka-carolyn-fredi-washington-harlem-video/ [Accessed 18 May 2019].

Courtney, S. (2005) *Hollywood Fantasies of Miscegenation*, Princeton: Princeton University Press.

Crane, C. (2012) *Imitation of Death*, New York: Kensington Books.

Davidson, K. (2011) *Torn Between Two Masters: Encouraging Teens to Live Authentically in a Celebrity-Obsessed World*, Portland: Tate Publishing.

Doherty, T. (1999) *Pre-Code Hollywood: Sex, Immorality and Insurrection in American Cinema 1930-1934*, New York: Columbia University Press.

Dyer, R. (2004) 'Lana: Four films of Lana Turner', in L. Fischer and M. Landy (eds) *Stars: The Film Reader*, New York: Routledge, pp 409–428.

Else, G.F. (1986) *Plato and Aristotle on Poetry*, edited by P. Burian, Chapel Hill: University of North Carolina Press.

Everson, W.K. (1976) *Claudette Colbert*, New York: Pyramid Publications.

Ferry, A. (2002) '"Anonymity": The literary history of a word', *New Literary History*, 33(2): 193–214.

Giles, D. (2000) *Illusions of Immortality: A Psychology of Fame and Celebrity*, London: Macmillan.

Goffman, E. (1959) *The Presentation of Self in Everyday Life*, New York: Anchor Books.

Gussow, M. (1996) 'Ross Hunter, film producer, is dead at 75', *New York Times*, 12 March. Available at: www.nytimes.com/1996/03/12/movies/ross [Accessed 19 May 2019].

Halliday, J. (1997) *Sirk on Sirk: Conversations with Jon Halliday* (revised edn), London: Faber & Faber.

Harrison-Kahan, L. (2011) *The White Negress: Literature, Minstrelsy, and the Black-Jewish Imaginary*, New Brunswick: Rutgers University Press.

Hiro, M. (2010) '"Tain't no tragedy unless you make it one": *Imitation of Life*, melodrama, and the mulatta', *Arizona Quarterly*, 66(4): 93–113.

Housman, A.E. (1956) *Collected Poems*, London: Penguin.

Hughes, L. (1963 [1935]) 'Mulatto', in W. Smally (ed) *Five Plays*, Bloomington: Indiana University Press, pp 1–36.

Hurst, F. (2004 [1933]) *Imitation of Life*, edited by Daniel Itzkovitz, Durham: Duke University Press.

Hurston, Z.N. (nd) *Glossary of Harlem Slang*, African American Literature Book Club. Available at: https://aaalbc.com/content.php?title=Glossry+of+Harlem+Slang [Accessed 19 May 2019].

Itzkovitz, D. (2004) 'Introduction', in F. Hurst, *Imitation of Life*, Durham: Duke University Press, pp vii–xlv.

Kempis, T.A. (1999 [1418–1427]) *The Imitation of Christ*, translated by W. Benham. Available at: www.gutenberg.org/cache/epub/1653 [Accessed 27 June 2019].

Maltby, J. and Day, L. (2011) 'Celebrity worship and incidence of elective cosmetic surgery: Evidence of a link among young adults', *Journal of Adolescent Health*, 49(5): 483–489.

McCutcheon, L.E., Lange, R. and Houran, J. (2002) 'Conceptualization and measurement of celebrity worship', *British Journal of Psychology*, 93(1): 67–87.

McCutcheon, L.E., Ashe, D.D., Houran, J. and Maltby, J. (2003) 'A cognitive profile of individuals who tend to worship celebrities', *Journal of Psychology Interdisciplinary and Applied*, 137(4): 309–322.

Mulvey, L. (2006) *Death 24x a Second*, London: Reaktion Books.

Piaget, J. (1991) *Play, Dreams and Imitation in Childhood*, London: Routledge.

Plato (1974) *The Republic*, translated by D. Lee, Harmondsworth: Penguin.

Rousseau, J.-J. (2004) 'Letter to D'Alembert', in *Letter to D'Alembert and Writings for the Theatre, vol. 10 of The Collected Writings of Rousseau*, translated and edited by A. Bloom, C. Butterworth and C. Kelly, Hanover: University of New England Press, pp 253–352.

Schwartz, H. (1996) *The Culture of the Copy*, Cambridge: Zone Books.

Stacey, J. (1994) *Star Gazing: Hollywood Cinema and Female Spectatorship*, London: Routledge.

Woodruff, P. (1992) 'Aristotle on mimesis', in A.O. Rorty (ed) *Essays on Aristotle's Poetics*, Princeton: Princeton University Press, pp 73–96.

Wordsworth, W. (1954 [1850]) 'The prelude; or, growth of a poet's mind', in *The Prelude, Selected Poems and Sonnets*, edited by C. Baker, New York: Holt, Rinehart and Winston, pp 203–438.

6

Natural Imposters? A Cuckoo View of Social Relations

Martin Abbott and Daniel Large

Introducing the common cuckoo: 'the quintessential cheat'?

Perched on a branch, a cuckoo hen watches in silence, waiting for a robin to leave her nest. Once the robin leaves to forage, the cuckoo approaches its nest, and crouches amid the robin's eggs. Within seconds the cuckoo lays its egg and flies off. In colour the cuckoo's egg approximates that of the robin. The robin soon returns to resume brooding her clutch. As far as we can tell, she does not react to the appearance of an additional egg now in her nest. On hatching, the robin raises the cuckoo nestling alongside her biological offspring. The cuckoo chick outcompetes some of the nestling robins. This results in the robin chicks' death by starvation, and the ejection of their corpses from the nest. Upon sexual maturity, the surviving offspring – cuckoo and robins alike – go on to repeat this ecological cycle.

Ecologists describe this behaviour as brood parasitism. Cuckoos are a diverse family of 141 species ranging across all continents bar Antarctica. Not all cuckoo species are brood parasites. Those that are lay their eggs in the nests of other birds. Their eggs typically exhibit morphological features resembling those of the host, as we described with the robin. Seemingly none the wiser, the host birds rear the cuckoos' offspring, often to the demise of their biological offspring. By mimicking host species egg morphology (and in some cases consuming the host bird's egg; Fromme, 2018) the cuckoo reproduces without engaging in parental care. In this way, cuckoo behaviour conjures the idea of imposture in the most intimate of contexts.

In the opening vignette, we tried to depict the cuckoo's actions in purely descriptive terms. But we found it took some considerable effort to avoid using terms like 'sneak', 'invade' or 'cheat' to refer to the events transpiring in the nest. The fact that purposive clinical detachment is difficult to maintain while writing about the nature of brood parasitism is noteworthy. As we discuss in this chapter, cuckoos' brood-parasitic lifestyles are entangled with human conceptions of imposture. Indeed, human culture is littered with a variety of potent references to cuckoos' nesting practices. Among these images, we are concerned with the notion of imposture that cuckoos' brood-parasitic lifestyles evoke.

For millennia, cuckoos' reproductive strategy has fascinated peoples around the world. Examples suggest that this recurrent fascination derives partly from general interest in the phenomenon, and from uncertainty about how to understand the bird's deviant nature. Four thousand years ago in India, Sanskrit literatures note the cuckoo was reared by other birds (Payne, 2005). Over two thousand years ago in Greece, Aristotle meditated on the common cuckoo's brood parasitism (Hett, 1936). Aristotle noted the cuckoo's 'cold nature' – which was commonly thought at the time to thwart the fertility and ability of the species to rear its own offspring – was redeemed by the bird's beguiling resourcefulness in reproduction (Armstrong, nd). While the cuckoo is now generally thought of as a parasite, in the Late Middle Ages a belief was that the female cuckoo copulated with the male foster species (Rothschild and Clay, 1952). These varying accounts concerning the cuckoo's reproductive behaviour reflect the inscrutable character of the bird that continues to grip the human imagination.[1]

In Western social contexts, descriptions of cuckoos' nesting behaviour, scientific or otherwise, appear thoroughly caught up

---

[1] More recently in music and cinema, too, the cuckoo appears over and over. American musician Cass McCombs alludes to the notion of deception in his song 'Sleeping Volcanoes' in which he describes 'cuckoo land' as 'home of the fake'. To cite another example, the English composer Benjamin Britten's whimsical tune 'Cuckoo!' was brought to new audiences in Wes Anderson's film, *Moonrise Kingdom*. Anderson (in Platt, 2012) hints that 'the color of the movie' comes from Britten's music.

with cultural and normative images of the nuclear family, caregiving, kinship and deceit (see for instance Acworth, 1946). By parasitizing the nest of another bird, the cuckoo performs a set of behaviours that offend conceptions of reproductive, familial and intimate conduct. In doing so, brood parasitism strikes a chord with 'matters of the heart', as Coopmans (Chapter 4, this volume) puts it. Such heartfelt matters give rise to deep-seated anxieties. These anxieties stir concerns regarding the ethics of reproduction, the identity of family members, and trust in intimate relations. For the same reason, vivid allusions to cuckoo brood parasitism also turn up frequently in folklore and works of literature.

A key aspect of cuckoo brood parasitism is the multiple questions it raises about the identity of the imposter. Is the cuckoo hen (and/ or her mate) the imposter? Is the egg, the nestling or the fledgling the imposter? Indeterminacy concerning identity, culpability, intent, instinct and deceit characterizes conceptions of who or what the imposter is and represents. The cuckoo nestling embodies this ambiguity. On the one hand, the cuckoo chick is hardly at fault when it hatches out of an egg laid in the robin's nest. On the other hand, the chick commits a violent act when it evicts the robin chicks from the nest (Davies, 2000). These multiple questions about intent, complicity and innocence remain unanswered. Ultimately, the uncertainty leaves the precise identity of the imposter contentious.

In this way, the cuckoo operates symbolically as an imposter figure that piques questions of identity, intent and malice that cannot be easily resolved. How the cuckoo-cum-imposter is interpreted – as hatchling or monster – matters because these contrasting conceptions elicit visceral reactions. Depending on one's vantage point, different ways of understanding the bird's lifestyle will be more salient. This in turn informs our speculations and emotional involvement concerning the identity of the imposter. The uncertain identity of the imposter, and apprehension about what the imposter will do, generates suspicion and fear in its audience. This makes the cuckoo's character unsettling and heightens the image of the imposter as a monstrous figure.

In order to analyse the entangled relations between the cuckoo and preoccupations about impostering, this chapter examines how their intermingling works in literary practices. This can better inform our understanding of how the cuckoo resonates so potently with the fear of being duped and cheated by an imposter. As we discuss in greater detail throughout this chapter, cuckoo brood parasitism has exerted a strong grip on the social imagination. Why should this bird and its reproductive habits function as such an evocative figure for stirring a preoccupation with impostering? The resonance and the appeal that the cuckoo sustains across an array of cultural examples strikes us as a remarkable puzzle.

To address this puzzle, we ask: How in practice are notions of the cuckoo and the imposter entangled, and with what dramatic effect? We consider how the cuckoo and the imposter function as a literary device by analysing texts that draw on this entanglement. In the following section, we focus on three spheres of social life in which parasitizing another's nest strikes a chord: reproductive, familial and intimate relations. We analyse select passages from *The Midwich Cuckoos* (Wyndham, 2000 [1957]), *Wuthering Heights* (Brontë, 2014 [1847]) and *Othello: The Moor of Venice* (Shakespeare, 2008 [ca. 1603]), each of which features representations that are exemplary of these matters of the heart. In each text, we examine the way in which the cuckoo is entangled with human preoccupations surrounding imposture. To understand the significance and function of these entanglements, we assess how they produce dramatic effects. In section three, we discuss our findings. We conclude that the figure of the cuckoo imposter is a potent force in disturbing and challenging social and moral orders.

The cuckoo in entanglement with heartfelt matters

A wide range of references to cuckoo brood parasitism intermingle with taken-for-granted conceptions about what is natural and normal and, especially, what is considered morally appropriate behaviour. Yet there is less detailed appreciation of how these entanglements function in practice. In this section, we look at the dynamics of these entanglements in three sources that mobilize the cuckoo in ways that heighten the moral stakes jeopardized by imposture.

Cuckoo brood parasitism elicits concerns and anxieties that are unsettling to matters of the heart. Central to this is the assumption that biological relations should trump those amongst non-biological kin. Of course, these relational matters are not always as straightforward as they are sometimes made out to be. Nevertheless, in cases where a surrogate mother births another parent's child and adoptive or foster parents raise non-biological offspring, knowledge of the 'transplant' in their midst provides some measure of confidence in the child's identity. Such knowledge informs and builds confidence in the relations between the parties and in the respective roles they are expected to play.

The cuckoo, therefore, seems to upend certain precepts underlying in Western heteronormative culture. In this chapter, we suggest that the task of disentangling cultural preoccupations with the cuckoo can be undertaken across three spheres of social life. Specifically, our analysis of literary works shows how brood parasitism challenges and disturbs social orders associated with reproductive, familial and intimate relations. If we return to the nest with which we began this chapter, the robin is (apparently) confident about the identity of her eggs, even if such confidence turns out to be misplaced. This way of disrupting the moral order of *reproduction* only succeeds if the imposter successfully evades detection – a good thing that birds can fly. By evading the notice of the host bird, the cuckoo ensures that their progeny will not only survive, but thrive, in order to repeat the cycle. The enactment of a scenario whereby a stranger surreptitiously leaves their child in the hands of unknown parents, who then unwittingly raise the imposter-child as their own – potentially at the cost of their own offspring – in turn disturbs the *familial* order. This moves us into territory that agitates moral sensibilities. The final matters that the cuckoo unsettles are basic assumptions about *intimate* relations and sexual codes of conduct.

Reproductive relations

The figure of the cuckoo disrupts the normal order of reproduction in John Wyndham's (2000 [1957]) classic science fiction novel, *The Midwich Cuckoos*. Disturbing as it may be to consider, the book elevates brood parasitism to a new level; the women of the nondescript English village of Midwich are impregnated by unknown brood-parasitic aliens. Wyndham (2000 [1957]: 54) builds on Thomas Huxley's notion of 'xenogenesis' – defined as the production of offspring permanently unlike their (host) parents – to generate consternation among the village's residents.

The concealment of these events outside of the village by government authorities for fear of public backlash fills the villagers of Midwich with indignation. In this strange context, pressure to conform with gender roles is a source of tension between the sexes. In particular, these matters arise in chapter 14, which is the focus of our analysis in this section. How the women of Midwich are treated highlights the role of gender in impostering, especially as it relates to reproduction and the stigmatization of motherhood.

During an event called the 'Dayout', an impenetrable, invisible dome covers Midwich for a 24-hour period. Put in place by aliens, the dome renders everyone inside unconscious, which enables the aliens surreptitiously to impregnate the women of Midwich. Almost all the women of 'child-bearing age' become surrogate mothers to 'golden-eyed children' who are described as, inter alia, 'intruders, changelings' and 'cuckoo-children' (Wyndham, 2000 [1957]: 92).

While the women who bear these children are central to the story, they are seldom given voice to discuss their welfare. Instead, village deliberations are dominated by the paternalistic concerns of male characters. Consequently, any mother who contemplates taking matters into her own hands or acts independently of male counsel, is slandered as 'hysterical'. Dr Margaret Haxby, for example, 'ditches' her cuckoo-child in Midwich and returns to London. Gordon Zellaby, an aristocratic character, questions the intentions of the highly educated researcher and her prerogative to abandon the cuckoo-child. Zellaby thus ponders the consequences of Haxby's departure without the cuckoo-baby: 'This could be awkward. I can see a pretty panic starting up among the other women who've taken these girls in. They'll all be throwing them out overnight before they get left in the cart, too. Can't we stall?' (Wyndham, 2000 [1957]: 94). Haxby's right to decide what is best for her own welfare, first as an individual and then a mother, is disregarded in favour of maintaining the pretence that no brood-parasitic aliens visited Midwich. Moreover, despite having been violated by the alien, Haxby is reproached for deviating from a strong reproductive norm: mothers are not 'supposed' to abandon their child, which is what makes the ditching so obscene. In Zellaby's anxious mind, were a mother to abandon her cuckoo-child, such an act might inspire further insubordination and disorder in the village.

In the following passage Wyndham invokes the image of a robin hen to normalize the 'hysterical' comportment of the women. In Wyndham's portrayal, the robin hen is said to satiate the unrelenting appetite of their cuckoo-chick at the cost of her own life. The insinuation is that women are 'naturally' unable to make rational decisions about their own welfare because they are hostage both to their maternal instincts and to the entreaties of their alien children. The robin standing in place of the mothers, as Zellaby once again reflects, is unable to overcome its 'natural' urge to protect cuckoo-children:

> That's why the poor hen works herself to death feeding the greedy cuckoo-chick. It's a form of confidence-trick, as I told you—the callous exploitation of a natural proclivity. The existence of such a proclivity is important to the

continuation of a species, but, after all, in a civilized society we cannot afford to give way to *all* the natural urges, can we? (Wyndham, 2000 [1957]: 93)

The cuckoo-children are portrayed not only as unnaturally 'evil' and 'determined survivors', but also as 'monsters' that take advantage of feminized weaknesses. Zellaby's recognition of the children's true colours resigns him to the inevitability that the mothers will 'nurture' these intruders 'and betray our own species' (Wyndham, 2000 [1957]: 98). In contrast to their surrogate mothers, whom Zellaby assumes have given in to their natural urges, the young children are assumed initially to be innocent. Subsequently, the cuckoo-children will stop at nothing to ensure their survival. They are described as a malevolent 'fifth column' working in concert with Mother Nature to overthrow human civilization. In doing so, the figure of the hysteric mother is described as entangled with this evil: 'There is no conception more fallacious than the sense of cosiness implied by "Mother Nature." *Each species must strive to survive, and that it will do, by every means in its power, however foul*' (emphasis ours; Wyndham, 2000 [1957]: 97).

The conclusion that Zellaby arrives at, unilaterally taking this matter of the heart into his hands again, is that kamikaze suicide is the only recourse once a 'nest is infested' by brood-parasitic 'monsters'. This extreme position reflects a patriarchal bias for moralizing self-destructive violence against others (that is, killing the cuckoo-children by blowing himself up) as opposed to less violent means. In the end, the cuckoo-children are deceived by Zellaby because they thought he was someone to be trusted. This juxtaposition creates ambiguity for the reader in determining precisely who best merits the appellation of imposter.

In a key passage, Wyndham addresses impostering directly. The author's insight is not necessarily as we expected. Zellaby argues that what matters is not necessarily the figure of – who or what is – the imposter, but anticipation of their next move:

> Now, the important thing about the cuckoo is not how the egg got into the nest, nor why that nest was chosen; the real matter for concern comes after it has been hatched—what, in fact, it will attempt to do *next*. And that, whatever it may be, will be motivated by its instinct for survival, an instinct characterized chiefly by utter ruthlessness. (emphasis ours; Wyndham, 2000 [1957]: 92)

Familial relations

Familial relations refer to the bonds and understandings shared between and among parents and children. These bonds and understandings involve assumptions concerning the inheritance of money, property and hereditary titles. Brood parasitism unsettles assumptions about the conveyance of these social and economic properties from parent to child. By enacting the image of a child being surreptitiously inserted into a household that is not its own, brood parasitism ruptures the familiar model of social order around family-economic relations.

In this section, we analyse how this enactment entangles taken-for-granted conceptions about what is natural and normal and, especially, what is considered appropriate behaviour between families and among family members. In particular, we are concerned with hereditary matters of economic wealth and status.

Emily Brontë's (2014 [1847]) novel *Wuthering Heights* provides a useful case to inform this analysis. Next, we excerpt passages from chapter four of the novel that pertain to one of the primary protagonists, Heathcliff Earnshaw. Heathcliff is an anti-hero character who is the subject of sustained description and representation as an ill-begotten cuckoo-child.

When asked about Heathcliff's adoptive background, Mrs Dean (the housekeeper of the manor over which Heathcliff had ultimately come to lord), described Heathcliff's history this way:

> It's a cuckoo's, sir – I know all about it: except where he was born, and who were his parents, and how he got his money at first. And Hareton has been cast out like an unfledged dunnock! The unfortunate lad is the only one in all this parish that does not guess how he has been cheated. (Brontë, 2014 [1847]: 36)

By equating Heathcliff's origin with a 'cuckoo's history' in this way, Mrs Dean raises several points about the nature and character of Heathcliff. From Mrs Dean's perspective, Heathcliff was not an innocent child in need of rescue and welcome to the home. Rather, he was an interloper who was not only illegitimate, but also aggravated relations by 'cheating' his adoptive brother out of his rightful station. Just as a parasitic cuckoo chick does in a robin's nest, Heathcliff prevents the biological offspring from claiming its inheritance. In the same vein, Mrs Dean's reference to 'money' is notable because it questions the legitimacy of Heathcliff's economic rise vis-à-vis Mr Earnshaw.

In a subsequent passage describing Mr Earnshaw's arrival at the manor with Heathcliff in tow, the author extends the comparison of Heathcliff to a cuckoo-chick. Brontë draws attention to Heathcliff's indignant and reluctant adoptive mother's suspicions of malicious intent. This highlights the stakes that would be risked if the cuckoo-child is admitted into the Earnshaw nest:

> See here, wife! I was never so beaten with anything in my life: but you must e'en *take it as a gift of God; though it's as dark almost as if it came from the devil.* We crowded round, and over Miss Cathy's head I had a peep at a dirty, ragged, black-haired child; big enough both to walk and talk: indeed, its face looked older than Catherine's; yet when it was set on its feet, it only stared round, and repeated over and over again some gibberish that nobody could understand. I was frightened, and Mrs. Earnshaw was ready to fling it out of doors: she did fly up, asking *how he could fashion to bring that gipsy brat into the house, when they had their own bairns to feed and fend for?* (emphasis ours; Brontë, 2014 [1847]: 38)

In English folklore, cuckoos were commonly associated with the devil (Roberson Wallace, 2016). Comparison with the devil, a superlatively evil character, captures the negativity and revulsion with which cuckoos were then regarded. Although the exact origins of this connection are not known, they may well have to do with the potent moral disgust that cuckoos elicited as 'homewreckers', cheating innocent host parents of the fruits of their labours. The reference to the Earnshaw's 'own bairns' reveals apprehensions about the potential for homewrecking, particularly the risk that adopting the child would come at the cost of their biological children. The implication is that the adoptive child is, by nature, less worthy of inheriting Earnshaw's wealth and status than his biological children.

Against the protest of family members for whom the figure of the 'dark', 'staring', 'foreign' Heathcliff aroused suspicion and dread, Mr Earnshaw is described as exhibiting unwavering commitment and affection in raising Heathcliff as his own son. The paternal attachment shown by Earnshaw is endearing, at least initially, despite the fear, anxiety and disdain of the other family members for the adoptive child. In the best of circumstances, adopting orphans is viewed as a virtuous act. The image of Earnshaw benefits in part from this perspective.

Nevertheless, the reader is left with a lingering sense that Mr Earnshaw was duped. This sense of deceit is heightened by Earnshaw's status as a wealthy man. In normal circumstances, it would be unreasonable to doubt the intentions of a young, innocent child. However, by portraying Heathcliff as a cuckoo-child, the boy is transformed into a precocious character poised to upset the rightful economic order. These circumstances give rise to feelings of both pity and resentment towards Earnshaw for having been so easily duped, at the cost of tarnishing his family's status and the purity of their aristocratic line, by a cuckoo. By depicting Heathcliff as a nefarious cuckoo-child, the adoptive Earnshaw becomes a 'parental victim'. Earnshaw is incapable of resisting his paternal instinct to nurture the child, whose malevolence is only obvious to other family members. As the lord of the manor, Earnshaw loses nothing from taking in Heathcliff. For the other family members, however, Heathcliff poses a threat to their inheritance.

Norms of the time dictated that the first-born son should inherit the estate (Beaumont, 2004). This order of succession was interrupted because Heathcliff grew to be his adoptive father's favourite son. As a result, Mr Earnshaw's first-born son grew to hate Heathcliff, seeing him as 'usurper of his parent's affections and his privileges'. This is why Mrs Dean describes the first-born son as having been 'cast out like an unfledged dunnock' who 'does not know how he has been cheated'.

By portraying Heathcliff as a cuckoo-child, he is cast as a monstrous figure that maliciously robs innocent family members of their rightful social and economic entitlements. This characterization is accentuated by reference to the issue of ethnicity and race. It is notable that both Mrs Dean and Mrs Earnshaw refer to the foreign-looking adoptive child in disparaging, xenophobic terms, describing him as speaking in 'gibberish that nobody could understand' and as a 'gipsy brat'. In this way, characterizing Heathcliff as an inscrutable, dark, ethnically alien character singles him out as 'other' and further imbues him with monstrous qualities. This image reflects the rightful family members fear of the cuckoo–cum–monster invading their nest.

Intimate relations

References to the cuckoo play a central role in disrupting intimate partner relations in William Shakespeare's tragedy, *Othello: The Moor of Venice* (2008 [ca. 1603]). In English folklore, the cuckoo hen is considered an adulteress because she lays her eggs in the nests of other birds. For this reason, the cuckoo hen is associated with unfaithful female romantic partners; providing the basis for the term 'cuckoldry'. As Desens (1994: 74) notes, 'references to cuckoldry are so widespread in English Renaissance drama as to suggest it was a subject of intense anxiety' (see also Maus, 1987). And indeed, matters of cuckoldry drive *Othello*'s heady, tragic plot.

Shakespeare, more than any of his peers, examined the theme of cuckoldry to the extent that his name, according to a contemporary, may have grown 'synonymous' with the concept itself (Collington, 1998). Cuckoldry features prominently in many of the playwright's works.[2] This recurrent usage points to the importance of cuckoldry in Renaissance drama which involves intimate romantic themes.

The tragedy that is *Othello* presents the playwright's 'most bleak account' and 'devastating effects' of imagined cuckoldry (Collington, 1998: 190). The play pivots around the marriage of Othello and Desdemona, a relationship lacking in trust and intimacy from the outset. As commander of the Venetian armies, Othello sets out from Venice for the island of Cyprus with Desdemona, Cassio (his lieutenant), Iago (his ensign) and Emilia (Iago's wife). Expecting to intercept an invading force of Turks, the party learns upon their arrival in Cyprus that the Turkish fleet has been destroyed in stormy seas en route to the island.

With the stage set for battle, rather than fight the enemy, the idle soldiers are instead manipulated by Iago and turn against one another. Iago schemes against Othello in this way for multiple reasons. The reason articulated initially is that Othello overlooked Iago for promotion. Underlying this, however, is Iago's fear that Othello has slept with his wife, Emilia. To take revenge, Iago insinuates infidelity on the part of Desdemona, to unsettle Othello's confidence in his wife and to ignite volatile sentiments of jealousy and betrayal that will lead Othello to ruin. The thought of Othello and Emilia betraying him in the bedroom is incendiary to Iago, as vividly illustrated in his soliloquy:

> I do suspect the lusty Moor
> Hath leap'd into my seat; the thought whereof
> Doth like a poisonous mineral gnaw my inwards;
> And nothing can or shall content my soul
> Till I am even'd with him, wife for wife. (Shakespeare, 2008 [ca. 1603]: 2.1)

---

[2] Cohen (2004: 6) notes that 'Shakespearean plots in a variety of genres [are] driven by the threat or fear of cuckoldry (*The Merry Wives of Windsor, Othello, The Winter's Tale, Cymbeline*), and others exploring the catastrophic familial and social consequences of infidelity (*Hamlet, Troilus and Cressida*)'. Plot references to cuckoldry appear also in a number of the playwright's romantic comedies: *The Merchant of Venice* (1596–97), *Much Ado about Nothing* (1598–99), *As You Like It* (1599) and *All's Well That Ends Well* (1602–3).

In this way, the fear of intimate imposters operates as a driving force to action among the male protagonists in the play. On the part of Othello, the idea of Desdemona cheating on him – having been insinuated by Iago – overwhelms Othello's insecurities. Convinced that Desdemona is an adulteress, Othello questions his wife's love and speaks of cuckoldry as a husband's inevitable destiny:

'Tis destiny unshunnable, like death:
Even then this forked plague is fated to us
When we do quicken. (Shakespeare, 2008 [ca. 1603]: 3, 3)

Tormented by this imagined cuckoldry, Othello murders Desdemona. Upon finally learning from Emilia of Iago's machinations, and realizing that Desdemona was no cheat, Othello wounds Iago and commits suicide, rather than face trial for murdering his wife. This tragic outcome highlights the stakes at play when intimate relations are invaded by suspicions of imposture. Iago is more than aware of the disruptive and malicious effects that cuckoldry can wreak on a marriage, to which his scheming against Othello attests. He describes the jealousy arising associated with cuckoldry as a monster:

O, beware, my lord, of jealousy; It is the *green-ey'd monster*
which doth mock
The meat it feeds on: that cuckold lives in bliss
Who, certain of his fate, loves not his wronger;
But O, what damned minutes tells he o'er
Who dotes, yet doubts, suspects, yet strongly loves!
(Shakespeare, 2008 [ca. 1603]: 3, 3)

Shakespeare's references to cuckoldry reflect how views concerning marital betrayal were structured by gender norms in 17th-century England. *Othello* explores the absurd and insecure imaginations of men, which Traub (1992: 1) argues reflected the 'dominant ideologies' of the period in which the play was written and performed. In Renaissance England, Kahn (in Falconer, 1998: 35) argues that love relationships were shaped by 'misogyny', 'double standards' and 'patriarchal marriage'.[3]

---

[3]   Kahn (in Falconer, 1998: 35) argues that love relationships were shaped by three main beliefs in Renaissance England: (1) 'misogyny' presumed all women were promiscuous and could not control their sexual desires; (2) 'double standards' ensured infidelity was acceptable for men and unconscionable for women; (3) 'patriarchal marriage' made women the property of their male partner.

The image of an intimate imposter in the nest (i.e. mind) of Othello and Iago hatches feelings of jealousy. This eruption of jealousy impairs both the characters' reasoning and incites unforgivable deeds and tragic consequences.

Encounters with cuckoldry on the Shakespearean stage should thus be understood as premised on a social order scandalized by acts – often imagined – of insubordination performed by a woman. A woman acting on her own desires, independently, would have been an 'affront to patriarchal social order' (Cohen, 2004: 6). With the patriarchy unsettled – based on fictional events – the wife is slandered as a whore because her act subverts the ruling order and transforms her male partner into a cuckolded 'victim' fearful of emasculation (Matthews Grieco, 2014; Alfar, 2017).

However, cuckoldry need not always be understood as immoral and thus unnatural. That which humiliates the male character is sometimes considered a source of sexual gratification. Contravening the treachery of intimate partner relations discussed thus far, cuckoldry is natural and healthy from the standpoint of some married couples ('Brian', 2019). In this other context, participants describe the thrill of acting as a 'voyeur'. This offers an alternative understanding of cuckoldry, in which one half of a couple gains pleasure from performing and being watched, and the other half derives enjoyment by watching this intimate act. While this act may no longer be on the Shakespearean stage, the acting out of such sexual preferences should be understood as no less performative.

The cuckoo: an imposter entangled?

We have shown the importance of the entanglement between the cuckoo and the imposter in three texts. In this section, we reflect on what the dynamics of entanglement that we identified in the foregoing analysis tell us about the social and moral preoccupation with imposture. To understand the function and effectiveness of these entanglements, we then assess the ways in which they produce dramatic effects.

Reproductive

In *The Midwich Cuckoos*, Wyndham uses cuckoo imagery to explore how matters of imposture can take on gendered dimensions. In the village of Midwich, brood-parasitic aliens upset villagers' assumptions about the role of men in reproduction by usurping their position as fathers. This mirrors widespread concerns and anxieties that arose amid shifting reproductive norms in mid-20th century England. Hubble contends that Wyndham, witnessing the shifts in traditional gender roles in England, envisioned 'the next generation as so different as to be alien' (2018: 42).

The entanglement of the cuckoo and the alien imposter also gives rise to contentions concerning ethics of reproduction (for example, the rights of men and women to plan parenthood for themselves). For example, when Dr Haxby ditches her cuckoo-child, she is reproached by Zellaby who worries that this act will lead to disorder by encouraging the other women to kick out their cuckoo-children onto the street. In this way, Wyndham transcends the cuckoo-as-imposter trope as pitting the villagers against aliens, and instead uses this entanglement dramatically to question the social conventions that structure reproductive relations between men and women.

At the same time, the way that Wyndham science-fictionally mobilizes the cuckoo with the imposter elicits discomfort and horror by staging an unfamiliar departure from the status quo. In sum, Wyndham's approach serves to interrogate aspects of reproductive relations that might not otherwise be subject to question except via the rupture that the cuckoo's brood parasitism produces. At stake are tasks of caring for and taking care of others paired with belonging and attachment or rejection and alienation.

Familial

In *Wuthering Heights*, Brontë's account depicts young master Heathcliff – adopted by Mr Earnshaw and introduced to the Wuthering Heights manor – as a cuckoo-child. By doing so, Brontë achieves a marked dramatic effect: the cuckoo metaphor casts Heathcliff as an evil and unpredictable character and equates him with a monster. These dark associations with the cuckoo were familiar to Brontë's contemporary audience (Roberson Wallace, 2016). This approach contributes to the broader character development of Heathcliff in later parts of the novel as a brooding anti-hero.

While the equation of Heathcliff as a cuckoo nestling makes him repulsive, he also attains a romantic status in the novel. This is suggested early on when Heathcliff is described as Mr Earnshaw's favourite son, leaving the reader to surmise that Heathcliff may have some redeeming qualities after all. Later, Heathcliff's romantic status is consolidated by his role as the subject of affection in an epic love affair. Despite his 'biologically-rightful' place in the family, with which Mrs Dean sympathizes, Hindley is also portrayed as a petulant brat. These plot lines raise questions about whom the reader should empathize with, unsettling taken-for-granted conceptions of familial order.

More generally, the arrival of Heathcliff-as-imposter in the Earnshaw family unsettles the 'rightful' modes of inheritance tied to money, property and status. The uncertainty of who the cuckoo-child's parents are, in turn elevates uncertainty regarding the identity of who the children and/or spouses are. For example, one could speculate whether Mr Earnshaw fathered Heathcliff with a mistress. Who among the characters occupies their rightful place in the family home? Might one of them be usurped by a competitor for their position? Will family members be betrayed by loved ones who turn out to be imposters? These identity issues generate dramatic intrigue because they pivot around questions of parental abandonment, sibling rivalry and the sense of a lurking imposter in the home, amid one's most intimate relations. Indeed, the stakes are heightened by the Earnshaw family's aristocratic status. In this case, imposture impinges on the transfer of substantial wealth and hereditary titles from parent to child.

Intimate

Discussions of intimate partner relations that depict cuckoldry as immoral remain commonplace. Take the *Oxford English Dictionary*'s (nd) definition of cuckoldry – 'the dishonouring of a husband by adultery with or on the part of his wife' – which appears to have changed little in more than 400 years. Cuckoldry is still widely considered, inter alia, adulterous, deceitful and unnatural. Cuckoldry, in this guise, is frowned upon because of male anxiety about looking foolish and impotent. Shakespeare uses these uncertainties and insecurities in *Othello* for dramatic effect: by staging fears of imposture at the hands of other men as unresolved and indeterminate.

Thinking with intimate imposters gets at the often dichotomous framing of human–cuckoo relations in moral and/or immoral terms. Monsters are mythical constructions that promote outcomes in such moral and/or immoral arenas. They are deployed ad hoc to steer us in particular directions to achieve desired outcomes. In *Othello*, Iago wields a monster – the fear of being cuckolded stirs jealousy, which in turn spurs action throughout the play.

At the same time, Othello and Iago can also be considered the monsters of this tragedy. They distrust their wives and view them as their material playthings. The question of who the imposter is operates on multiple levels in the play. Othello is a successful outsider who is naive to the customs and values of the upper echelons of Venetian society. The fact that Othello is so easily manipulated by Iago attests to Othello's imposter-like status. Despite this naivety, Othello remains the tragic hero of the play. In contrast, Iago is variously described as a 'viper', 'villain' and the 'devil'. For these characters, honour, pride and masculinity are at stake.

Summary

We have thus seen that the cuckoo features as a central trope in various English cultural expressions. In each of the foregoing examples, entangling characters with notions of the cuckoo and imposture unsettles aspects of social and moral orders. Where this entanglement becomes particularly effective is in revealing tensions and ambiguities facing reproductive, familial and intimate relationships. By playing on these deep-seated anxieties concerning these matters of the heart, the cuckoo-as-imposter produces dramatic intrigue.

We identify action(s) performed in the name of and/or response to the imposter as a specific dramatic effect that is produced by entanglement with the cuckoo. This point emphasizes how these entanglements are a call to action. Irrespective of whether imposters are real, imagined or fake, it is what humans and birds do in apprehension of the imposter's next move, which is informed by their *expectation* of 'what, in fact the imposter will attempt to do *next*' – to use Wyndham's turn of phrase – that drives the progression of the drama. In this sense, the figure of the imposter is both a multifaceted engine of uncertainty, and a generator of social activity. Part of the potency of the imposter, therefore, derives from its capacity to provoke unguarded reactions from others.

In addition, we can say that the issues of identity, culpability, intention, instinct, will and deceit that cuckoo brood parasitism raises, all serve as potent plot devices. Building on this perspective, we understand cuckoos as actors possessing 'lively' characteristics. This liveliness imbues them with properties that make their intersection in society hard to predict. For example, the cuckoo may disrupt or subvert comfortable, established assumptions about the natural order of things. This is particularly true given that brood parasitism blurs the social categories that we erect to moralize orderly behaviour: cuckoo behaviour is 'disorderly', yet at the same time understood as 'natural'. In the way that the cuckoo 'naturally' bridges these social boundaries, the cuckoo also legitimates rupture. For this reason, we find that cuckoo brood parasitism blends order and disorder in a way that subverts or twists taken-for-granted conceptions about what constitutes the natural or normal social order.

An analytical commitment of our study derives from the concept of positionality. We understand positionality as key to analysing impostering as an act performed before an audience. In the case of the cuckoo, when, how, and by whom imposture registers as natural or unnatural depends on the situatedness of the 'audience member' who is implicated. In these three texts, the cuckoo–imposter entanglement orders (and thus moralizes) the relations among the different actors. This set-up then affords different participants different kinds of judgements. Importantly, these judgements are not always the same, but differ from case to case.

A final concept raised repeatedly in the three texts is the allusion of the cuckoo to a 'monster'. The agency of the cuckoo-cum-monster is increased by the social ambiguity of the imposter. Given that cuckoos and imposters are disturbing or frightening, it is useful to consider these monstrous categories in terms of their social construction (Law, 1991). In this regard, Warner (1994: 28) suggests that the 'main philosophical argument' of Mary Shelley's *Frankenstein* is often overlooked: the monster's violence is learned, and not endogenous to the (human) body. In other words, monsters are produced at the hands of society. Monsters do not exist naturally; they are fabricated unnaturally, and tangled within society. Society, in turn, may be beholden to the monster it created if they are not dismantled. How and who dismantles these monsters is contested. This demonstrates that monsters are at once both alien yet intimately familiar, much like the sympathies garnered by the golden-eyed cuckoo-children, Heathcliff and Othello.

Cuckoo! The upshot of unsettling reproductive, familial and intimate relations

What then can the case of the cuckoo tell us about imposture? As a literary device, the cuckoo plays on matters of the heart in reproductive, familial and intimate relations. These matters are constitutive of social and moral orders that are taken for granted by the characters (and the audience) in each of the three texts. By operating at the level of deeper social and moral orders, the case of the cuckoo as an imposter articulates with the broader theoretical questions raised by this volume; namely, that 'impostering practices are actually necessary to generating and sustaining social relations and cultural forms' (Vogel et al, Chapter 1, this volume).

In an immediate sense, invoking the cuckoo brood-parasite as a literary device accentuates the deviation from order that has occurred. At a deeper level, however, invoking the cuckoo as a specific case of disorder also implicitly or explicitly exposes the inverse, underlying, tacit values and commitments that structure the identification of so-called deviancies in the first place. By performing this move, the disorder that the imposturous cuckoo enacts strains the established social order. This staging implicitly sets up an uncertain proposition. When the cuckoo imposter arrives, how will characters react, and in what way will the social context be affected? The cuckoo–imposter entanglement unsettles social relations in particular ways in different cases.

In Wyndham's village of Midwich, the arrival of the imposter heightens tensions in gender roles. The author uses brood-parasitic aliens impregnating the women of the village to reveal how patriarchy orders relations between women and men. In Brontë's *Wuthering Heights*, the arrival of Heathcliff as a cuckoo-child permanently disturbs familial relations, unseating Hindley as rightful heir and installing a dark brooding anti-hero as lord of the manor. In Othello, the insinuation of cuckoldry devastates Iago, leading to the tragic murder of Desdemona and to Othello's demise. In each case, the cuckoo ruptures the reproductive, familial and intimate matters that the characters hold close to their hearts.

A critical element that contributes to how the cuckoo unsettles social order lies in the inherent indeterminacy of brood parasitism. Through the multiple and uncertain guises in which brood-parasitic actors masquerade and are interpreted, the cuckoo evades clear indictment of who has committed the act of imposture; whether the imposture can be considered malicious or a happenstance; and finally, whether the hosts of the cuckoo's intrusion can be considered victims, or should instead be understood as less than innocent, or even complicit in some way. This indeterminate quality challenges the established social order and can leave the resolution to these challenges open-ended. The complexity and flexibility built into the entanglement of the cuckoo with the imposter contribute to the bird's utility as a literary device.

We conclude that cuckoos and impostering serve as an evocative and enduring literary device by virtue of how their entanglement unsettles established social and moral orders. This entanglement casts established values and commitments in a new light. The opportunity to envision social relations through a bird's eyes produces a dramatic tension. By ejecting matters of the heart outside the nest of established order, thinking with the cuckoo–imposter entanglement creates an opportunity to re-envision social and moral relations in an alien, yet natural, context.

Acknowledgements

Many thanks to the people who generously gave their time to read our chapter. We appreciate their insightful comments, suggestions and counsel. These readers include Amy Cheatle, Cecelia Madsen, Marty Miller, Aftab Mirzaei, Trishna Senapaty and Malte Ziewitz.

References

Acworth, B. (1946) *The Cuckoo and Other Bird Mysteries* (2nd edn), London: Eyre & Spottiswoode.

Alfar, C.L. (2017) *Women and Shakespeare's Cuckoldry Plays: Shifting Narratives of Marital Betrayal*, Abingdon: Routledge.

Armstrong, R. (nd) 'No. 2231: Cuckoos and cuckoldry', *Engines of Our Ingenuity*, [online]. Available at: https://uh.edu/engines/epi2231.htm [Accessed 25 June 2020].

Beaumont, M. (2004) 'Heathcliff's great hunger: The cannibal other in Wuthering Heights', *Journal of Victorian Culture*, 9(2): 137–163.

'Brian' (2019) 'Watching whenever', *The Casual Sex Project*, [online] 2 November. Available at: https://casualsexproject.com/watching-whenever/ [Accessed 25 June 2020].

Brontë, E. (2014 [1847]) *Wuthering Heights*, Minneapolis: Lerner Publishing Group.

Cohen, S. (2004) 'No assembly but horn-beast: The politics of cuckoldry in Shakespeare's romantic comedies', *Journal for Early Modern Cultural Studies*, 4(2): 5–34.

Collington, P.D. (1998) *O Word of Fear, Imaginary Cuckoldry in Shakespeare's Plays*, Toronto: University of Toronto.

Davies, N.B. (2000) *Cuckoos, Cowbirds, and Other Cheats*, London: T & A D Poyser Ltd.

Desens, M.C. (1994) *The Bed-Trick in English Renaissance Drama: Explorations in Gender, Sexuality, and Power*, Newark: University of Delaware Press.

Falconer, E. (1998) '"My life upon her faith": Love, relationships and cuckoldry in Othello and Much Ado', *Articulāte*, 3: 35–39.

Fromme, A. (2018) 'This baby bird is a mother's nightmare', *National Geographic*, [online]. Available at: https://www.nationalgeographic.com/magazine/2018/01/explore-bird-cuckoo-brood-parasitism/ [Accessed 25 June 2020].

Hett, V.S. (1936) *Aristotle: Minor Works. On Marvellous Songs Heard*, London: Heinemann.

Hubble, N. (2018) '"The choices of Master Samwise": Rethinking 1950s fiction', in N. Hubble, A. Ferrebe and N. Bentley (eds) *The 1950s: A Decade of Modern British Fiction*, London: Bloomsbury, pp 19–51.

Law, J. (ed) (1991) *A Sociology of Monsters: Essays on Power, Technology, and Domination*, London: Routledge.

Matthews Grieco, S.F. (ed) (2014) *Cuckoldry, Impotence and Adultery in Europe (15th–17th Century)*, Farnham: Ashgate.

Maus, K.E. (1987) 'Horns of dilemma: Jealousy, gender, and spectatorship in English Renaissance drama', *ELH*, 54(3): 561–583.

Oxford English Dictionary (nd) [online] Available at: https://www-oed-com.proxy.library.cornell.edu/ [Accessed 25 June 2020].

Payne, R.B. (2005) *The Cuckoos*, Oxford: Oxford University Press.

Platt, R. (2012) 'Benjamin Britten's "Moonrise Kingdom"', *The New Yorker*, [online] 6 August. Available at: https://www.newyorker.com/culture/culture-desk/benjamin-brittens-moonrise-kingdom [Accessed 25 June 2020].

Roberson Wallace, E. (2016) 'Caged eagles, songsters and carrion-seekers: Birds in Jane Eyre and Wuthering Heights', *Brontë Studies*, 41(3): 249–260.

Rothschild, M. and Clay T. (1952) *Fleas, Flukes & Cuckoos: A Study of Bird Parasites*, London: Collins.

Shakespeare, W. (2008 [ca 1603]) *Othello: The Moor of Venice*, Waiheke Island: The Floating Press.

Traub, V. (1992) *Desire and Anxiety: Circulations of Sexuality in Shakespearean Drama*, London: Routledge.

Warner, M. (1994) *Managing Monsters: Six Myths of Our Time: The 1994 Reith Lectures*, London: Vintage.

Wyndham, J. (2000 [1957]) *The Midwich Cuckoos*, New York: Rosetta Books.

7

Conjuring Imposters: The Extraordinary Illusions of Mundanity

Brian Rappert

A conjuror ... is an actor playing the part of a magician.
(Robert-Houdin, 2011 [1868]: 43)

Please, consider if you will the intrigues that the figure of the magician holds for thinking about the imposter.[1] Doubtless you appreciate that entertainment magic (aka 'modern conjuring') involves reciprocally recognized and monitored deception. While magicians might proffer all sorts of verbal and non-verbal explanations for their feats, audiences are aware that both can function as techniques of subterfuge. Magicians, in turn, craft their performances by anticipating that many eyes and ears are primed to detect tell-tale signs of chicanery. Robert-Houdin's much-quoted characterization of the conjuror suggests still another level of pretence; while the conjuror takes on the guise of a magician, this semblance is only an outward show that obscures the real role: actor.

---

[1] This chapter is the result of my interactions with many individuals. My special thanks to the editors, and to Olga Restrepo Forero and Malcolm Ashmore for their considered comments on an earlier draft. I would also like to thank all those that participated in the magic sessions analysed here because, after all, what is a magician without an audience? And, of course, thanks to 'Emma' for, well, quite a bit.

In what follows, I want to set the performance of magic and the act of impostering next to each other in order to appreciate how they can mutually inform one another. In doing so, my intent is to examine how imposters and their audiences co-constitute each another. I am going to do so from an unconventional tack. Rather than approaching conjuring from the lofty heights of internationally renowned performers – the David Copperfields, Penn & Tellers, David Blaines and so on of this world – my attention is mainly with the stuttering forays of a beginner with little claim to skill. Namely, me. It is from a position of comparative ignorance and inability that I want to voice certain appreciations about co-constitution.

Motley impostures

Let's start with some basics about persona as this can serve as a portal into many other issues. As part of their performance, conjurors frequently adopt a range of guises. Maybe you have seen one play the part of a psychic, mad scientist, clown, streetwise hustler, psychologist or, perhaps the most devilish of all, 'themselves'. Each of these guises is able to be enacted in ways more or less suggestive that the performer does (or does not) possess extraordinary abilities. Each persona can also more or less serve as a ruse for dissimulation.

Within the history of magic, an extreme example of impostering was realized by the American performer William Robinson. At the start of the 20th century, Robinson came to adopt the stage presence of Chung Ling Soo, copying the act of a contemporaneous peer named Ching Ling Foo. Pulling off this part required making various alterations in how Robinson looked, dressed and spoke (which became almost not at all) on stage. It also required a backstory of his familial lineage to throw off the scent of the curious. But Robinson (or is it Soo?) did not stop there. A persona opportunistically adopted because of a cabaret opening would become a consuming double life. When interviewed by reporters Robinson appeared as Soo speaking through translators so as to not give away his identity. In addition to turning his mistress (Olive 'Dot' Path) into his Chinese stage assistant wife, he found Asian children to pose in family photos. By the time he died on stage in 1918 during a bullet-catching trick that went horribly wrong, few outside the circuits of professional conjurors had any inkling that Soo was Robinson.[2] While

---

[2]   A life told in Steinmeyer (2005).

pretence is commonplace in magic, the case of William Robinson suggests how a performer can become their performance.

In other respects, some conjurors have warned against impostering. For instance, the manner in which supernatural powers might be evoked or gestured towards as part of performances has brought questions about when conjurors slide into what Simon During labelled as their 'dangerous doubles' (During, 2002: 81). With the vanishingly thin line between the skills necessary for illusionist shows and spiritualist séances, for instance, it is perhaps unsurprising that magic and forms of spiritual belief in the West have closely intertwined pasts. Within the histories told about this intertwining, conjurors have been variously positioned. In the past as today, some entertainment magicians have made strident efforts to debunk pretenders who are deemed as promulgating the wrong kind of misinformation – psychics, clairvoyants, spiritualists and the like.[3] In the past as today, on occasion such debunkers have even claimed superior abilities compared to research scientists because the latter's professional training makes them poorly suited to exposing wilful deception.[4]

And yet, the positioning of conjurors has not always been favourable. Concerns that the deceptions of entertainment magic might inadvertently foster mystic-making, or that practitioners might dabble in a bit of both, have been recurring themes for hundreds of years (Mangan, 2007). Too, the 'will to out' has hardly been a danger-free exercise. For instance, through his performances and writings, the illusionist Derren Brown has repeatedly sought to unmask superstition and New Age spiritualism.[5] A frequent means by which he accomplishes this debunking is to offer psychological (that is, non-paranormal) explanations for his conjuring feats (for example, the use of subliminal messaging). However, as critics have argued, such explanations have themselves often been forms of misdirection that have served to promulgate pseudo-beliefs. This is so as some of his effects are actually 'little more than clever magic tricks' (Singh, 2003). The form of magic Brown is popularly associated with – mentalism – is often regarded as ethically problematic, not least by its practitioners, because of the way it refers to mindreading powers claimed by others to be real. Thus mentalists

---

[3]   As in, for instance, Penn and Teller (1992).

[4]   For past and more recent arguments along these lines, see Pettit (2013: chapter 3) as well as Measom and Weinstein (2014), the latter being a documentary of the magician with the show name: 'The Amazing Randi'.

[5]   Not to mention beliefs about religion, alternative medicine, and environmentalism. For instance, see Brown (2006: 14–15).

hazard inducing the belief that such abilities are genuine (even if the fine print of their performance programmes disclaim otherwise).[6]

What conjurors should and should not appear to be has been taken up along other dimensions. Consider one example. Citations of Robert-Houdin's (2011 [1868]: 43) characterization of the conjuror as an actor invariably leave out a connecting clause. The original quotation reads as 'A conjuror is not a juggler; he is an actor playing the part of a magician'. By this, Robert-Houdin intended to disassociate magic proper from the plebeian performances of the market fair. In its place, he fashioned a persona for himself as a gentleman of society dressed in evening wear performing within the refined space of his classy theatre. If you have seen a magician in a top hat and a tailcoat, you have witnessed the enduring legacy of Robert-Houdin.

In other respects, those involved in conjuring have counselled one another on the hazards of the identities they assume. Henning Nelms warned young conjurors not to imitate their idols. Such strategies, he pointed out, were bound to fall flat since each performer must work with their own unique personality (Nelms, 2000 [1969]). Derren Brown (2003: 42) advised his peers against playing the 'implausible impostor' – namely, someone able to pull off feats that audiences know cannot really be done (such as floating a banknote in mid-air). Instead of playing a demigod, a bill-floater or some other farfetched figure, the conjuror should be a 'hero' who, through an intriguing personality:

> offers a connection to a secondary world of wonder, which will shine through momentarily if circumstances are arranged correctly here in the world which we experience. Let us make this trick have real meaning for the spectator, and let us give them a little cathartic journey with it that will not revolve around the mundane question of 'How did he do that?' (Brown, 2003: 42).

As Brown went on to contend, what makes magic is the ability of the performer to create drama and enchantment through the crafting of characters and scenes. As he argued elsewhere, 'magicians starting out are prone to falling victim to a rather convincing trick being played upon them: that they are better than they are' (Brown, 2010: 81). Whereas anyone can perform a shop-bought trick and call themselves

---

[6] For a discussion of these points, listen to Maven (2019).

a magician, neither the ability to pull off certain tricks alone nor the technical skills learnt define individuals as conjurors proper. As Brown lamented, it is all too tempting for dabblers in the craft with some purview of the occulted to imagine themselves able to offer criticisms of seasoned professionals (Brown, 2010: 83).

Many others have sought to define pretence against genuineness. Conjurors have long debated the rights and wrongs of employing stooge participants that pretend to be ordinary members of the public, shills hidden in the crowd that try to affect audience responses, as well as professional informers that glean the secrets of competitors.

Just as *who* should take part in magic has been a topic of dispute, so too has *what*. Robert-Houdin (2011 [1868]: 29) dismissed contemporaries of his day that relied on 'a stock of apparatus[7] that kind in which of itself works the trick'. He likened such performers to musicians that produce 'a tune by turning the handle of a barrel-organ' (Robert-Houdin, 2011 [1868]: 29). As he also contended, 'to succeed as a conjuror, three things are essential – first, dexterity; second, dexterity; and third, dexterity' (Robert-Houdin, 2011 [1868]: 29). And yet, as examined by Mangan (2007: chapter 6) among others, Robert-Houdin did not always rely on dexterity. Whatever his handiness with cards and coins, many of the signature acts he became known for – such as the Marvellous Orange Tree act in which oranges blossomed on a tree with butterflies dancing in the air – were mechanical automatons (Lamont and Steinmeyer, 2018). In general, Robert-Houdin's stage performances often utilized electromagnetic or mechanical contraptions purposefully hidden from public view in order to bolster his onstage claim to extra-ordinary abilities (Smith, 2015). Today, a whole cottage industry exists to service professional magicians, much of it striving to develop props and gimmicks that eliminate the need for dexterity, dexterity or dexterity. The ready availability (for purchase) of not only gadgetry but of performance scripts has led to the charge that many contemporary conjurors are little more than rock cover bands reproducing (without the audience's awareness) art devised by others (Greenbaum, 2019).

Skilful deceptions?

Impostering then has a complex relation with conjuring. Conjurors might usefully feign the presentation of themselves and the objects

---

7   Meaning specially designed props, gimmicks and other artefacts.

of their trade, but the scope for simulation offers a plethora of disagreeable outcomes.

The previous section did more than draw some lines between conjuring and impostering. Indirectly, it hinted at some of the commonplace commitments in how conjurors are portrayed. To expand: like some popular histories of great inventors, accounts of magic often treat conjurors as larger-than-life individuals. These are figures who by their ingenuity, craft and tenacity are able to take audiences 'by the hand ... to deceive themselves' (Steinmeyer, 2003: 17). In line with this approach, histories of magic are commonly told through recounting the actions of prominent professionals.[8] These individuals are positioned onto asymmetrical vantage points from which they manipulate perceptions.[9] In this spirit, it has been said that crucial 'to the magician's art is the ability to control audiences' interpretations of what they perceive' (Villalobos et al, 2014: 638). Or, stated coarsely by Dariel Fitzkee (2013 [1943]: 71) decades ago but in a spirit with continuing relevance for ways of thinking prevalent today:

> The intended dupe of the magician's wiles is, of course, the spectator. He is the objective. All of the performer's endeavour is aimed at deceiving him. He is the obstacle the magician encounters. In him are combined the formidable barriers the deceiver must breach and the very weaknesses that make him vulnerable.

So there you go ....

As a contrast to attempting to see from great heights, let me indulge you by recounting my experience as a novice. In December 2017, I began learning card magic. Prior to that time I had no familiarity with the mechanisms of card tricks beyond a couple of crude ones I had learnt in my childhood. My learning consisted of a structured process of first reading 'how to' books, then following video instructions and then receiving personal instruction.[10] By the time of writing (Spring 2020), I have put on over a dozen public shows.

---

[8] As in, for instance, Steinmeyer (2003), Mangan (2007) and During (2002).

[9] As illustrated in Nardi (1984) and as critiqued in Rolfe (2014).

[10] For a recounting of my experiences, see https://brianrappert.net/magic

Back in early 2018, I started to video small group performances as part of my self-study. There were 30 sessions in total, across four different routines. The background and rationale for each of these routines has been elaborated in detail elsewhere.[11] In terms of what we are sharing together now, though, let me just point out that they were run in the mode of small focus groups, each combining roughly nine tricks with interspersed pre-formulated questions. I modified the questions and overall composition of the sessions on an ongoing basis in order to make my emerging reflections and lessons into topics of conversation. The 69 participants were largely university faculty, academic visitors or PhD students who, at the time, were located in universities in the UK or Sweden.[12]

I am going to focus on my first routine consisting of 13 sessions. In crafting and revising it, impostering figured as a recurring theme. While I initially toyed with the idea of claiming mental powers of 'mindreading', in the end I cast my 'powers' in these sessions as relating to 'body reading'. That entailed overtly claiming I was identifying cards on the basis of individuals' eye movement, voice, facial expressions, and so on. I decided to proffer this kind of verbal explanation (what those in the trade call 'patter') because I thought it would ambiguously (and thereby productively) resonate for many academic participants with notions of 'embodiment'. I started boning up on body language and other literature in order to weave in the appropriate parlance.[13]

In doing so my expectation was that participants would reject any suggestion that I could identify their cards because I read their bodies, while still recognizing that how we gazed, spoke and otherwise behaved bodily mattered for how the magic got accomplished overall. Looking back now, it is in no way astonishing that the ironic dissonance which I sought led to expressions of confusion about what kind of tack I was taking towards

---

[11] See Rappert (2021) as well as the 'Going On' entries at https://brianrappert.net/magic/performances.

[12] In six cases their non-academic partners attended. In 16 of the sessions my academic wife (designated as 'Emma') attended.

[13] Shortly after setting off on this path, I read Simon During's (2002) *Modern Enchantments*, and found my positioning of body reading had a historical parallel. In the late 19th century, so-called 'mind readers' neither claimed to be in contact with spirits nor to employ sleight of hand techniques. Instead, they positioned themselves as accomplishing all sorts of bizarre feats on the basis of poorly understood mind–body interactions that were becoming known through the then emerging of field of 'psychology'. Nearly a year after reading about this historical parallel, I became aware that the use of 'neuro-linguistic programing' to read facial expressions and the like is also a positioned as a means of undertaking the inexplicable. See Brown (2006: 173–188).

embodiment. What was surprising then (and remains so today) was that, on not a few occasions, participants stated that they thought I had indeed identified cards through attending to their eye movements, voice, facial expressions, and so on. My reactions to hearing such statements were complex and I felt a sense of unease. Towards the end of the sessions in the first routine, I toned down the 'body reading' mumbo jumbo. Instead of explicitly saying that I was reading participants' body language, I visibly attended to their eye movements, voice, facial expressions and so on through my eye movements, voice, facial expressions and so on in order to encourage them to infer that body reading was afoot. It is hard to be sure, but, if anything, this suggestive approach seemed to be a more persuasive basis for playing a 'body-reader' than overtly bluffing abilities. As I see it, bluntly claiming body reading facilities left participants with choices: challenge me or not, believe me or not, and so on. Enacting the abilities through embodied action without mentioning them enabled participants to draw their own inferences. As every rhetorician and advertiser knows, getting the audience to make connections for themselves goes a long way towards persuasion.

Brian-the-conjuror: cultural beliefs and the constitution of skill

Let me elaborate on another aspect of impostering you might find more relevant. Consider the following exchange. It was prompted by asking participants how they thought the tricks had been done:

1	Paula:	Well, attention is being directed.
2	Brian:	By who?
3	Paula:	By you. Yeah, yeah.
4	Brian:	How have you felt me directing your attention?
5	Paula:	Well … because it's a contract and we are here to be
6		entertained and in order to be entertained we know
7		we have to play along with the rules and you are the
8		person that is providing the rules and so you are saying
9		things like … um, check these cards, now have a good
10		look at them.
11	Brian:	Uhmm.
12	Paula:	And it is impossible for us to do that while also
13		paying lots of attention to you.
14	Brian:	Yeah.

15	Paula:	So we are having our attention drawn away from
16		where the action is going on.
17	Brian:	OK, OK.
18	Paula:	That's how I have seen, and that is how whenever
19		I have seen anything about magic that it has been
20		explained that it is just amazing that you can
21	Alex:	Draw attention.
22	Paula:	Draw attention away from what you are doing.
23		[...]
24	Paula:	Some of it is physical manipulation and you
25		having a chance to look at cards and re-arrange
26		cards in interesting ways. But in order to do that
27		surreptitiously our attention has to be elsewhere.
28		And you have to have a lot of physical dexterity.
29		And it's like playing a musical instrument, singing,
30		or, you have got to do more than one thing at a
31		time. So you have to get the patter going.
32	Brian:	Yes, yes, yes.
33	Paula:	And sound really confident as well as the fact that you
34		are surreptitiously looking at what the bottom card is
35		because in a lot of these tricks, sorry am I saying too
36		much?
37	Brian:	No, no, no, it is fine. Whatever.
38	Paula:	But a lot of tricks, what is happening is that the
39		card is either being placed on the top or the bottom
40		but seems to be concealed, but is in a prime
41		place and that means as long as you have enough
42		dexterity, you can make sure you know roughly
43		where it is.
44	Alex:	But he needs to know more than roughly.
45	Paula:	But we need to then not be distracted, we need to
46		be distracted. In some of the tricks it is easier to see
47		that happening than others.

Herein, the participants and I unfolded a sense of the scene together through the exchanges. What Paula perceived in our encounter was spoken to through reference to a prior familiarization with card tricks. In lines 5–10 she suggested that performances entailed a contract between audiences and magicians in which the former play to the rules of the latter. In lines 18–20 she compared this experience to previous encounters with magic, and grounded her statements on the basis of past exposure to explanations. Such utterances functioned in a real-time manner to develop a sense of the scene at hand and the identities of those in it. They framed our interactions in terms of distinct roles; authorized me to act as the performer; gave a gloss of our previous interactions as in line with conventional roles; accounted for limitations in being able to specify the detailed mechanisms of the tricks; and provided a resource for making sense of later interactions.[14] Subsequently in this exchange (lines 33–34), Paula would offer a general-level description for the mechanisms for tricks – an act that transgressed the proper audience role spoken to in lines 5–10. In subsequently seeking approval for this action (lines 35–36), she sought to repair any perceived transgressions. In this way, both a sense of the specific scene as well as the nature of magic as an activity was worked up through the exchange.

What I want to do now though is to consider how our moment-to-moment doings together constituted notions of identity. In particular, I want to attend to how something of an imposter was realized – namely, Brian-as-skilled-conjuror. By way of elaborating how, let me reveal one critical matter I have not mentioned yet. This initial routine did not rely on sleights of hand techniques such as palming, forcing, backslipping or passing. The tricks all fell under the label of 'self-working'.[15] Instead of deft movements, they were based on other principles. In other words, my tricks were not at all what Robert-Houdin (2011 [1868]: 29) had in mind in his requirement that conjuring entails 'first, dexterity; second, dexterity; and third, dexterity'. Rather, they are the modern card magic equivalent of the 'stock apparatus' he dismissed.[16] While

---

[14]   For an analysis of the manner in which norms and situations are mutually constituted, see Wieder (1974).

[15]   The term 'self-working' is used in magic to designate tricks that don't require much fine motor or other physical skills. The first book I learnt magic from stated that such tricks require 'no skill' (Fulves, 1976: v) though such evaluations are often disputed by magicians. In the case of my tricks in the first routine, one did require pushing a card out of the deck to glimpse it and another entailed covertly turning over a deck, acts which required some dexterity, but it would seem difficult to regard them as entailing anything like 'a lot'.

[16]   While appropriateness of the label of 'self-working' is disputed, the status of tricks as self-working can matter. For instance, at the time of writing I am seeking to gain entry into a professional society for conjurors called the Magic Circle. As part of this, apprentices must pass a performance examination. While the act can include self-working tricks, applicants are advised that 'an act consisting entirely of a succession of standard selfworking dealer tricks is unlikely to earn you sufficient marks' (see Magic Circle, 2017).

I would go on to learn classic sleights, in the first routine, I simply did not know any and, in general, could not make claim to possessing any refined abilities as a magician because I was a novice (for instance, the ability to direct attention).

And yet, in the exchange presented earlier as well as others, in their efforts to determine 'how the tricks were done', participants offered various stock explanations echoing popular understandings of the mechanisms of magic. In keeping with prevalent cultural understandings of how magic is done, they frequently attributed me with some sort of noteworthy talents. This included the ability for complex probabilistic reasoning in counting cards as well as the agility to control cards covertly in combination with distraction (for example, lines 24–26, 40–43).

Although it is not evident from the transcript, my brief responses to such suggestions ("OK", "yes") were intended as mere acknowledgement tokens of what was said, rather than as clear affirmations. Through doing so, I positioned myself as someone I had no claim to be: a technically skilled performer.

In relation to the temporal unfolding of the tricks, it is noteworthy that such explanations simultaneously supported a retrospective characterization of what had taken place, and offered a prospective basis for interpreting future actions. Any subsequent inability to detect, say, dexterous card manipulations (and as I say there was almost none of this at play) presumably stood as grounds for my abilities. In this way a kind of self-ensnarement trap was set: the more audiences attempted to discern the basis for tricks by attending to the manual details of the performance, the more evidence was established in support of my deftness – again a deftness I cannot see any basis for claiming I possessed at the time.

After the 30 recorded sessions with academics during 2018–19, I would go on to offer a series of similarly formatted public shows and related impromptu performances.[17] Again and again across these different settings, audiences would offer accounts investing me with capabilities, intentions, plans and skills I would have made no claim on. Perhaps most tellingly, this took place when I errored. On a number of occasions when the cards handling went wrong for some reason, participants offered verbal accounts that attributed some sort of skilful agency to me for this outcome. That took the form, for instance, of them reporting that they thought I deliberately intended the trick to go wrong; including because the mess-up was itself part

---

[17] See https://brianrappert.net/magic/public-shows.

of the forthcoming revelation of a greater effect. On some occasions I needed to make extended explanations that the mess-up was, well, just a mess-up and please could we move on now .... Darwin Ortiz (1994: 108), a prominent theorist of magic, spoke to this theme in relation to his experience:

> I find that if I've built up enough prestige and audience rapport, people will simply refuse to believe that I can make a mistake. If something does go wrong, as long as I can come up with any half-way credible 'out,' they will cheerfully conclude that it was 'all part of the act'.

While, as an early learner, I have often experienced a 'skilling up', professional magicians have recounted how cultural beliefs and situational expectations can result in a 'skilling down'. Darwin Ortiz (1994: 100), for instance, likewise spoke to an occasion when:

> A few of us went over to a Denny's about 2 o'clock in the morning. One of the group, a topnotch magician, insisted on doing a trick for the waitress. The trick was a true miracle in which a rubber ball penetrated the bottom of a drinking glass he'd borrowed from her. She looked at him skeptically and said, 'Do that again.' ... He did it again. She scowled and said, 'You're throwing it in the glass somehow.' End of discussion. I'm sure that if he'd done the same trick in a professional performance where people knew he was being paid a great deal of money to appear, the reaction would have been tremendous. But, to the waitress, he was just some clown at Denny's. As far as laymen are concerned, great magicians don't hang out at Denny's at 2 A.M. (If they only knew!)

Brian-the-conjuror: scientific insights and the constitution of perception

As I mentioned before, the second, third and fourth routines did utilize sleight of hand. As part of these routines, there was a converse example to the one elaborated in the previous sub-section.

But before I get into that let me take a step back and first note that there is a whole field of research called 'the Science of Magic'. In a nutshell, psychologists and others use tricks as experimental stimuli for examining cognitive heuristics and visual perception. The kinds of conclusion they have reached are pretty stark. The short of it, as Kuhn (2019: 109) has argued, is that 'all perception turns out to be an illusion'. We can miss changes that take place in visual scenes over time (even by just blinking), we can miss what is in our direct line of sight, we are constantly filling in gaps in our sensory experience with guess work, and so on. Sleight of hand techniques make use of such fallibilities to create a sense of the inexplicable.

What is interesting in light of such findings is that in only two of my sessions was there anything like a suggestion from participants that their perceptions might be significantly fallible. As reported in the quoted exchange, while participants repeatedly acknowledged that sight can be (mis)directed this way or that, what was nearly completely absent were wider claims that visual perception was meaningfully fallible or, more generally, that what was observed was contingent on the means of observing. Instead, the scene was rendered 'mundane'. I mean this in the spirit of Melvin Pollner's (1987) use of the term; the world was regarded as 'out there' and appreciation of it was delivered by the senses. Thus, rather than accounting for the magic through references to our shared limited human capabilities, attempts to explain it generally attended to *my* will, abilities, plans, doings, and so on as the one directing sight lines.

To turn the screw further, though, the two references to the limits of perception both related to 'perceptual' or 'inattentive' blindness; that is the possibility that an object in plain sight can be rendered hidden because attention is focused on other objects in our field of vision. In both, the iconic example of the 'invisible gorilla' psychological experiment was cited by participants.[18] While in one of the sessions this experiment was only mentioned in passing, in the other it figured as a recurring reference point. This latter audience consisted of three philosophers of mind, all versed in the science of human perception and, thus, the fallibility of perception. In this session, one trick entailed a participant signing a selected card. Later on that card was selected again (sleight magic!) and this time I signed it and returned it to the deck. Presto! Several minutes later the card signed by both us appeared inside a capped water bottle on the table. As part of recounting this trick, the following exchange ensued after Ben mentioned he had recently re-watched one version of the gorilla experiments:

---

[18] As recounted in Simons and Chabris (1999). If you are reading this because you don't know the gorilla experiment then you must visit http://www.theinvisiblegorilla.com/gorilla_experiment.html at this very moment and very carefully count the number of passes.

Chiara: … but this that kind of a trick, if you focus on that maybe you are a little … but with this kind of thing it makes me feel, oh crazy, because … it … there is a lot of time you still have to, it takes—

Emma: Yeah.

Chiara gestures toward bottle; then makes opening bottle gestures.

Chiara: to do this, to open.

Emma: Yeah, yeah.

Chiara: You have to

Emma: Yeah, yeah.

     [Multiple voices talking over each other]

Ben: That's how inattentive we were. That's how inattentive we were.

Emma: I mean it is good that he pulls the card and then signs.

Chiara: Maybe it was there since the beginning.

Emma: No. I don't know.

Chiara: When we started.

Ben: But he could have easily taken the bottle down like from the side.

Ben gestures with right hand moving down over the edge of the table.

Chiara: No, no, no.

Brian energetically simulates twisting a bottle cap open at the centre of the table

Brian: I did it right here in the middle of the table. Was this your card?

Ben: Really. Really. If you like played the tape and that is what happened I won't, I would not be surprised.

     [Group laughter]

Brian simulates pushing card into a bottle at the center of the table.

Brian:	I push the card down.
Ben:	I was, I was so inattentive. I was like so into shuffling it. [Laughter] You could have put on a gorilla costume.
Emma:	Yeah, yeah, yeah. Who would have noticed? [Laughter] He is naked. [Group laughter]

In this unfolding interaction, a sense of what happened was reconstructed. Inattentive blindness became an explanation that not only provided a general basis for explaining the trick, but a sense of participants' (flawed) perceptual capabilities.

OK, now let me again reveal something important: this performance did *not* rely on inattentive blindness.[19] Well, I should say not in any significant sense. The card-in-the-bottle was only readily visible on the table to the participants for several seconds before I explicitly directed their attention to it. Even if they had seen it at the start of this period, the trick would have had its intended effect. Funnily enough, the 'Card in Bottle' instructions I learnt *had* suggested making use of inattentive blindness by prescribing the card-in-the-bottle be placed in view for a lengthy period (DaOrtiz, 2018). I did not take this path though. As a beginner I adopted a far more cautious strategy for getting the card in the bottle that meant the card-in-the-bottle was only able to be seen for a short time.

In making the concept of 'inattentive blindness' relevant to our interactions, these participants thereby created a sense of what was going on and the identities and capabilities of those involved. I wish I could say these investments had been made as a consequence of an intentional plan on my part to get credit for more than I could do. I wish I could say what is frequently said by many conjurors: that

---

[19] As those seeking to act skilfully with deception, even in peer discussions, conjurors often anticipate that others might be sceptical about their claims. Attend a magic convention, and you are bound to hear meta-disclaimers like: "I will tell you the truth …", "This is not patter now …" or "Hand on heart …". So, for what it is worth, hand on heart, my performance of this trick did not rely on inattentive blindness (in any significant sense). Since you have followed the argument this far, I might as well reveal that the card-in-the-bottle was obscured by other objects on the table until moments before I brought it out into the open and then shortly afterward drew attention to it. The real question is 'How did the card get in the bottle in the first place?'. But I am not telling … (though you could read the Dani DaOrtiz text if you really wanted to know how this might be done).

they deliberately use the (typically realist, conventional) pictures of the world lodged in the heads of the audience to achieve their effects (Nardi, 1984; Hill, 2010: 142–149). I would even be happy with a more modest claim that I was able to rise to the opportunity that presented itself in the moment because of my intense preparedness.[20] Instead, though, Chiara, Ben and Emma discursively co-performed notions of their and mine capabilities.

To acknowledge these possibilities is to envision the conjuror not as a sculptor finely shaping audiences' perceptions and understandings, but rather as a surfer moving along deep currents that cannot fully be under control.

To recap then. While I did not don the attire of a fanciful character claiming mysterious powers gleaned through trials of character in the Orient, impostering was afoot nonetheless in the interactions. In telling you about some of my experiences, I wanted to bring attention to how both participants and I were mutually implicated in bringing each other into being in a type of activity typically understood as highly asymmetrical if not downright manipulative. This has not been a story of the performer from on high able to 'control' audiences. Nor have I wanted to suggest our identities simply come into being because we each played our respective roles. The situation has been far more complex and interesting. Statements about our roles and competences – what they are, what it means to follow them, how relevant they are in a specific instance – were marshalled as part of our situated ... and unfolding interactions to ... provide a sense of our identities and abilities in ways which ...

Emma:	OK, do you want comments on this now or later?
Brian:	Sorry?
Emma:	You asked me for comments, so do you want comments now or later?
Brian:	Well they seem a little inappropriate now, I am just coming to some conclusions.
Emma:	You paused.
Brian:	Yes, I was struggling for where to go with this point, so, yeah, OK, maybe comments now.
Emma:	There is lot going on here. Maybe too much. For starters, I was struck by your language; the conversational voice

---

[20] As in the example of Dynamo (2012: 270). Compare and contrast this example with the power attributed to anticipations of magic given on page 230.

you have for speaking directly to the readers. 'As you can clearly see, dear reader …' and the like.

Brian: Yes.

Emma: As far as I can tell, you are doing this to run a parallel with magic. Magicians—

Brian: OK, let's say conjurors just to be consistent and avoid confusion.

Emma: Conjurors then, invite spectators to peek behind the curtain, and you are doing much the same thing through your analysis. 'Oh you thought you knew what was going on, well look again …'. I see the parallel, it is kind of fun as a change, but what's the point?

Brian: Well, so one of the motivations is to alert readers to the conventions of social research. In one way or another I think a good deal of analysis positions itself as a form of revelation. Whereas once the world was understood in this way or that, research gets under the surface to show the actual situation to be otherwise, more complicated, counter-intuitive, etc. The trope of revealing is pervasive, even in works that might distance themselves from its common realist undertones. I wanted to bring that trope to the fore by using language that plainly entailed a revelatory move. And yes, in doing so I wanted to point to the rhetorical similarities between the performance of magic and the performance of scholarship.

Emma: I don't understand why you don't make that argument directly. Just say it. A danger is that readers will be drawn towards the superficialities of the rhetorical style and thereby away from your argument. In conjuring parlance, I guess you would call that misdirection. But I don't think you are aiming to display competency with that type of misdirection.

Brian: I see what you are saying. I could have just limited myself to attending to conjuring or maybe footnoted a few sentences that raised questions about the relevance of the notion of imposter for social research. What I am aiming for though is a kind of questioning sensibility. Let's say I am trying to evoke a curiosity born from the manner the reader is positioned and stirred into engagement. What is this? What's going on now? And now? That's that kind of spirit I want to engender.

I think this is in line with the editors' stated intention in the Introduction to develop a set of sensitivities through this volume.

Emma: I understand but I imagine the parallels you want to draw are simply going to be missed. You mentioned before with your first routine that people didn't cotton on to the irony of it. I think the same is going to happen here unless you undertake some massive revisions.

Brian: Sure. I can try to find some ways to state what I am seeking to accomplish. Unconventional forms of writing tend to divide opinion, sometimes even among those writing in this form.[21] In relation to magic, though, playing around with writing forms has particular justifications. Being seen to be a bit crafty in writing can serve to mark the dissimulation skills learnt from conjuring. Lots of conjurors have choreographed very, shall I say, roguish autobiographies in which they cast doubt about the truthfulness of their own accounts. In doing so they illustrate, and not just recount, their skilfulness in working with concealment and disclosure. If Penn & Teller can milk that they are being less than straight with readers,[22] maybe I can utilize something similar. In this respect, writing unconventionally could be a way of strengthening, not calling into doubt, my authority. Or at least the authority of Brian-the-conjuror, which is something I need to pull off here too in addition to the normal demands for academic credibility. No, wait, actually there is more than that at stake. My academic credibility in this case rests on my credibility as a magician. And vice versa even too. They don't exist apart.

Emma: Funny. However, all ways of seeing are ways of unseeing.[23] I am interested in how being upfront about some aspects of your constitution as a conjuror has simultaneously pushed other matters into the shadows.

Brian: Such as?

---

[21] For what might (or might not) be an instance of this see Woolgar and Ashmore (1988).

[22] See, for instance, Penn and Teller (1989).

[23] For an analysis of how this pertains to constructivist sociological analyses, see Woolgar and Pawluch (1985).

Emma:	Well, what about that trick in the second routine where you left the room and did a card identification over the speakerphone. I did all the physical jiggery-pokery with the cards for that, but you took the credit for possessing mindreading abilities.
Brian:	Yes, that's right. Actually I forgot about that.
Emma:	Given what you have told me about the history of conjuring, it's probably not the most notorious instance of a husband-magician taking credit for his wife-assistant's labour. This particular trick seems worth a mention if you are purporting to tell how you got formed in the eyes of others.
Brian:	Yes, but for me that was not your most significant contribution. Having you at some of the sessions—
Emma:	Most.
Brian:	Yeah, most of the sessions. As I gained experience I began to recount to participants how previous participants acted. I did that to get those present to reflect on their own actions. That got them self-consciously minding what they were doing. That served to take their attention away from me. But moreover, I also I wanted to get them thinking about how my characterizations of others related to what they were doing there and then. To achieve both I needed people to take that aspect of my patter seriously and not just treat it as so much waffle. Your presence there as a witness implicitly served to warrant my accounts.
Emma:	Sort of the academic equivalent of the DVDs we watched where Derren Brown has lots of spectators on the street reacting to his tricks in order to warrant that the outcomes were not the result of some fancy video manipulation.
Brian:	Exactly.
Emma:	If I indirectly validated your arguments then, I suppose I am doing the same thing now.
Brian:	Probably.
Emma:	Thinking about this then, even if I started directly challenging what you were saying, that would function to reinforce the overall warrant for your argument.
Brian:	Well I guess in practice there would be some point at which challenge would spill into undermining. But as a starting point I would imagine most onlookers

would expect you to contest some points. So some challenge might even be a necessity. Stage conjurors and escape artists have long appreciated the importance of introducing challenge into their performances. Some have planted people in the audience whose job it is to raise set-up objections that the performer can then brush away. In fact, such challenges can be part of how the skulduggery is accomplished. 'I want to check to make sure his chains are tight', someone might interject just as a bound Houdini is about to be lowered, head first, into a cell of water. And they walk up and slip him the key to open the locks under the guise of ensuring to onlookers that Houdini is tightly bound. There is no end to this sort of thing. In this sense, disruption and disordering are not well conceived as the opposites of fixing and ordering.

Emma: This work is insidious.

Brian: Maybe, no, yeah, it is. But so are a lot of things. My reading of Erving Goffman[24] is that he regarded the strategies used to corroborate our trustworthiness in everyday face to face interactions as the very same ones we use to dissimulate. This condition does not give way just because one moves from a magic performance to a piece of social research to an everyday conversation to a fictionalized storyline.

Emma: Is that one of your points, how much can become grist to the mill for stories of individuals and their achievements?

Brian: But in my case my reliance on others goes much further. A great thing with performing the recorded session with academics is that not only do I learn from their reactions, they also brought insights from their fields. Despite having a PhD in sociology, I didn't know that much about Goffman. I learned all sorts of things through these sessions though. Others have helped me think about the ethics of care, the notion of communities of practice, the meaning, no, I should say the thousand and one meanings of intersubjectivity.

Emma: I suggested examining magic as a form of distributed agency. I think the person you named 'Ben' had already suggested participatory sense making too.

---

[24]   See Goffman (1956).

Brian: The list would go on and on. I have had various follow-up emails proposing this or that line of thinking. Whatever conclusions I come to at the end of this work will be a product of those interactions, interactions that lasted well beyond the time I pulled a participant's card out of a deck. Or a bottle of water.

Emma: That raises lots of questions too. Academic ideas are typically born and raised within communities even if in the end authorship gets accredited to specific individuals. You need to find some way into acknowledging the contributions of others.

Brian: Hmm.

Emma: I think you are going to struggle. Trying to break or even draw attention to academic conventions is thorny undertaking. Even now, bringing me into the discussion seems highly questionable. Who is going to believe my contributions aren't anything more than a creation of your mind crafted to suit your divined purposes? At least for anyone that knows me, as soon as they read these words, I am sure they would think, 'Well that isn't the so-called Emma I know'.

Brian: Yes. It is a struggle. I want to find ways of performing conjuring and researching conjuring that pose questions about the centring and de-centring. But how to evoke the spirit of this?

Emma: But that is just the beginning of your troubles. As I have understood it, you want to both mobilize and deconstruct a sense of the facts-of-the-matter. You are aiming to draw on revelatory tropes and caution about their ubiquity. You want readers to witness phenomena for themselves but then call in question how each of us does so. What else? Oh yes, and establish your credibility to likely academic-type readers while questioning certain conventions for credibility in the academe.

Brian: I suppose that, in my mind, what is really important is to ensure I speak to how actors and their audiences co-constitute one another. In doing magic I was playing the role of the performer to an audience. In so many ways though the audiences performed me. I need to find ways to evoke the subtleties of such co-constitutions. It is not as simple as saying we shift between distinct roles or perspectives. The dynamics are definitely more … umm … intimate, even if they aren't unmediated.

Emma: But how are you going to do all that? There is already too much going on. Worse, the more we discuss this, the more loose ends appear that need to be tied down. Really, the more that is said, the more becomes notable as unsaid.

Brian: That is the trick. I cannot see this happening in any other way that the ends tie themselves. The troubles in performing conjuring need to be informed by and inform the troubles of understanding magic as an interactional activity. And both of these topics need to be informed by and inform the conditions of analysis in-general. Like a compelling magic trick, I will need to use play between what is revealed and concealed to nurture, but then sustain, puzzles and troubles. Again, call the ultimate aim the development of a sensitivity. A sensitivity towards the kaleidoscope phenomena that is impostering.

Emma: I await the revision because I definitely want to read it over. Also, that sensitivity needs to felt, not just analytically dissected. So you are going to need to achieve that affect-effect for others too.

Brian: Yes, I don't see any easy answers. How to pull all that off skilfully? I have some thoughts.[25] I am certainly going to try at least to point towards something fundamental to impostering.

---

[25] Readers are asked to speculate how this outcome of the dialogue compares to the recommendations given by two reviewers to an earlier draft of this chapter. Against various points of criticism regarding the clarity of the aims of the argument, the quality of the transcriptions and their analysis, my use of revelation tropes, and more besides, these reviewers proposed that the text be left more or less as it is. But that in the dialogue I should claim a similar skill-free position with regard to my writing as I did with regard to conjuring. The very assertion of his lack of writing skills, suggested the reviewers, would have the magical reflexive effect (cf Ashmore (1989): 'the reflexive move is double edged') of convincing his readers of 'the irresistible essence of [my] clever, and accomplished (and fully reflexive) writing. **All tricks will be found to have been pulled off faultlessly!**' (bold in original). I did not opt for a 'skill-free' position in relation to my writing but, inspired by these comments, I doubled down my efforts to hold the hauntings of writing and conjuring together in a manner that informs the understanding of both, while simultaneously trying to maintain open-endedness.

Emma: How we make it happen together.

Brian: How we make it happen together.

References

Ashmore, M. (1989) *The Reflexive Thesis*, Chicago: University of Chicago Press.

Brown, D. (2003) *Absolute Magic* (2nd edn), London: H&R Magic Books.

Brown, D. (2006) *Tricks of the Mind*, London: Channel 4 Books.

Brown, D. (2010) *Confessions of a Conjuror*, London: Channel 4 Books.

DaOrtiz, D. (2018) *Card in Bottle*, Malaga: Grupokaps.

During, S. (2002) *Modern Enchantments*, London: Harvard University Press.

Dynamo (2012) *Nothing is Impossible*, London: Ebury Press.

Fitzkee, D. (2013 [1943]) 'Showmanship for magicians', in J. Jay (ed) *Magic in Mind*, Sacramento: Vanishing Inc., pp 97–102.

Fulves, K. (1976) *Self-Working Card Tricks*, New York: Dover.

Goffman, E. (1956) *The Presentation of Self in Everyday Life*, New York: Doubleday.

Greenbaum, H. (2019) *The Insider*, podcast, 18 November [online]. Available at: https://podcasts.apple.com/gb/podcast/the-insider/id1437930783 [Accessed 12 January 2020].

Hill, A. (2010) *Paranormal Media*, London: Routledge.

Kuhn, G. (2019) *Experiencing the Impossible*, London: MIT Press.

Lamont, P. and Steinmeyer, J. (2018) *The Secret History of Magic*, London: Tarcherperigee.

Magic Circle (2017) *Guide to Examinations*, London: Magic Circle. Available at: https://themagiccircle.co.uk/images/The-Magic-Circle-guide-to-examinations.pdf [Accessed 1 May 2020].

Mangan, M. (2007) *Performing Dark Arts*, Bristol: Intellect.

Maven, M. (2019) *The Insider*, podcast, 20 May [online]. Available at: https://podcasts.apple.com/gb/podcast/the-insider/id1437930783 [Accessed 12 January 2020].

Measom, T. and Weinstein, J. (2014) *An Honest Liar*, Left Turn Films.

Ortiz, D. (1994) *Strong Magic,* Washington, DC: Kaufman and Company.

Nardi, P.M. (1984) 'Toward a social psychology of entertainment magic (conjuring)', *Symbolic Interaction*, 7(1): 25–42.

Nelms, H. (2000 [1969]) *Magic and Showmanship*, Mineola: Dover.

Penn, J. and Teller (1992) *The Unpleasant Book of Penn & Teller or How to Play with your Food*, London: Pavilion.

Pettit, M. (2013) *The Science of Deception*, London: University of Chicago Press.

Pollner, M. (1987) *Mundane Reason*, Cambridge: Cambridge University Press.

Rappert, B. (2021) 'Pick a card, any card', *Secrecy & Society*, 2(2), https://scholarworks.sjsu.edu/secrecyandsociety/vol2/iss2/8.

Robert-Houdin, J.E. (2011 [1868]) *The Secrets of Conjuring and Magic*, translated by Louis Hoffman, Cambridge: Cambridge University Press.

Rolfe, C. (2014) 'A conceptual outline of contemporary magic practice', *Environment and Planning A*, 46: 1601–1619.

Simons, D.J. and Chabris, C.F. (1999) 'Gorillas in our midst: Sustained inattentional blindness for dynamic events', *Perception*, 28: 1059–1074.

Singh, S. (2003) 'I'll bet £1,000 that Derren can't read my mind', *The Daily Telegraph*, 10 June. Available at: https://www.telegraph.co.uk/news/science/science-news/3309267/Ill-bet-1000-that-Derren-cant-read-my-mind.html [Accessed 10 December 2019].

Smith, W. (2015) 'Technologies of stage magic', *Social Studies of Science*, 45(3): 319–343.

Steinmeyer, J. (2003) *Hiding the Elephant*, London: William Heinemann.

Steinmeyer, J. (2005) *The Glorious Deception*, New York: Carroll & Graf Publishers.

Villalobos, J.G., Ogundimu, O.O. and Davis, D. (2014) 'Magic tricks', in T.R. Levine (ed) *Encyclopaedia of Deception*, Thousand Oaks: Sage, pp 636–640.

Wieder, D.L. (1974) 'Telling the code', in R. Turner (ed) *Ethnomethodology*, Harmondsworth: Penguin: pp 144–172.

Woolgar, S. and Pawluch, D. (1985) 'Ontological gerrymandering', *Social Problems*, 32: 214–227.

Woolgar, S. and Ashmore, M. (1988) 'Introduction to the reflexive project', in S. Woolgar (ed) *Knowledge and Reflexivity*, London: Sage: 1–11.

8

States of Imposture: Scroungerphobia and the Choreography of Suspicion

James Kaufman

Introduction

Recent decades have witnessed the emergence of an international agenda to reorganize welfare states and their social security provisions around the 'activation' of benefit recipients, often in the form of 'welfare-to-work' programmes. In the United Kingdom this agenda has been pursued through an emphasis on 'behavioural conditionality', whereby entitlement to benefits is made conditional upon the 'active job-seeking' or 'work preparation' efforts of claimants, whose conduct becomes the object of ongoing monitoring. Controversially, failure to meet these conditions results in punitive and potentially lengthy benefit sanctions – the withdrawal of benefit payments. In the context of pervasive political and media narratives about 'benefit scroungers', questions about entitlement frequently become freighted with concerns about fraud and imposture and, for claimants, with an urgent need to appear genuine and deserving.

This chapter explores the choreography of suspicion in welfare-to-work services. It begins by situating these contemporary social policies with respect to 'scroungerphobia' and long-standing anxieties about skivers and benefit cheats. It then describes these policies and their recent development in more detail, with a particular focus on behavioural conditionality. This section underlines the way policy discourse has often drawn a distinction between practices of support and practices of sanction. Based on fieldwork, the next two sections elaborate the everyday experience, practice and management of suspicion in welfare-to-work services. They show how uncertainty is produced and maintained, and how suspicion is administered and made durable through bureaucratic organization and practice. Here the chapter underscores how this suspicious mode of disordering permeates the field. It is not simply the claimant who is rendered suspect, their status indeterminate, but the meaningfulness of welfare-to-work 'support' and those who provide it. The final section considers moments when the administration of suspicion fails or breaks down, and the state and its agents become the fakes and frauds.

Scroungerphobia and the anxieties of imposition

Like the other contributions to this volume this chapter uses 'the imposter' as an analytic device. This perspective is sensitive to the sorts of anxiety about status and entitlement that frequently attend the welfare state, and which policies of behavioural conditionality and sanctions simultaneously heighten and promise to assuage. The word imposter is rarely encountered in this context, however, and imposter dramas instead revolve around the figure of the undeserving claimant or 'benefit scrounger'. This figure can stand for different forms of imposture. It might imply unambiguously fraudulent claims and the cunning deception of bureaucratic authority. Or it can suggest more ambiguous forms of imposture as imposition – opportunistic freeloading or the indolent 'shirking' of responsibility. Enduring anxieties about imposture in this context can be harnessed, heightened and deployed by those seeking political consent and legitimacy, particularly when advancing controversial policies or during moments of political crisis. This heightened and crafted anxiety about welfare imposters is referred to as 'scroungerphobia' (Golding and Middleton, 1983; Shildrick and MacDonald, 2013).

Anxieties about cheating and freeloading pre-date the modern welfare state, dating back at least to the Elizabethan Poor Laws and their arrangements of outdoor relief. Suspicions about deservingness and entitlement have long been tangled up with moral economies of exchange, dependency and the work ethic (Fraser and Gordon, 1994). Howe (1990) suggests that notions of the deserving and undeserving poor are integral ideological components of liberal individualism, gaining prominence with the development of industrial capitalism and waged labour. Shilliam (2018) has shown how these ideas are imbricated with the colonial foundations of capitalist development, and have been racialized and deployed in processes of racialization from the outset. Tyler (2013, 2020) has explored the contemporary cultural political economy of these ideas, in particular their role in processes of stigmatization and as part of 'the machinery of inequality'. Anxieties about undeserving 'scroungers' are thus long-standing and symptomatic of some of the deeper and more durable (dis)orders of the societies in which they are found.

These enduring anxieties about cheats and freeloaders can be crafted, harnessed and deployed in more local contexts for particular political ends. Narratives of imposture mobilize powerful tropes and motifs for those seeking political legitimacy and consent. In the US the 'welfare queen' condenses various misogynist and racist stereotypes and has been recurrently deployed to justify processes of welfare retrenchment from the 1980s onwards (Hancock, 2003) – processes which have served as model and inspiration for British policy makers. The UK expansion of welfare-to-work and behavioural conditionality in the early 2000s was accompanied by a series of 'public information campaigns' that sought to raise the profile of benefit fraud (Connor, 2007). Television and billboard adverts portrayed benefit fraudsters as a hidden but menacing threat to society and encouraged people to be vigilant and report their friends and neighbours using a telephone hotline. Adverts addressed putative fraudsters directly with the phrase "We're onto you". More recently, seeking consent for post-2008 crisis austerity measures, the Conservative chancellor George Osborne sought to 'weaponize' welfare policy (Jensen and Tyler, 2015) by talking up the problem of 'skivers' who avoided responsibility by claiming benefits while others went out to work. Drawing its dividing line simply between those who work and those who do not, the rhetoric of skivers helped shift attention towards those claiming sickness and disability benefits – people who previously tended to be considered among the genuine and most 'deserving' of claimants but who are now also subject to stringent behavioural conditionality (Garthwaite, 2011; Byrne and McEnhill, 2014).

Scroungerphobia is not confined to political discourse but is part of a broader cultural political economy in which television and the tabloid press are key conduits. The post-2008 austerity period witnessed a flourishing of the 'poverty porn' television genre, represented by shows such as *Britain on the Fiddle* and the notorious *Benefits Street*. The latter claimed to be an 'observational documentary' but presented highly stereotyped images of poverty, ideally crafted to enter into a 'fast media' circuit and its affective economy of sensation and outrage (Jensen, 2014; Jensen and Tyler, 2015). *Benefits Street*'s depiction of widespread and casual benefit fraud led to questions in parliament and on the flagship BBC topical debate show *Question Time*, and for a period framed much public political discussion of social security and welfare issues. Among other things, scroungerphobia refers to the way that distorted and highly manipulated images of poverty and need can be used to shape political action and policy.

Welfare-to-work and behavioural conditionality

While suspicions about imposture are long-standing in this context, the specific conditions and criteria regulating claims for benefits in the UK have undergone significant change. For much of the second half of the 20th century the social security system focused on tests to establish peoples' circumstances, to assess their ability to work, or to categorize their needs according to relevant bureaucratic criteria. Critiques of this settlement argued that it was too 'passive' and fostered long-term economic inactivity, dependency and demoralization (Mead, 1997; Murray et al, 1999). These critiques, though widely discredited, have nonetheless been politically influential and since the 1990s there has been an increasing emphasis on the 'activation' of supposedly inactive claimants (van Berkel and Valkenburg, 2007). In terms of bureaucratic criteria of eligibility and entitlement this has been accompanied by a shift from conditions of category and circumstance (for example, whether one is unemployed or happens to be living on a low income) to 'conditions of conduct' or 'behavioural conditionality' (Clasen and Clegg, 2007). This form of conditionality, extended and intensified across various domains of the welfare state (Dwyer, 2004; Dwyer and Wright, 2014), involves the ongoing regulation of entitlement based on the actions and conduct of those making a claim. One way of understanding this development is in terms of a dissolution of trust and solidarity, and the institutionalization of suspicion.

In practical terms this has meant that, from the 1980s onwards, the monitoring of claimants' conduct and work search activity has steadily increased. So too has the power of bureaucrats to suspend or stop benefit claims when suspicions arise about the authenticity of a claimant's efforts to find work. While this shift towards monitoring claimant conduct was initially confined to unemployed claimants, it has since been extended to those claiming on the basis of parental responsibilities, disability and ill-health. The replacement of Incapacity Benefit (IB) with Employment and Support Allowance (ESA) in 2008 introduced requirements that those claiming on the grounds of health or disability demonstrate their efforts to 'prepare' for a return to work by attending regular 'work-focused interviews' (WFIs). Behavioural conditionality of this sort has also been extended to those on low incomes claiming in-work benefits, who are now required to evidence their ongoing efforts to secure more or better paid work (Dwyer and Wright, 2014). Failure to comply can have immense consequences. A recent report by the UN Special Rapporteur for extreme poverty and human rights identified the UK's punitive sanctions regime as a key driver behind rising poverty and an explosion of homelessness, rough sleeping and foodbank use (Alston, 2019).

The increasing use of behavioural conditionality as a social policy tool has made understanding its everyday dynamics an important task, and it has been the subject of extensive debate among academic social policy researchers. There is a broad critical consensus that this agenda is ill-conceived and deleterious in its effects on those seeking support from the welfare system. Ill-conceived because it is based on empirically unsupported and widely discredited concepts: the underclass, cultures of dependency, narrowly calculative theories of agency (Wright, 2012, 2016). Harmful because it forces people (further) into poverty. However, there remains a tendency within policy debate, which necessarily criss-crosses various discursive fields (that is, between practice, research and politics), for certain broad categories of description and understanding to organize discussion and disagreement. Debates about conditionality have often been framed as a question of finding the appropriate balance between 'support' and 'sanctions', holding these two terms apart as if they named qualitatively distinct and easily distinguishable phenomenon (Dwyer, 2016).

Questions of balance have been a recurrent feature of the political rhetoric used to justify more stringent conditionality, such as New Labour's much-repeated rhetoric about balancing 'rights with responsibilities'. When introducing tougher conditions or a harsher sanctions regime, politicians have often been keen to suggest that these are really about strengthening genuine rights of entitlement. Greater compulsion and sanctions are accompanied by an emphasis on the support available to genuine claimants. The suggestion is that those who fall foul of the system's more punitive provisions must, in some sense, really deserve it. One of the arguments of this chapter is that 'sanctions' and 'support' are not always experientially distinct features of conditional social security. This has to do with how the 'support' on offer is more often than not experienced as a test designed to identify imposture – a test that is always received under the shadow of suspicion and the threat of sanction. If welfare-to-work sustains a mode of disordering in which claims to entitlement are rendered indeterminate and persistently suspect, in practice this also has the effect of blurring any clear distinction between sanction and support.

Street-level ethnography

This chapter is based on a street-level ethnography of welfare-to-work services. It draws on 30 interviews with benefit claimants and job coaches, as well as my own participation in this field as a job coach – someone responsible for helping people find work, but also for monitoring their job search and 'work preparation' activity. This involvement – for nine months during 2008–9 – played an important role in this research, not least in terms of facilitating access to coaches, but also in terms of an embodied understanding of the context in which they worked, with its particular pressures and dilemmas. The fieldwork interviews on which much of this chapter is based were conducted in 2014–15 as part of ESRC-funded doctoral research exploring the street-level implementation of behavioral conditionality in welfare-to-work services.

The street-level approach to policy analysis sees the interactions between citizens and low-level employees of the state (or state-funded services) as a site of policy making (Lipsky, 2010). The focus on interaction and practice distinguishes it from more discursive forms of policy analysis, or analysis focused on official policy. This also distinguishes it from more top-down approaches to policy analysis, which have tended to view what goes on in places such as welfare-to-work offices in terms of either compliance or deviance. Rather differently, street-level analysis focuses on coping. In articulating the case for street-level research, Lipsky drew attention to the various ways that street-level workers must, of necessity, make important decisions in the course of their work which cannot be envisaged or legislated for in advance. In doing so they mediate a variety of frequently conflicting demands and policy goals, and the coping strategies they invent in the process effectively become the policies they carry out. They are also the form of policy as it is encountered by citizens. However, one consequence of adopting the imposter analytic in this context is to draw attention to those occasions when coping fails. In this way the imposter analytic opens street-level analysis, not simply to coping and the achievement of order, but to modes of disordering that sustain policy ambiguity and beyond formal settings and into the sites where these are usually considered to be resolved.

'Scroungerphobia' and the uncertain drama
of suspicion

The drama of imposture, as it played out in welfare-to-work services, was often intensely personal. Suspicions arose between individuals in face to face encounters, and involved the interpretation of words, tone of voice, expression and body language. They might be founded on interpersonal absences such as missed appointments and unanswered calls, or on small and seemingly incongruous observed details of someone's life. There was also something reflexive or recursive about suspicion in this context – being suspect was, for claimants, often to suspect the suspicions of others. One consequence of scroungerphobic policy and practice was to make claimants self-consciously preoccupied with how they appeared to others (Howe, 1998).

For their part, welfare-to-work organizations did not discourage – and in many respects actively cultivated – a suspicious attitude towards claimants. This is something that one participant, Alexandria, recalled as particularly striking on her first day at work as a job coach. Taking the role because she was interested in the idea of coaching, she also imagined it would be a good way to put her education in psychology to good use. Expecting to find an organization staffed by similarly earnest and enthusiastic people, she was instead

> '[S]urprised by the [other advisors], they were very jaded and cynical about every single person who walked in the door, and there was an assumption, one of the managers said on the day, just assume that the majority of people are already working and your job is just to legitimize what they're doing ... cash in hand, claiming benefits on the side. So assume that there's some kind of trickery going on.' (Alexandria, work coach)

Many of Alexandria's ideas about the sorts of problem addressed by welfare-to-work programmes were congruent with the broader field and political terrain. She agreed with many of its paternalist assumptions – that economic inactivity and unemployment might be problems of psychology and culture, of demoralization and dependency. However, her assumptions and preconceptions diverged from the field in one important respect. Inclined to believe that most people were claiming honestly and in good faith, Alexandria encountered a working environment where fraudulence and imposture were assumed to be widespread, and where a suspicious attitude was part of the expected professional disposition.

On the other side of the desk, claimants were intensely aware of the doubt and suspicion with which they were regarded. For those unaccustomed to making social security claims this could be a surprising and discomfiting experience (although others assumed and anticipated a degree of doubt, hostility, and suspicion). Jill was a young unemployed person who had until recently been homeless, sleeping on different friends' sofas. For Jill, bureaucratic encounters with unhelpful and sceptical functionaries were part of the problem of being unemployed and homeless. In the past Jill had experienced particular difficulty making her housing situation understood, and this led to various problems with her claim and payments. These difficulties made her increasingly reluctant to explain her situation unless she felt certain to be believed and understood:

> 'The way they look at you, you feel quite bad and they make you feel quite small, like a bit of an idiot and it's quite embarrassing. I don't tell them everything because they'll be thinking [I'm] just lying about that or [I'm] just saying that. If there's somebody at the Job Centre who's friendly and they know you, then you can explain to them ... If the person was not nice I would worry [about being suspected of lying] and then I would end up thinking, oh god, I'm just not going to go in so I don't get in trouble about it.'
> (Jill, Jobseeker's Allowance (JSA) claimant)

Of the different feelings provoked by the experience of being doubted or disbelieved, Jill describes embarrassment but also humiliation. Other people felt anger, indignation or frustration. In either case it was often difficult for people to know what to do with these feelings. The risk of incurring sanctions meant that people were keen to maintain good relations, and so expressions of irritation, anger or simple assertiveness carried clear risks. For Jill, feelings of humiliation and embarrassment produced a strong desire to withdraw and avoid future encounters – another risky response given the requirement to attend. Feeling doubted or disbelieved was an uncomfortable experience, but one that was very difficult to avoid or correct. Such experiences didn't entirely exclude other kinds of encounter, however, and people often mentioned workers who were more inclined to listen, assume honesty, and deal in good faith. Yet this unpredictability, and the uncertainty it produced, itself contributed to a general and diffuse sense of threat and worry.

Uncertainty, more generally, was a common theme in conversations with people claiming benefits. Although claimants usually had a very clear idea about the consequences of breaking the rules, they were often less certain about the rules themselves – what they actually meant, or how they might be interpreted by different workers. Changing circumstances and changing administrative requirements also brought its own uncertainty, since new and different imposter tests created new and different risks of being identified as an imposter. Bridget had been out of work for nearly ten years, had problems with her mental health, and although still entitled to claim Employment and Support Allowance (ESA) on health grounds, had made a voluntary decision to transfer to JSA because she was keen to start working again. This transfer, once made, was not easy to reverse, but she thought that as a JSA claimant she would receive more help to find the right job. Changing benefits changed the conditions attached to her claim, which became stricter. Whereas ESA carried no requirement to apply for or accept jobs, claims to JSA were conditional on making a defined number of applications per week, following the directions of her job coach, and accepting any 'reasonable' offer of employment. Although Bridget's health condition remained unchanged on transition to the new benefit, recognition of its limiting impact on her ability to undertake certain kinds of work became a discretionary rather than defining feature of her claim.

This change rendered Bridget a potential imposter in a way she hadn't anticipated. Having made the decision to move to JSA voluntarily, she assumed that her job coaches would continue to recognize the legitimacy of her health needs and allow her to retain control over the sorts of work she pursued. However, as time went on and advisors came and went, Bridget found that her assertion of health needs was as likely to be interpreted as evidence of 'shirking' her responsibilities as it was to be believed and taken seriously. Her health problems, when officially recognized, had secured moderations to conditionality and a loosening of bureaucratic strictures. Yet when she stepped into a new bureaucratic regime the articulation of those same problems rendered her differently, not as someone describing a health condition, but as someone making excuses – a potential imposter. Bridget found that different advisors saw and treated her as very different types of person, and she was as a result constantly uncertain about where she stood and what was expected of her:

> 'If I go up to my advisor, this is the one that's always wanting to sanction me, and if I says to her, I can just leave that and I'm going to leave it, she'd probably say to me, you've

got to do it. And then you go to another advisor that'd say you don't have to do it, [but] *you have*, and they sanction you. So you're caught, caught in the middle. I don't know if that can happen, I don't know.' (Bridget, JSA claimant)

For Bridget, as for other claimants, trying to anticipate how her situation might be perceived and interpreted was exhausting. It was, in addition, anxiety-provoking given the severe financial penalties for getting it wrong. The regime of conditionality thus created an atmosphere of fear and uncertainty even – and perhaps especially – for those keen to observe the rules of the game. This situation had the quality of a trap – something in which Bridget found herself caught. This was one version of a double bind articulated by different claimants, in which they felt compelled to focus on how they appeared to others at the expense of how they seemed to themselves. Not enough to merely be or experience oneself as being genuine, one also had to convince others of this. Paradoxically, the need and attempt to appear genuine itself became a form of dissimulation. This seemed like an impossible situation because the people, processes and procedures used to verify authenticity seemed intent on identifying and producing its opposite – pretence.

It was also an impossible situation because the bureaucratic demands of conditionality could never be satisfied. They did not end unless you found work or were sanctioned for failing the attempt. Elizabeth, a JSA claimant, told me about a time when she needed to attend a hospital appointment which clashed with her welfare-to-work appointment. On the occasion of her clashing appointments, Elizabeth had called ahead to notify her work coach that she would be at the hospital. Some days later she received a letter from the Department for Work and Pensions (DWP). Her message had not been passed on. The automated letter advised her that missing a mandatory appointment could result in a sanction. Elizabeth was both furious and upset. The process of ongoing testing was experienced as an ongoing accusation – one that magnified existing stigma – and as an impossible demand:

'You know a lot of that are on these [television] programmes, I feel like they classify everyone under, you know, the shoplifting and things. But there's a lot of people that aren't like that, and with the government cutting certain benefits, making it push, push, push. I can understand doing that, but there's people that are pushing themselves, you know? Genuinely trying. They're going – what else can I do? And they say you're not doing enough, not doing enough, not doing enough. It feels a bit … that's what's demoralising. You're going, I am doing that! You know? I'm doing everything, what else can I …? Tell me another way to do it or something else and I'll do it, but I don't know what else I can do.' (Elizabeth, JSA claimant)

Compliance with the regime of behavioural conditionality was overdetermined: it resulted from the need to avoid sanction and maintain an income, but also from the need to avoid stigma and maintain a sense of one's worth and integrity. It resulted from claimants' need to be seen as 'genuine' in various contexts. Discussing his own experiences of participating in such programmes, the writer Ivor Southwood coins the term 'non-stop inertia' to describe the energy that must be expended to get precisely nowhere (Southwood, 2011). The drama of imposture, if it wasn't to end with a sanction, often seemed interminable.

Managerialism, verification and the administration of suspicion

Bureaucratic practices significantly shaped the drama of imposture in welfare-to-work services. They played an important role in sustaining a mode of disordering in which claims to identity and entitlement were always held in doubt. Alongside the explicitly suspicious discourse of scroungerphobia, bureaucracy had its own tendency to produce uncertainty and suspicion. This was only partly linked to the assessment and verification of status claims. Other practices, unrelated to explicit imposter tests, also played an important role. These included practices concerned with performance management and the surveillance of workers. Alongside behavioural conditionality, bureaucratization was a consequence of new public management, contracting-out, and the quasi-marketization of public services. It was this general bureaucratization which, for claimants and workers alike, often gave rise to a sense that the drama of imposture was interminable and unavoidable.

Claim-making has, since the turn of the millennium, become increasingly bureaucratized and complex. Beside the ongoing monitoring associated with behavioural conditionality – regular appointments, action plans, job search diaries, and so on – a tightening of eligibility and entitlement criteria has brought new tests and checks. Claim-making also involves multiple organizations and actors from the outset. When Bridget first made a claim for sickness benefit – then Incapacity Benefit (IB) – she was able to do so on the basis of a simple note from her GP. When IB became ESA in 2008, the criteria became stricter and the process more complex. A new claim for ESA required applicants to undergo a notoriously inflexible and often humiliating points-based assessment, conducted by dubiously qualified individuals working for outsourced 'service providers'. The 'work capability assessment' (WCA), to give it its proper name, has been the subject of much controversy and provides the inciting incident for the film *I, Daniel Blake* in which a man is declared fit for work despite a heart condition which, by the end of the film, proves fatal.

Once successfully claiming – either JSA, ESA or now Universal Credit – claimants dealt with the DWP via the public agency Job Centre Plus. From the point of view of claimants, this agency appears more like a collection of disparate silos. Claimants needed to communicate with different divisions online, over the phone, and in person at their local Job Centre Plus office. Enquiries regarding administrative matters could not be dealt with in person, however, and when claimants had such problems (for example missing payments) they were instructed to contact the call centre. Finding the appropriate person to deal with a problem could often take considerable time, effort and resources (time spent on hold on the telephone was expensive). Different divisions often seemed to hold different and sometimes conflicting information about claimants. Referral to outsourced providers of welfare-to-work services further increased the different agencies involved in a claim. It multiplied the points of potential communication, but also of miscommunication. Stories about appointment letters that never arrived or were for a date that had already passed were so common as to be mundane, were it not for the devastating impact such mistakes could have when, for example, a much-needed benefit payment mysteriously stopped. Like fortifications with the potential to harm, the bureaucratic complexity of claim-making was a deterrent that sent a clear message to claimants. It implied the scale of the perceived imposter threat, and the seriousness with which it was taken.

At the same time as claim-making has become increasingly bureaucratized, so too has frontline work. The intensive monitoring and documentation of claimants' activity means that such work occupies much of street-level workers' time and attention. In addition welfare services have, since the 2000s, undergone successive waves of marketization and the extension of marketized governance via the use of payment-by-results contracting arrangements. Managerial practices of performance management and individual target setting have spread and intensified throughout the sector. The frontline workers I spoke to were overwhelmingly preoccupied with the monitoring of their own performance against an array of targets and expectations:

> 'There is a lot of monitoring, an awful lot of monitoring. It's called an MSL [minimum service level] list – there's a list of about 15 MSLs. We've got to make sure we're compliant to 95 per cent every day. So it's things like, the MSL 4 is to ensure that every client on your caseload has a future appointment. MSL 11 is you have meaningful contact with every client on your caseload within the last two weeks. And if your MSLs drop below 95 per cent, well, it's just we can't, you know?' (Judy, work coach)

The same recording and monitoring practices were often involved in more than one bureaucratic regime: the monitoring of claimants, of workers, and of organizations involved different uses of the same data. The recording of claimant attendance, activity and outcomes took place on casework management software, with a mix of drop-down menus, text boxes and automated prompts. The MSLs that Judy described, and against which her individual performance was judged, related directly to expectations set by the DWP – the purchaser of welfare-to-work services. Aggregate data gathered from across the organization was used for overall contract monitoring, but this data could also be accessed by managers at various organizational levels, and was used in the performance management of regions, teams, but also individual workers. As Judy intimates, noncompliance with MSLs was not an option. The power of MSLs derived from the threat of contract withdrawal or termination – failure to perform meant that organizations might lose their contracts, individual workers their jobs.

At the same time, the process of gathering this data formed a significant part of workers' interactions with claimants and was integral to the implementation of behavioural conditionality. Interactions with frontline staff often involved answering questions and watching as a job coach worked through various screens and online forms. In trying to understand what the point of all this might be, many claimants came to the conclusion that it was evidence gathered in an ongoing case of potential prosecution. As one ESA claimant, Sybil, told me, "In my experience they don't do anything at all apart from keep an eye once a fortnight on their claimants. You know, and report back to the government if they see anything that they consider indicative of someone being a benefit scrounger". Alongside the anxiety associated with the threat of sanctions, claimants also complained about the pointlessness and stupidity of their involvement with welfare-to-work services – that it was a waste of time. With repetition, pointless interactions gave rise to a sense that such encounters were designed to catch them out, even when ostensibly about the provision of help and support to find work. In this vein, Greg describes a typical interaction with his work coach:

> 'If you're on the Universal Jobmatch and you've applied for jobs, you've done your job search history. They'll check out what you've been doing all week before and then they'll go "There's a vacancy there". They'll turn the computer round to you, and I'll say, "I've applied for that". "When did you apply for it?" I say "Tuesday, check my job search, you should already know". And they've went through the job search and went, "Oh aye, there, it's there".' (Greg, JSA claimant)

In conveying the tedious pointlessness of such interactions, Greg also suggests his work coach was more concerned with monitoring than the provision of useful help to find work. Alongside their anxieties about being judged an imposter, claimants were often wearied by the inanity of their welfare-to-work encounters. Sometimes this took the form of an accusation that welfare services were not doing what they were supposed to – what they advertised themselves as doing – namely, helping people with their problems. Such moments were a mode of disordering in which the positions were reversed, and where the state and its agents became the ones guilty of pretense.

States of imposture

In March 2015 six employees of the private welfare-to-work provider Action 4 Employment (A4E) were jailed for fraud (BBC News, 2015). Specifically, they had made up files and forged signatures to falsely claim that they had helped their clients into jobs. Under payment-by-results contracting arrangements with the DWP, service providers needed to submit evidence of successful job outcomes before they received payment. Persuading former claimants and their new employers to sign an 'evidence sheet' to this effect was a crucial – but rarely straightforward – part of the work coach role. Without the evidence there would be no payment. The reporting of this story highlighted the total figure fraudulently claimed from the DWP as approaching £300,000, from a total of 167 false claims. Less is said about the context for the fraud, or the precise nature of the benefit obtained. Whereas some welfare-to-work providers did offer modest incentives for workers to reach their job outcome targets – often in the form of retail vouchers, rather than bonuses – more often than not the incentive was simply to keep the job. Outcome targets were counted, and the tally reset, each month. It isn't clear, from the reporting, how things operated within the provider A4E, but it is certainly the case that any financial benefit obtained by the individuals involved would have been a tiny fraction of the overall sum, which would have gone to their employer.

This case offers a clear and seemingly unambiguous reversal, where street-level agents are the frauds and cheats. The idea that street-level job coaches were fakes and charlatans was fairly common among claimants. In one version of this idea, work coaches were seen as well-meaning but ineffectual, constantly needing to cover for their lack of knowledge and expertise. This helped explain the welter of pointless bureaucratic activity and bad advice. In another version, the motives and intentions of coaches were suspect, and claimants were wary and mistrustful of the frequently professed desire to help. This offered an explanation for the inconsistent attitudes and actions of advisors, and the prevalence of punitive practice including but not limited to sanctions.

That street-level workers were not who they seemed, presented themselves to be, or to be trusted was, perhaps surprisingly, also an idea voiced by coaches about themselves and their colleagues. This idea was less a conclusion, and more of a question – am I, are we, the frauds? – and one which job coaches inhabited as a conflict or dilemma. Simone, a former work coach, expressed it this way:

> 'It's like a split personality there actually. I actually often felt like that when I was there. I was like, who am I? Am I a sales person? Or am I a caring, more support workery, interested in people with mental health person? Am I a people person? And we got given the role, yeah, very much this is more sales.' (Simone, work coach)

Experienced as an existential question of personal identity, Simone also intimates that this question about her true working identity was also about the conflicts and antagonisms present in the broader situation. This was a multifaceted conflict between the personal identities and dispositions of work coaches, the commercial demands of the target regime, the ostensible and advertised aims of the service, and the needs and situations of claimants. As Simone elsewhere put it:

> 'You had your job security on the one side, and also your ego as well, I mean to do well, all that kind of stuff. But equally there's people who have got loads of, who just aren't ready or willing to work, or have issues and family, you know, loads of issues, and you just – sometimes, if I'm really honest, I probably pushed it too far. Kind of pushed and pushed and pushed. And I felt guilty after. But that's because you have the target in the back of your head. So you have to put pressure on someone to work when they weren't actually really ready.' (Simone, work coach)

While some job coaches were indeed open to (or invested in) the idea of identifying cheats or meeting targets, others had taken the role believing it would revolve around forms of helping – that it was akin to other forms of advice giving or support work. The discovery that the role was overwhelmingly driven by commercial pressures and the need to attain job outcome targets was disillusioning, but also potentially revelatory. It was an experience that changed some people's understanding of the situation and their place within it. This was a mode of disordering that had a restructuring effect on those who experienced it. As another advisor, Emily, put it to me, "I guess that was the reason I left, because I thought actually, I don't like the person I'm becoming". Here the possibility of becoming or continuing as an imposter led people to quit the situation. However, in contrast to the sorts of public revelation and restructuring described by Derksen (Chapter 3, this volume), these individual exits left the overall situation unchanged. Revelation and its restructuring effect remained a private and individual affair.

Nor does this reversal, in which work coaches came to see themselves as suspect, offer any symmetry. That workers were able to so misread and misjudge the situation, and find themselves so out of place, was partly a consequence of the Janus-faced discourse of welfare-to-work – 'rights and responsibilities' – and its rhetoric of balance. But work coaches like Simone and Emily came to understand that this was not a balanced situation. Emily, in particular, felt this very strongly, telling me that behavioural conditionality "was essentially a weapon of, if you do what we want and get a job, we will give you back your autonomy. Until then you belong to us". Although it was difficult for work coaches to quit their jobs, it remained a decision they were able to make, and in their own time. For the vast majority of claimants, it was much more difficult and costly to simply withdraw from the situation, even if this is what they might have wished.

Lipsky argues that street-level bureaucracies such as Job Centre offices and welfare-to-work providers teach the public 'political lessons contributing to future political expectations' (Lipsky, 1984: 9) and 'socialize citizens to expectations of government services and a place in the political community' (Lipsky, 2010: 4). They can do so, in part, because such services are often ones that people have little choice but to accept. To make a claim for benefits was to learn how and what it meant to be a claimant. The lesson almost everyone learnt was that to be a claimant was to be doubted and regarded with suspicion, and for their needs to be suspect and considered illegitimate. Yet although at times threatening and punitive, there was also something ridiculous and absurd about this situation. Here Jill describes being questioned about her date of birth:

> 'I made an appointment which had got cancelled so I had to go in on Christmas Eve, and the woman was arguing with me saying we can't find you on the system anywhere. I went, "I'm on the system", and she was talking down to me. She was like, "What's your date of birth?". I went, "1993", and she went, "Are you sure it's not 1996?" I was like, "I'm pretty sure what year I was born in". They could find a Jill for 1996 but not for '93. She was quite rude as if I don't know my own age.' (Jill, JSA claimant)

Such absurdities are a common theme of both academic and popular accounts of street-level bureaucracies (Vohnsen, 2017). For Graeber (2015), such 'stupid' interactions result from the inherently unequal power relations of bureaucratic encounters, and an unequal division of interpretive labour: it is the applicant who must demonstrate how they fit with the relevant categories and criteria, and who must do the interpretive work of understanding what the bureaucracy really wants from them. Here the labour of the imposter test falls on the suspect. For Graeber there is no such impetus for officials to attempt a similar imaginative understanding of applicants and their situation – street-level officials are much more concerned with understanding what their superiors want from them. Bureaucratic encounters thus become 'dead zones of the imagination' (Graeber, 2012) with bureaucratic officials accustomed to relying on the interpretive efforts of those who appear before them, while they preoccupy themselves with making everyday reality fit with the available bureaucratic representations.

This analysis draws attention to the power asymmetries of the bureaucratic encounter, which rest on the bureaucrat's ability to inflict harm by stopping someone's income. This asymmetry underwrote and pervaded welfare-to-work encounters, even when they seemed to be going well. For claimants there was always the worry that, even if this coach seemed understanding, the next one may not. The work coaches who felt their own imposture most keenly were those who both recognized this and disliked it. The problem for coaches like Simone and Emily was that this situation undermined their sense of themselves as helpful people. They wanted to establish genuine relations of support, but it was impossible for them to tell the difference between cooperation and coerced compliance. Whether they liked it or not, wanted to or not, they embodied a threat. Behavioural conditionality, as a mode of disordering that held the claimant's status permanently in question, also placed in question its own claims to status as a form of help and support.

Disordering the welfare state

Ostensibly designed to settle questions about genuine entitlement (or imposture), in effect welfare-to-work programmes continually reposed such questions, deferring their resolution. This chapter has shown how behavioural conditionality and its associated bureaucratic practices sustained a particular mode of disordering in which status claims – claims of entitlement – were always indeterminate and potentially suspect. It has described how the services associated with support to find work and conduct monitoring became sites of pervasive uncertainty and suspicion, and where imposture was often assumed until proven otherwise. As a consequence many claimants felt the need to 'passively' comply with 'activation' requirements, or to stereotypically perform their deservingness in order to avoid sanction. Here the imposter regime itself elicited a mild form of impostering. This suspicious mode of disordering also permeated the field. In the process of subjecting to claimants to relentless scrutiny, the meaningfulness of welfare-to-work services and those who provide them were themselves rendered suspect.

References
Alston, P. (2019) 'Report of the Special Rapporteur on extreme poverty and human rights on his visit to the United Kingdom of Great Britain and Northern Ireland', Geneva: United Nations.

BBC News (2015) 'A4e staff jailed for training fraud', BBC News. Available at: https://www.bbc.com/news/uk-england-32139244 [Accessed 13 June 2020].

Byrne, V. and McEnhill, L. (2014) '"Beat the cheat": Portrayals of disability benefit claimants in print media', *The Journal of Poverty and Social Justice*, 22(2): 99–110.

Clasen, J. and Clegg, D. (2007) 'Levels and levers of conditionality: Measuring change within welfare states', in J. Clasen and N. Siegel (eds) *Investigating Welfare State Change: The 'Dependent Variable Problem' in Comparative Analysis*, Cheltenham: Edward Elgar, pp 166–197.

Connor, S. (2007) 'We're onto you: A critical examination of the Department for Work and Pensions' "Targeting Benefit Fraud" campaign', *Critical Social Policy*, 27(2): 231–252.

Dwyer, P. (2004) 'Creeping conditionality in the UK: From welfare rights to conditional entitlements?', *The Canadian Journal of Sociology*, 29(2): 265–287.

Dwyer, P. (2016) 'Citizenship, conduct and conditionality: Sanction and support in the 21st century UK welfare state', *Social Policy Review*, 28: 41–62.

Dwyer, P. and Wright, S. (2014) 'Universal Credit, ubiquitous conditionality and its implications for social citizenship', *Journal of Poverty and Social Justice*, 22(1): 27–35.

Fraser, N. and Gordon, L. (1994) '"Dependency" demystified: Inscriptions of power in a keyword of the welfare state', *Social Politics: International Studies in Gender, State & Society*, 1(1): 4–31.

Garthwaite, K. (2011) '"The language of shirkers and scroungers?" Talking about illness, disability and coalition welfare reform', *Disability & Society*, 26(3): 369–372.

Golding, P. and Middleton, S. (1983) *Images of Welfare: Press and Public Attitudes to Poverty*, Oxford: Blackwell.

Graeber, D. (2012) 'Dead zones of the imagination: On violence, bureaucracy, and interpretive labor. The 2006 Malinowski Memorial Lecture', *HAU: Journal of Ethnographic Theory*, 2(2): 105–128.

Graeber, D. (2015) *The Utopia of Rules: On Technology, Stupidity and the Secret Joys of Bureaucracy*, Brooklyn/London: Melville House.

Hancock, A.-M. (2003) 'Contemporary welfare reform and the public identity of the "welfare queen"', *Race, Gender & Class*, 10(1): 31–59.

Howe, L. (1990) *Being Unemployed in Northern Ireland: An Ethnographic Study*, Cambridge: Cambridge University Press.

Howe, L. (1998) 'Scrounger, worker, beggarman, cheat: The dynamics of unemployment and the politics of resistance in Belfast', *The Journal of the Royal Anthropological Institute*, 4(3): 531–550.

Jensen, T. (2014) 'Welfare commonsense, poverty porn and doxosophy', *Sociological Research Online*, 19(3). Available from: https://www.socresonline.org.uk/19/3/3.html [Accessed 15 October 2015].

Jensen, T. and Tyler, I. (2015) '"Benefits broods": The cultural and political crafting of anti-welfare commonsense', *Critical Social Policy*, 35(4): 470–491.

Lipsky, M. (1984) 'Bureaucratic disentitlement in social welfare programs', *Social Service Review*, 58(1): 3–27.

Lipsky, M. (2010) *Street-Level Bureaucracy, 30th Ann. Ed.: Dilemmas of the Individual in Public Service*, New York: Russell Sage Foundation.

Mead, L.M. (1997) *The New Paternalism: Supervisory Approaches to Poverty*, Washington, DC: Brookings Institution Press.

Murray, C., Phillips, M. and Lister, R. (1999) *Charles Murray and the Underclass: The Developing Debate*, London: IEA Health and Welfare Unit.

Shildrick, T. and MacDonald, R. (2013) 'Poverty talk: How people experiencing poverty deny their poverty and why they blame "the poor"', *The Sociological Review*, 61(2): 285–303.

Shilliam, R. (2018) *Race and the Undeserving Poor: From Abolition to Brexit*, Newcastle upon Tyne: Agenda Publishing.

Southwood, I. (2011) *Non-Stop Inertia*, Winchester: Zero Books.

Tyler, D.I. (2020) *Stigma: The Machinery of Inequality*, London: Zed Books.

Tyler, I. (2013) *Revolting Subjects: Social Abjection and Resistance in Neoliberal Britain*, London: Zed Books.

van Berkel, R. and Valkenburg, B. (2007) *Making it Personal: Individualising Activation Services in the EU*, Bristol: Policy Press.

Vohnsen, N.H. (2017) *The Absurdity of Bureaucracy: How Implementation Works*, Manchester: Manchester University Press.

Wright, S. (2012) 'Welfare-to-work, agency and personal responsibility', *Journal of Social Policy*, 41(2): 309–328.

Wright, S. (2016) 'Conceptualising the active welfare subject: Welfare reform in discourse, policy and lived experience', *Policy & Politics*, 44(2): 235–252.

9

The Face of 'the Other': Biometric Facial Recognition, Imposters and the Art of Outplaying Them

Kristina Grünenberg

Introduction

On a Thursday in December 2018 in Leipzig, Germany, the self-proclaimed hacker and computer scientist Jan Krissler, alias 'starbug', and his colleague Julian showed the assembled crowd at the Chaos Communication Congress[1] how they could fool a hand-vein biometric sensor with the use of a fake hand. The hand was made by capturing their palm vein patterns with a digital camera from which the infrared filter had been removed, mounting the patterns on a wooden hand and covering it with a layer of wax skin (Burt, 2019) (see Figure 9.1).[2]

---

[1] The Chaos conference is the yearly gathering of the German 'Chaos Computer Club', the largest association of hackers in Europe, which was founded in 1981. According to their webpage, they have 5,500 registered members (https://www.ccc.de/en/).

[2] It is possible to 'capture' the veins under the skin and turn them into a biometric identifier with the use of infrared light. Haemoglobin in the blood contains oxygen when it is transported from the lungs to the tissues in the body by the arteries. When the blood flows back to the heart, this oxygen has been released. Deoxygenated haemoglobin absorbs infrared light and makes the vein patterns visible as dark lines. This was why 'starbug' and Albrecht had to remove the infrared filter that normally filters the light and colours in a digital camera.

Figure 9.1: Slides from the researchers' talk showing the manufactured hand

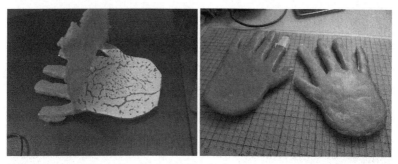

Source: Image courtesy of 'starbug' and Albrecht.

Before this event, vein biometrics was considered one of the most secure biometric technologies. Vein patterns are located under the skin and were therefore considered more difficult if not impossible to fake.

Biometric technologies are digital technologies developed to register, recognize and distinguish individual bodies. Fingertips, faces, eyes, veins and other body parts enrolled for biometrics are considered unique in their dimensions, textures and patterns. These technologies are increasingly being discussed as their use proliferates across contexts. Fingerprint-, facial- and to some extent iris recognition are widely used in border control and consumer electronics (computers, smartphones, cars), for access to restricted sites, for identification of beneficiaries in health or social systems, in refugee camps, and increasingly for identification in banking (Bonneau et al 2018; Jacobsen, 2017, 2019; Grünenberg, 2020a). Biometric technologies are often talked about in terms of convenience and/or security by industry representatives. They are convenient because they make it unnecessary for individuals to remember passwords. As one biometric researcher argued, "You can forget your password, but you can't forget your body" (interview, Peter, professor of biometrics). Furthermore, according to researchers, they potentially enable more seamless interactions: doors that open as you approach, car seats that auto regulate to the particular body of a driver by, for example, registering a fingerprint. Biometric technologies are conceived of as security-enhancing devices because ideally they make it possible to distinguish between individuals who may legitimately cross a boundary, whether a national border or a more prosaic perimeter such as the entrance to an office, and those who may not.

In biometric research the notion of 'spoofing' appears to identify a particular form of impostering. The example of palm-vein spoofing mentioned at the opening of this chapter is a case in point. Here a fake hand with vein patterns produced from images of the hand veins of 'starbug' and Albrecht respectively was recognized by a biometric vein sensor *as* theirs. The point made by the 'spoofer duo' was that anyone with the creative skills, knowledge and a month on their hands to experiment could impersonate another individual whose veins they had photographed. According to 'starbug', even a photo taken at a distance of five metres would suffice to recreate a particular set of veins.[3]

This chapter examines the line of biometric research that aims to prevent 'spoofing', that is, attempts by individuals to circumvent biometric systems through the presentation of artefacts such as 'fake fingers', 'fake irides' or facial masks. Through these fake body parts, spoofers attempt to hide their 'true identity' by pretending to be someone else, whether a particular individual (known as impostering) or an unknown other (known as obfuscation) (Bhattacharjee et al, 2018). The chapter takes the reader into a laboratory of biometric research and focuses on the work with 'anti-spoofing', or, in formal biometric vocabulary, 'presentation attack detection',[4] which focuses on continuously coming up with new ways to circumvent biometric sensors and thus beat spoofers at their own game.

What, then, is at stake in working on anti-spoofing in biometrics laboratories? How is the spoofer or imposter configured inside and outside the laboratory? What is the relationship between spoofers and anti-spoofing research, and what does this version of the imposter tell us about contemporary social relations and cultural forms? The chapter explores these questions by drawing on Alfred Gell's work on traps and hunting (1996) and Taussig's studies (1993, 1999) of mimesis and defacement. In this chapter, I argue that the work with facial masks that takes place in the biometric laboratory and that is aimed at the anticipatory unmasking of imaginary imposters implies the researchers becoming a type of imposter themselves. Researchers, in other words, have to *impersonate* potential imposters in order to

---

[3] 'starbug' initially became famous for using photos of the fingerprints of the German Minister of Defence and now head of the EU Commission, Ursula von der Leyen, to reverse engineer her prints and spoof a fingerprint sensor (Hern, 2014).

[4] For more on this difference, see Grünenberg (2020d).

develop what I have called 'algorithmic traps' to detect them.[5] In this context, rather than being simply antagonists, researchers, public spoofers and hackers like 'starbug', as well as imagined spoofers, enter into an endless self-perpetuating cycle, a form of symbiotic 'mimesis' (Taussig 2018 [1993]), a never to be resolved process of impostering that becomes productive in shoring up the funding for new biometric research and that further legitimizes the continuous search for what I call safer, more secure and seamless 'bio-machines of security'.[6] This search for seamless and spoof-proof security measures takes place in and around biometric laboratories, for example, in the daily work with custom-made facial masks and other spoofing practices, which is the focus of this chapter.

Just as the public use of biometric technologies has proliferated, so too has social science research on the topic. Over the past ten years, an increasing number of scholars have focused particularly on the problematic aspects of biometrics in relation to a number of issues: privacy and state surveillance (Haggerty and Ericson, 2000; Rao, 2018; Rao and Vijayankar, 2019); mechanisms of filtering and exclusion through, for example, the gender and racial biases that are inherent in algorithms and software systems (Magnet, 2011; M'charek et al, 2014; Boulamwini and Gebru, 2018); the potential abuse of biometric databases for political and other motives (Finn, 2005); biometric systems as expressions of the securitization of the state (Lyon, 2008; Andersson, 2016; Aradau and Blanke, 2017); and how such systems of surveillance and filtering came into being through the confluence of particular historical, political, commercial and economic developments (Feldman 2011; Amoore, 2013). The initial important scholarly work on biometric technologies was tied to the disciplines of philosophy, law, media studies and political science and focused mainly on the implications for individuals subjected to biometric technologies and surveillance (Van der Ploeg, 1999; Magnet, 2011; Amoore, 2013). More recent work has been based on ethnographic studies of the world of biometrics, including the role of

---

[5]   Briefly speaking, an algorithm is an ideal mathematical recipe or model of how best to solve a problem. In biometric research and practice, different algorithms are used for different tasks during the processing of bodies for biometric systems. Algorithms cannot stand alone, but need to be operationalized by mathematical code(s) and the scripture of different types of computer programs (cf Dourish, 2016).

[6]   I coin the term 'bio-machine of security' here to highlight the way in which biometric technologies are said to join together life/live (bio) and algorithms (machine) in a sort of 'objective truth-saying complex'.

commercial biometric companies, service-providers, case-handlers, biometric researchers and developers (see, for instance, Feldman, 2011; Jacobsen and Rao, 2018; Olwig et al, 2020; Grünenberg et al, 2020e). Ethnographic studies of biometric laboratories, however, are still relatively rare (see Bourne et al, 2015; Grünenberg, 2020a). Furthermore, whereas there has recently been an increased interest in hacking (see Coleman, 2013, 2014; Horstmann, 2020), to my knowledge the present chapter is the first ethnographic study of biometric spoofing and anti-spoofing research (see also Grünenberg, 2020d).

The messy, self-perpetuating configuration of biometrics

Biometric technologies may be new in their present form and entanglements, but identifying particular individuals on the basis of their bodies is not a new trend. The story of biometrics can be usefully characterized by ideas about the 'evolution' of scientific knowledge and practice developed by Michel Serres more generally. Serres argues that, rather than constituting a

> continuous and cumulative knowledge or a sequence of sudden turning-points, discoveries, inventions and revolutions plunging a suddenly outmoded past into instant oblivion, the history of science runs backwards and forwards over a complex network of paths which overlap and cross, forming nodes, peaks and crossroads, interchanges which bifurcate into two or several routes. (Serres, 1989, quoted in Brown, 2002)

Similarly, and contrary to how they are often portrayed in science literature, biometric technologies are the outcome not simply of inevitable scientific progress, but also of chance encounters, mobilized support, political and commercial choices and priorities, public events and research agendas. According to several social science researchers, for example, the recent proliferation of biometric technologies are tied to the attacks on 9/11 and the subsequent search for flawless security measures, in order to prevent such attacks from being repeated (Lyon, 2008; Gates, 2011; Magnet, 2011; Amoore, 2013; cf Grünenberg, 2020b).

Nonetheless, the enhanced contemporary focus on biometric technologies is only possible against the backdrop of already historically long-rehearsed ideas of employing bodily characteristics and features for identification. In 19th-century Europe, the French policeman-cum-physical anthropologist Alphonse Bertillon, the Italian professor of psychiatry Cesare Lombroso, the British colonial administrator William Herschel in India, the Scottish missionary and physician Henry Faulds and the English mathematician and psychologist Francis Galton, among many others, set out, each in their own way, to explore how faces, fingers and other body parts could reveal truths about people. Bodies, they found, could be used to identify individuals and possibly reveal their likely intentions and inclinations.[7] Yet even at this time, the technologies that were developed were contested, developed further and/or abandoned due to imposter practices and 'real-life challenges'. Cole (2002), for instance, cites the example of an identification clerk with a US police force who in the 1920s faked a fingerprint in order to frame a criminal and harvest the reward for his capture (Cole, 2002: 276). Another example of challenges to biometrics was the abandonment of 'Bertillonage', a widely adopted system of body measurement developed by Alphonse Bertillon. Bertillonage was cast aside as a consequence of the identical body measurements of a pair of identical twins. The twins, who had been taken into criminal custody years apart and for different reasons, spurred great confusion among prison officials, who found an exact match between the body measurement ID of a prison inmate in the prison archive and a newly arrived prisoner (see also Grünenberg, 2020c). This case led to the widespread adoption of fingerprint technology, which was formally developed around the same time as Bertillon's system (Farebrother and Champkin, 2014).

In the 19th century, the concerns attached to ID technologies, such as fingerprinting and facial and cranial measurements, mainly revolved around the identification of criminals and the signs on criminal bodies. Today the use of identification technologies has been extended to the identification of people defined respectively as desirable VIP travellers and unwanted border-crossing migrants and terrorists, as well as for the commercial and health-related purposes noted earlier. The recognition of body prints (hand, palm, finger), facial features, iris shapes and so on have also been automated, in contrast to the cumbersome and time-consuming registration, recognition and storing of body imprints in the large physical archives of the 19th century. Body prints and characteristics are converted into digital identity markers that are stored in large databases, which take up little space and can, at least in principle, be searched algorithmically by computer

---

[7]   The emergence of the discipline of anthropometry and the sub-discipline of phrenology were also part of this development.

processors and programs developed by researchers and engineers in biometric laboratories (see also Finn, 2009). However, what is often overlooked in the celebration of these digital technologies is the equally cumbersome and time-consuming work they imply. This includes writing algorithms, codes and programs; developing sensors and physical installations; training and retraining, for example, airport staff to manage these installations; and 'training' travellers to stand still between automated glass doors while they are being registered and/ or compared to their passport photos and fingerprints (Olwig et al, 2020). Moreover, it is only when these technologies 'work' and can communicate across different systems that they can enable a speedier determination of who is allowed or denied access to particular places (Olwig et al, 2020). Contrary to how they are often portrayed, notably by biometric vendors and some politicians, the idea of these technologies as inherently objective and reliable security infrastructures is not a given either. Instead, their level of security is the subject of negotiation between different stakeholders and will vary depending on the context of the biometric application. Typically consumer biometrics like facial recognition installed in computers, iPhones or PlayStations will be easier to spoof, since the threshold that decides how much a person should 'look like him- or herself' is set relatively lower (see also Møhl, 2019).[8] Every time a new biometric system is invented, biometric researchers have to clarify the optimal balance between on the one hand the *security* it provides by filtering out whoever is defined as a trespasser or illegitimate user, and on the other hand the *convenience* it should also imply for those defined as legitimate users of the system. In the jargon of biometrics, this is about calibrating 'false acceptance' and 'false rejection rates'.[9] Nonetheless, it is up to the

---

[8] Both in the labs and at the conferences I attended I was told that the setting of such thresholds, even in high-security systems, is not fixed, but is instead the outcome of negotiations between different stakeholders in the field at any given time. Furthermore, according to one researcher I spoke to, sometimes the large international databases that locally captured data are checked against operate with their own separate thresholds, making the whole operation of recognition even more complex. Møhl (2019) has also shown how at Copenhagen airport these stakeholders are commercial actors, airport security and national security representatives.

[9] These two terms refer to the probabilities that are used to assess biometric systems and algorithms accurately. False acceptance or match rates, also known as false positives, refer to the probability that a system will accept someone who is *not* a legitimate user. False rejection rates, also known as false negatives, refer to the probability that a legitimate user is not recognized by the system. The criteria for who the legitimate user actually is depends how a 'legitimate user' is defined by the stakeholders implementing the system.

companies, customers and other stakeholders to choose how secure, and thus how 'spoof-proof', their systems should be. This, of course also means that potential spoofers can spoof some biometric systems with greater ease and that the security potential – that is, the extent to which the systems might actually reliably recognize and differentiate individuals based on the mathematical models on which they rest – does not necessarily match their security performance in practice. Getting these systems to work, in other words, is a rather fragile achievement.

Identification and recognition based on faces, fingerprints and irides are the most frequently used biometric modalities at present, having been officially approved, among other things, for use at national borders and in passports. Many other biometric modalities are currently in use, including vein biometrics, voice and gait biometrics and heartbeat biometrics, and yet others which have been developed in the labs, but not yet tested. While research, development and the use of biometrics have intensified, so too has the amount of work directed at securing them against potential spoofing and other attacks.

Spoofing and anti-spoofing: impersonating the imposter

It is only within the past decade that research into spoofing has become a prominent field of research. The first handbook on biometric spoofing and anti-spoofing was published in 2014, this being the first concerted effort at compiling anti-spoofing research. In the foreword, written by professor of computer science Kevin Bowyer, several references are made to sci-fi films and their 'biometric spoofing scenes', which, even when they are not technically accurate, according to the author, point to the importance of the field (Bowyer, 2014: v). Bowyer commends the handbook's authors for 'anticipating an important emerging need' for anti-spoofing measures given the deployment of large-scale biometric installations around the world and the consequent expectation that the intensity and frequency of spoofing attacks will increase (Bowyer, 2014: vi).

Spoofing and anti-spoofing research thus deals more with the potential and future risks of spoofing than with actual known spoofing attempts and spoofers. This is probably one reason why in the biometric lab, apart from building on previous scientific experiments and their own imaginations, researchers also draw on sci-fi films and books as sources of inspiration in their work on anti-spoofing. Furthermore, everything from videos on YouTube to hackathons, hacker conferences like the

'Chaos Communication Congress' described in the opening paragraph, and other similar public events where self-declared hackers, hacktivists and/or digital and visual artists publicly display the circumvention of digital security devices are also used as source materials for imagining the spoofers. Some of these public events were championed by hacker and computer scientist 'starbug' mentioned in the introduction, but several visual artists have also shown publicly how biometric technologies, such as face detection- and recognition systems, may be spoofed (Harvey, 2018; Selvaggio, 2018). All these public spoofs have served as inspirational material for researchers working on anti-spoofing, who are positioned as the 'good guys', in contrast to the definition of spoofers as 'criminals', 'villains' and 'bad guys' in the literature and at the biometric conferences and meetings I attended. Bowyer himself addresses this distinction in the anti-spoofing handbook: 'spoofing attacks tend to attract the attention and publicity, but it is the anti-spoofing methods that are more important to "the good guys"' (Bowyer, 2014: vi). This is of course not surprising, given the real threat that particularly internet hackers and spoofers potentially pose to biometrically based ID systems (for example, in the form of ID theft) as well as to public infrastructure in an increasingly digitalized public sector.[10] Nonetheless, nobody knows the extent to which biometric installations or sensors are actually being circumvented through spoofing. While there, to my knowledge, are only a few known dramatic and failed examples of attempts at biometric mask-spoofing, these examples appear repeatedly in teaching and talks on biometrics (see also Grünenberg, 2020d). But as Merck argues (Chapter 5) in this book, even when imposter imitations fail they still advance the plot, as we shall see later (see also Chapter 1 in this volume). Now, however, I shall move on to describe the biometric laboratory and its work.

The laboratory

> The biometric laboratory where I did my fieldwork is located in a rather small town in a new building in the middle of an otherwise picturesque setting of green mountains and flowers. It is spring, and the air is cool and crisp. The building, with its large hall from which you can see the three upper floors,

---

[10] Recent ransomeware attacks that also paralysed part of the British national healthcare system are a case in point. Although not a case of spoofing, these examples testify to the weaknesses of digital/computer based systems and the possibilities of attacking or circumventing them.

houses several research facilities, research groups, laboratories, offices and even a hotel and a restaurant in the same building, which explains the apparent discrepancy between the immaculately dressed mainly young people (clerks, receptionists, maids, housekeepers, waiters and waitresses) on the ground floor and the casually dressed researchers who inhabit the other three floors. Here AI-based and algorithmic applications such as robotics, speech and audio-processing techniques, biometrics and bio-signalling are being worked upon. The biometric laboratory is located on the building's second floor, and just by the entrance to what otherwise appears as a relatively standard office is a shelf with at least 15 3D facial masks of different materials and qualities. Some of the masks are custom-made soft and skin-coloured silicone masks, while others are made of a type of hard plastic or wax or from folded paper. (Excerpt from fieldnotes)

I embarked upon my fieldwork with a focus on the daily working practices of researchers involved with body images, algorithms and code. Through arduous work, meticulous attention to detail, constant processes of trial and error, experimentation, complicated mathematics/statistics and computer coding and programming, the researchers in the lab worked on different biometric modalities (for example voice, finger veins, face recognition) and on anti-spoofing technologies. The researchers aimed to create ever more secure systems and provide seamless experiences. Other reasons they worked with biometrics included the creativity involved in experimenting with, and developing new technologies, and being at the forefront of science (see also Grünenberg 2020a). I was particularly interested in their work on systems to detect potential attempts to circumvent systems by using facial masks and other ways of hiding the face or 'putting on' the face of someone else. My interest on research on spoofing and anti-spoofing, particularly with face-masks, had been spurred during my attendance at lectures on biometrics and 'liveness detection' at a Danish university.[11] From there I traced particularly work on *face anti-spoofing* in the laboratory, with the kind assistance of Martin, the head of the laboratory, and the researchers there.

---

[11] The concept of 'Liveness detection' is interesting in itself and worthy of a separate analysis. Here it implies the ability of biometric systems to recognize 'life' from 'non-life', that is, that the body part captured or scanned by the biometric sensor is an extension of an authorized, live person who is actually present.

Face fascination

The perception of faces as a locus of identity and a site of expression of *what is really going on* 'below the surface' has a long track record. The Italian criminologist and physician Cesare Lombroso's measurements and analyses of facial traits as indicative of deviant behaviour in his book *Criminal Man* from 1876 is a case in point. In everyday life, the face seems to constitute the corporal surface where we search for clues as to who the other may be and what he or she may be thinking (Edkins, 2015: 1). With reference to Levinas, Edkins argues that the face is seen as a source of ethics and ontology and is furthermore 'hugely powerful in contemporary politics' (2015: 3). In their apparent corporal materiality, faces should therefore be available for inspection and scrutiny. Unsurprisingly, then, face masks and masked faces, aside from their multiple uses in rituals described by anthropologists, are often viewed with suspicion and even fear in countless literary works and films (see also Grünenberg, 2020d), as well as in public policies. A good example of the importance placed on the face in public is the Danish government's 'covering ban' (*tildækningsforbud*), passed into law in 2018. This bill bans any covering of the face, including 'fabric, helmets, body suits, face masks and fake beards' (Danish Ministry of Justice, 2018). The motivation as stated in the bill is:

> to make explicit that, in the government's view, it is incompatible with the values of and social cohesion in Danish society and the respect for our community to hide one's face in public. The face is the basis of recognition between people and makes it possible to read other people's signals and feelings. The face has a crucial role in social interaction. Those who cover their faces, seen from the government's perspective, are expressing their unwillingness to become part of Danish society. (Danish Ministry of Justice, 2018)

It is further argued that the law is being introduced in order to ensure that 'we in Denmark meet each other with trust and respect, face to face' (Danish Ministry of Justice, 2018).[12] Whereas this bill was passed

---

[12] Translation of quotes by author.

mainly in order to prevent the wearing of burkas and niqabs in the public sphere, it is revealing for its depiction of faces as individual 'telling surfaces' and the way in which 'accessible faces' are understood as important for trust.

The face in this account, as in many others, is understood as a stable, taken-for-granted index of the person. As social scientists and philosophers have pointed out, however, faces are not merely biologically stable facts located on individual bodies, but rather mutable, historical, political and cultural accomplishments. According to Deleuze and Guattari in their work on 'faciality', the face is in fact an outcome of specific assemblages of power and politics (Deleuze and Guattari, 1987, in Edkins, 2015; see also Gates, 2011). And as Taussig argues, 'the face itself is a contingency, at the magical crossroads of mask and window to the soul, one of the better-kept public secrets essential to everyday life' (1999: 3; also Edkins, 2015).

Nevertheless, faces and the apparent human ability to recognize something as a face, as well as to distinguish between different faces, have long fascinated researchers in behavioural sciences such as psychology. According to Black (2011), for example, human and primate infants are able to recognize even crude representations of key facial attributes within hours or days of birth, as well as to mimic other people's facial expressions without ever having seen their own (Black, 2011: 1). Biometric research on face recognition shares this fascination with the face, and automated facial recognition is an attempt to make the 'bio-machine' mimic what is considered to be the human capacity for recognizing faces under different circumstances. The widespread fascination with the face, the way the face is supposed to reflect one's 'true' intentions and identity, and the fear of and fascination with concealed, particularly masked faces could explain why the biometric laboratory working with face anti-spoofing has received a lot of public attention in the form of newspaper articles and TV reports. Yet perhaps the public attention to and general fascination with faces and their concealment were also among the reasons why working on face masks became so prominent in the lab in the first place.

> There are ten to twelve researchers and only one fully funded permanent head of research working in the lab, depending on the available projects and funding. They are all working on different projects related to spoofing, biometric security and template protection (i.e. how to protect user data in different types of database).

Whereas the office floor, apart from the shelf with masks, looks rather like any communal workspace, with computers and tables placed along the walls and a counter in the middle across which the researchers have their daily coffee breaks, the basement is where the work with facemasks takes place. The laboratory at the end of a dark concrete corridor is a small square room with no windows. This is Anna's domain, the place where the different types of props and disguises that the researchers might come up with are gathered, where spoofing tests and spoofing-presentation attack databases are assembled, and ultimately where the spoofer is materialized. The lab space is a small room with a mobile screen and multiple lamps making it possible to create a wide range of 'environmental conditions' (such as lighting and background) that might have implications for the automated detection of a potential spoofer. Then there is what for our purposes is fittingly known as the 'capturing device'. This consists of several types of image-recording devices built together as one by one of the researchers to enable RGB, depth, infra-red and thermal imaging. It is used to make different types of images of the spoofer (See also Heusch et.al 2020).[13]

Different types of facial masks are lined up on the red shelves; some of them are generic masks that do not share their features with specific individuals, while others are plastic transparent or coloured masks of famous people or politicians, custom-made silicone masks or relatively thin and folded paper masks that contain a facial image printed from a face photo database [see Figure 9.2].

'It took me twelve hours to make this mask', one of the researchers, Arthur, tells me, as he shows me how the paper mask has to be folded meticulously in order to obtain the shape of the particular face printed on its paper surface. The 'jewels in the crown', however, are the

---

[13] RGB (Red, Green, Blue) refers to the standard way of producing colour images; depth makes it possible to see the shape of the face, while the infrared makes it possible to see the heat omitted from the body, heat that is not usually visible (several animals and insects, like snakes, bed-bugs and mosquitos, use this ability to find their prey). Thermal technology also makes it possible to see heat, including the distribution of heat from different parts of the face.

Figure 9.2: Author wearing paper mask

Source: Image courtesy of Idiap Research Institute and author.

silicone masks. These masks are made by a particular mask-maker with a special effects company who specializes in life-like masks. 'It took a long time and a lot of research to find the right person to produce these masks, someone who knows his craft', Martin, the head of research, tells me. This is expensive equipment, he explains. The more intricate the details and texture of the masks, the more expensive they become. Masks like these are mostly made for cinematic film productions. They are made on the basis of meticulous measurements of facial attributes, or 'land marks' as they are also known (nose, eyes, mouth etc.), of the distance between these landmarks,[14] as well as on 2- and 3D images. These are masks that will mould to your face, due to their flexible material, but, as Martin continues: 'If you wear your own mask, you look like you have gained ten kilos' he laughs, showing me some images to prove the point.

---

[14] Such measurements also form the basis of systems of face recognition.

Figure 9.3: The author's silicone face

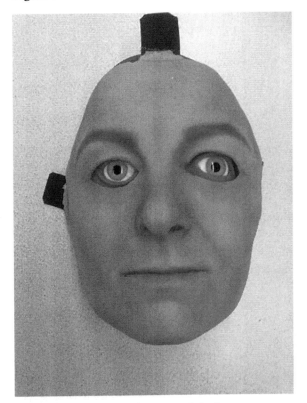

Source: Image courtesy of Idiap Research Institute.

Anna is the person who organizes the capturing of spoofing or presentation attacks for the database and who takes care of the masks. 'So you are the mask caretaker?' I ask. Anna laughs, 'Yes, I guess so. I am also trying to find them a storage place. I wipe the masks every now and then, to keep them clean, but it also works when they lose their stickiness.' Anna is holding my mask inside out on her hand. 'Mm', she says approvingly. 'It's working.' She touches the inside of the mask repeatedly with her index finger and exclaims: 'Yeah – YES! Finally! That's how the mask should be.' The mask is now sticky inside, and as she gently wipes my face with alcohol and places the soft sticky mask on my face, I imagine that this must be what it feels like to carry a large slug on my face. We start capturing images of myself and my superimposed alter ego [see Figure 9.3]. Later we continue with other types of masks, glasses, wigs and make-up from different angles using different light sources and having heated up the mask with a hair-dryer, and then doing so without heating the mask up, since that determines whether or not the thermal technology can detect the presence of a mask or not.[15] In between she presses multiple buttons, changes the light settings and background, and ensures the right positioning of the diverse spoofer headgear. Anna has made a script with the different types of attacks that the research group has decided to capture and train their systems on. The script serves to categorize, document and archive the different types of spoofing attacks. However, during the course of the project new spoofing attacks are invented and new 'spoofers' are staged, so Anna also keeps a log which serves to document the attacks that the research group comes up with along the way. (Excerpt from fieldnotes; see Figure 9.4)

The images that Anna captures in the basement are in this case partly based on a project commissioned by a US intelligence agency and partly on the researchers' own efforts to keep inventing new forms of disguise that could be used by spoofers. The images are stored in a database and are used to train and develop algorithms and test hardware anti-spoofing, such as the use of the thermal application. The next step, then, is to work on the detection algorithms that will make it possible to distinguish between a real and a fake face and thus enable the system to catch spoofers. Rather than a linear process, this is a process of trial and error, of tacking back and forth between different types of masked-face spoofing attacks, and of the elaboration of different types of algorithms, codes and devices for purposes of detection. It is Igor who develops and tests these systems and their ability to detect fakes on the computer in the upstairs office. However, it is not only the systems, but also its humans that need to be 'trained'.

---

[15] Even custom-made silicone masks are relatively easy to spot with a thermal camera since they remain colder than the average temperature of a face. They do, however, warm up once you wear them for longer. The hairdryer was used to preheat the masks and thus make it more difficult for the system to detect them.

Figure 9.4: Anna's script and log

Source: Image courtesy of Idiap Research Institute.

Recognizing real from fake faces, spoof recognition and visual enskilment

Perhaps surprisingly, the need for the biometric researchers themselves to be able to distinguish real from fake faces with their own eyes is important. Biometric systems never stand alone; rather, the researchers' visions, theoretical models and assumptions are folded into them. Without the ability to distinguish fake from real, it is much harder for the researchers to tune and train their algorithms to classify and distinguish the 'real' from the 'fake' or the entitled individual from the spoofer. This ability is also based on an ongoing process of 'visual enskilment' (Grasseni, 2007), that is, the ongoing 'apprenticeship of the eye' (Grasseni, 2007: 42) and the training to pay attention to the subtle differences that mark 'real' from 'fake'.[16]

> This morning I am sitting beside Omid in front of his computer in the lab. The screen is showing multiple facial images. I am trying to understand how Omid recognizes 'real' from 'fake' face images, how he produces a database for anti-spoofing, and how he also uses his knowledge

---

[16] Cf Palsson (1994) and Ingold (2000) for more on enskilment.

about real and fake faces, as well as his imagination, to produce face images that can be used to spoof biometric installations. We look through the images to see whether I can detect the fakes from the real. This proves really difficult for my untrained eye and has to do with detecting the subtle differences in colour, glare, background, texture, image-borders etc. compared to what the researcher would consider a real facial image or video to look like. The system is then trained to detect such subtle differences. This is not about facial recognition: Omid's vision is not attuned to the 150 IDs, that is, to the specific faces in his database. In fact he hardly sees them, since they are simply not of interest to him or to the work he does. Working with face anti-spoofing is not about recognizing individuals, neither those printed on the mask nor those beneath it, but about recognizing the signs of real and fake faces. Omid's vision is attuned to identifying spoofs, and this is what he sees in his database. (Excerpt from fieldnotes)

As Igor, who works upstairs with the images captured by Anna, points out:

> Usually systems that work for face recognition don't work for PAD [Presentation Attack Detection] because here we are trying to work with other problems. In face recognition we want to understand the features – what your eyes look like, how your nose is positioned, your mouth, and things like that. But in PAD we want to be independent of identity. We don't care what the person looks like. We want to pick up the specific [presentation/spoofing attack] moments or elements. It is even 'dangerous' if we focus on identity. [For example] if we train our system using you in the database, we take your image, we show it to the system, we train the system with you as an attacker, then the system will be biased to identity. You come to the system, and it says – 'Hey! I've seen you in an attack!' (Igor laughs). (Excerpt from fieldnotes)

Igor's explanation makes it easier for me to understand Omid's seemingly uncomfortable silence when, during my attempts to tell a fake from a real face, I exclaim: "Hey! That's Pekka!" as I recognize a person in his database from an event in Finland that I participated in six

Figure 9.5: Different face captures

Source: Image courtesy of Idiap Research Institute.

months earlier. Face anti-spoofing then is about identifying the signs of impostering, in this case signs located on the sur-face (skin texture, for example) or signs emanating from the face (the temperature of the face, the movements of lips and eyes).

The process performed by Omid is a good example of how initially, through their own processes of trial and error and experimentation, the researchers need to acquire the skill to see in new ways. Through this skilled vision, they are able to teach their systems of algorithms, codes and programs to catch different kinds of imposters (see Figure 9.5).[17] This is what sets the scene for developing what I call 'the algorithmic trap'.

In his essay 'Vogel's Net' (1996), Alfred Gell addresses the notion of traps in game-hunting. Referring to a particular central African arrow-trap, Gell argued that the trap set by the hunter in the process of hunting signified both the hunter and the prey. Traps, once they are set and left to do their job, initially communicate the absence of both the hunter and the prey. In this context of absence, the trap, although

---

[17] How and where in the process the researchers' acquired knowledge is engaged depends on the type of anti-spoofing approach that they are using or developing.

designed to merge into the background, becomes a form of surrogate hunter, one that does the work of the hunter. The trap, Gell argues, 'Is in fact an automaton or robot, whose design epitomizes the design of its maker', just as it is a model of its victim (Gell, 1996: 27). In this context, the trap may include the contours of the prey animal in the design, but it may also 'more subtly or abstractly represent parameters of the animal's natural behaviour, which are subverted in order to entrap it' (Gell, 1996: 27). Gell mentions a rat trap that recreates the narrow spaces that rats like to poke around in, or a giraffe trap, which is made of the negative contours of the lower half of the giraffe – a hole fitting its lower body. In this sense, we may also say, with reference to Taussig's work on mimesis, that the hunter 'mimics' the prey (Taussig, 2018 1993).

In the laboratory, the work with masks of different colours, textures and materials, captured with different backgrounds, lighting and contrasts, the capturing logs displaying and numbering all the different types of attacks captured by Anna, and the algorithms developed, enhanced and modified by Igor on the upstairs computer, all together result in the crafting and setting of traps in Gell's sense, or more precisely in this case what I call 'algorithmic traps', in order to catch potential spoofers. Whereas the hunter's traps in Gell's accounts are normally crafted after lengthy observations of the animals' behaviour, environment and habitual responses, the lack of actual and known spoofing examples makes it necessary for the researchers to learn about and become a type of imposter themselves. As mentioned they do this by drawing on scientific literature, their own imagination and experiments, as well as public spoofing events, such as those performed by 'starbug' mentioned at the start of this chapter, and on science-fiction literature and films.[18] The algorithmic traps, however, are more than simply models of the anti-spoofing researchers and the spoofer as two separate entities. Gell's description of how the animal trap binds together hunter and prey also fits the algorithmic trap: 'The trap embodies a scenario, which is the dramatic nexus that binds these two protagonists together, and which aligns them in time and space' (Gell, 1996: 27). The algorithmic trap in other words is where the knowledge about spoofer tricks, intentions and ability – and researcher skills, knowledge and foresight – come together and are aligned. Like the hunter's trap described by Gell, once the algorithmic trap has

---

[18] My use of the pronoun 'he' is not coincidental. In the biometric events I attended during my fieldwork, the spoofer was always depicted or addressed as male.

been made the researcher's skills and knowledge are located in the trap in objectified form, otherwise it would not be 'readable' to others (Gell, 1996: 27). Newly conceived algorithms by implications can often be used for a range of purposes and in different settings. The process of imagining the spoofer that goes into the production of the algorithmic trap could be made intelligible by referring to Taussig's concept of mimesis.

In his book *Mimesis and Alterity* (1993), Michael Taussig argues that, through their mimetic practices (mimetic faculty) and production of figurines resembling the Spanish invaders, the indigenous Cuna in present-day Panama were able to grasp 'the other' and thus acquire some form of power over them (Taussig, 2018 [1993]: 35; see also Stone 2018). Although used in a very different way and in an entirely different context, Taussig's work on mimesis, together with Gell's thinking on traps, may serve as an inspiration for understanding what is going on in this type of anti-spoofing research. In other words, the biometric researchers and the potential spoofers, as well as self-proclaimed public hackers such as 'starbug', come to form a symbiotic-mimetic process. They are bound together by the crafting of the algorithmic trap and all the nitty gritty details of impostering practices involving face-masks, make-up, wigs and glasses, images and video replays. In these practices, the researchers attempt to mimic the imagined actions of potential spoofers by performing as imposters and thus predicting their next moves.

At the same time, public hackers like 'starbug' mirror these efforts in their continuous public displays of biometric weaknesses, which also take place through the mimicking of the researcher's work, a type of work that really requires the same skills and knowledge as that possessed by biometric researchers. This veritable 'arms race' or endless cycle of innovation and subversion (Von Schnitzler, 2013) resonates with Taussig's appeal to the sensitivity to circulation: 'we must be sensitive to the crucial circulation of imageric power between these sorts of selves and these sorts of anti-selves, their ominous need for and their feeding off each other's correspondence—interlocking dream-images guiding the reproduction of social life no less than the production of sacred powers' (Taussig, 2018 [1993]: 48). The images of public and unknown spoofers and researchers that circulate at biometric conferences, in the laboratories and between the laboratories and the public social-hactivists also feed off one another. They serve to reproduce a whole range of socio-material practices that come together to establish biometric technologies in general and anti-spoofing research in particular as an important field of research and guarantor of security.

All these processes take place in spite of the fact that no one *really* *knows* the magnitude of this type of problem, nor whether there *are* actually people out there who are attempting to spoof biometric technologies by wearing facial masks. As a researcher from another biometric laboratory I encountered drily commented about the work with facial masks: "It is still a whole lot easier to get hold of a fake passport than to go through the trouble of fabricating expensive facial masks." Asked about this view, Martin, the head of the lab, responded that yes, that might be the case right now, but with the falling prices of 3D printers and other materials, custom masks could easily become affordable in the future. Furthermore, inspired by Taussig's work on defacement, I would also argue that it is not so important whether or not the researchers believe in the reality of spoofers or not – what is important is their ability to determine what is true or false, real or fake, in a specific social field 'in which knowledge is power and the reality of the illusion [of the spoofer] serves the social contract' (Taussig, 1999: 104) – in this case a social contract based on a promise of security that positions biometric technologies as the solution to security problems. Furthermore, the laboratory attempts to 'unmask' the spoofer rather than simply anticipating and preventing impostering may actually 'add to the reality' of the masked imposter (cf Taussig, 1999: 59). The concrete attempts to 'unmask' the spoofer in this sense become a way of enacting him.

By the time anti-spoofing software in the form of algorithms and codes have been invented, the spoofer has become an entity that the researcher has imbued not only with certain surface properties, such as silicone, but also with possible intentions (Grünenberg, 2020d). By then, however, the facial masks are no longer merely tokens of deception, but also artefacts that hold together and co-produce a range of practices that are all part of impostering, as well as, it seems, partly holding together the biometric industry itself (see also Chapter, 1, this volume). It is in this sense that we can draw on Merck's insights about how, even when they fail, impostering practices – and, I would add, potential ways of entrapping them – still advance a specific plot (see Chapter 5, this volume).

Conclusion

The proliferating concern with spoofing, circumvention and fraud, and the consequently enhanced engagement with spoofing or presentation attack detection as a line of research within biometric research, is rooted in an acute search for objective and accurate security technologies and an apparent confluence of political, economic and researcher interests in the field of biometrics.

The particular type of imposter at issue here, 'the spoofer', and his potential practices of circumvention destabilize the dreams of biometric technologies as 100 per cent safe security measures. However, rather than merely destabilizing the ideas of biometric technologies as an adequate response to insecurity, the imposter in this chapter on the one hand ends up reaffirming the need for more investments in and a greater focus on biometric research and biometric technologies that prevent imposters from threatening the longed-for security attributed to these technologies, which 'just need a bit more work'. On the other hand, and by extension, the imposter serves to organize a particular field of research and research practices rooted in the expectations and predictions of spoofers' actions and the setting of 'algorithmic traps' based on the researchers' imaginations. The spoofer-imposter is in other words a key player in the ongoing generation of order and disorder in a particular field of research, as well as being important in configuring notions of digital security (see Chapter 1, this volume). Furthermore, in practice the attempts to detect potential imposters and predict their actions creates new socio-material relations as they are being carried out in the laboratory. Through staged mimetic play-acting involving researchers, facial masks, capturing devices, algorithms, hairdryers, operators, mask-makers, sci-fi films, artists and public hackers, the proximity of the otherwise distant and unknown spoofer is facilitated, by making him 'come to life' in what is a never-ending mimetic process.

To sum up, what may we learn from this example of what we could call 'counter impostering'?

Firstly, 'counter-impostering' is not necessarily a response to the actual practices of imposters, but may also rest upon imagined and speculative ideas about their probability.

Secondly, a related point, whether or not imposters exist as 'real villains' out there is not always important. The mere probability of unknown, ill-intentioned imposters who may use any number of tricks in order to disturb a sense or order or security is enough to set in motion a whole range of socio-material practices and hold together particular fields of interest.

Acknowledgements

I would like to thank the biometric researchers who generously allowed me to participate in their lab work and shared their thoughts and expertise with me, the editors of this book for great feedback and collaborative spirit, my colleagues in the Border World project, and not least the Velux Foundation for the support for the 'The Biometric Border World' project.

References

Amoore, L. (2013) *The Politics of Possibility: Risk and Security Beyond Probability*, Durham: Duke University Press.

Andersson, R. (2016) 'Hardwiring the frontier? The politics of security technology in Europe's "fight against illegal migration"', *Security Dialogue*, 47(1): 22–39.

Aradau, C. and Blanke, T. (2017) 'Politics of prediction: Security and the time/space of governmentality in the age of big data', *European Journal of Social Theory*, 20(3): 373–391.

Bhattacharjee, S., Mohammadi, A. and Marcel, S. (2018) 'Spoofing deep face recognition with custom silicone masks', IEEE 9th International Conference on Biometrics Theory, Applications and Systems (BTAS), Redondo Beach, CA, USA, pp 1–7.

Black, D. (2011) 'What is a face?', *Body and Society*, 17(4): 1–25.

Bonneau, V., Probst, L. and Lefebvre, V. (2018) *Biometric Technologies: a key enabler for future digital services*, report for the EU commission. Available at: https://ec.europa.eu/growth/tools–databases/dem/monitor/sites/default8files/Biometrics%20technologies_v2.pdf [Accessed 6 August 2020].

Bourne, M., Johnson, H. and Lisle, D. (2015) 'Laboratizing the border: The production, translation and anticipation of security technologies', *Security Dialogue*, 46(4): 307–325.

Bowyer, K.W. (2014) 'Foreword', in S. Marcel, M.S. Nixon and S.Z. Li (eds) *Handbook of Biometric Anti-spoofing: Trusted Biometrics under Spoofing attacks*, London: Springer, pp v–vii.

Brown, S. (2002) 'Michel Serres, science, translation and the logic of the parasite', *Theory, Culture & Society*, 19(3): 1–27.

Burt, C. (2019) 'Researchers spoof biometric palm vein recognition system with inexpensive fake', Biometric Update.com. Available at: https://www.biometricupdate.com/201901/researchers-spoof-biometric-palm-vein-recognition-system-with-inexpensive-fake [Accessed 24 June 2020].

Cole, S.A. (2002) *Suspect Identities: A History of Fingerprinting and Criminal Identification*, Harvard: Harvard University Press.

Coleman, G. (2013) 'Anonymous in Context: The Politics and Power behind the Mask', *Internet Governance papers 3*, The Centre for International Governance Innovation. Available at: https://www.cigionline.org/sites/default/files/no3_8.pdf.

Coleman, G. (2014) *Hacker, hoaxer, whistleblower, spy: The many faces of anonymous*, London: Verso.

Danish Ministry of Justice (2018) Forslag til lov om ændring af straffeloven (Tildækningsforbud). Available at: https://www.retsinformation.dk/eli/ft/201712L00219 [Accessed 24 June 2020].

Dourish, P. (2016) 'Algorithms and their others: Algorithmic culture in context', *Big Data & Society*, 3(2): doi: 10.1177/2053951716665128.

Edkins, J. (2015) *Face Politics*, New York: Routledge.

Farebrother, R. and Champkin, J. (2014) 'Alphonse Bertillon and the measure of man: More expert than Sherlock Holmes', *Significance*, 11: 36–39.

Feldman, G. (2011) *The Migration Apparatus: Security, Labor, and Policymaking in the European Union*, Stanford: Stanford University Press.

Finn, J. (2005) 'Photographing fingerprints: Data collection and state surveillance', *Surveillance and Society*, 3(1): 21–44.

Finn, J. (2009) *Capturing the Criminal Image: From Mug Shot to Surveillance Society*, Minneapolis: University of Minnesota Press.

Gates, K. (2011) *Our Biometric Future: Facial Recognition Technology and the Culture of Surveillance*, New York: New York University Press.

Gebru, T. (2018) 'Gender shades: Intersectional accuracy disparities in commercial gender classification', *Proceedings of Machine Learning Research*, 81: 1–15.

Gell, A. (1996) 'Vogel's net: Traps as artworks and artworks as traps', *Journal of Material Culture*, 1(1): 15–38.

Grasseni, C. (2007) *Skilled Visions: Between Apprenticeship and Standards*, Oxford: Berghahn Books.

Grünenberg, K. (2020a) 'Body cartographers: Mapping bodies and borders in the laboratory', in K.F. Olwig, K. Grünenberg, P. Møhl and A. Simonsen, *The Biometric Border World: Technologies, Bodies and Identities on the Move*, London: Routledge, pp 34–53.

Grünenberg, K. (2020b) 'The biometric community: Friends, foes, and the political economy of biometrics', in K.F. Olwig, K. Grünenberg, P. Møhl and A. Simonsen, *The Biometric Border World: Technologies, Bodies and Identities on the Move*, London: Routledge, pp 53–69.

Grünenberg, K. (2020c) 'Epilogue', in K.F. Olwig, K. Grünenberg, P. Møhl and A. Simonsen, *The Biometric Border World: Technologies, Bodies and Identities on the Move*, London: Routledge, pp 69–70.

Grünenberg, K. (2020d) 'Wearing Someone Else's Face: Biometric Technologies, Anti-spoofing and the Fear of the Unknown', *Ethnos*, doi: 10.1080/00141844.2019.1705869.

Grünenberg, K., Møhl, P., Fog Olwig, K. and Simonsen, A. (2020e) 'Issue introduction: IDentities and identity: Biometric technologies, borders and migration', *Ethnos*, doi: 10.1080/00141844.2020.1743336.

Haggerty, K.D. and Ericson, R.V. (2000) 'The surveillant assemblage', *The British Journal of Sociology*, 51: 605–622.

Harvey, A. (2018) 'CV dazzle', *Art Projects about Privacy, Computer Vision, and Surveillance*. Available at: https://ahprojects.com/projects/cvdazzle/ [Accessed 24 June 2020].

Hern, A. (2014) 'Hacker fakes German minister's fingerprints using photos of her hands', *The Guardian*, 30 December. Available from: https://www.theguardian.com/technology/2014/dec/30/hacker-fakes-german-ministers-fingerprints-using-photos-of-her-hands [Accessed 24 June 2020].

Horstmann, N.D. (2020) 'The power to selectively reveal oneself: Privacy protection among hacker-activists', *Ethnos*, doi: 10.1080/00141844.2020.1721549.

Ingold, T. (2000) *The Perception of the Environment: Essays on Livelihood, Dwelling and Skill*, London: Routledge.

Jacobsen, K.L. (2017) 'On humanitarian refugee biometrics and new forms of intervention', *Journal of Intervention and Statebuilding*, 11(4): 529–551.

Jacobsen, K.L. (2019) 'New forms of intervention: The case of humanitarian refugee biometrics', in N. Lemay-Hébert (ed) *Handbook on Intervention and Statebuilding*, Cheltenham: Edward Elgar, pp 282–293.

Jacobsen, K.U. and Rao, U. (2018) 'The truth of the error: Making identity and security through biometric discrimination', in M. Maguire, U. Rao and N. Zurawski (eds) *Bodies as Evidence: Security, Knowledge, and Power*, Durham: Duke University Press, pp 24–42.

Lombroso, C. (1876) *Criminal Man*, Durham: Duke University Press.

Lyon, D. (2008) 'Biometrics, identification and surveillance', *Bioethics*, 22(9): 499–508.

M'charek, A., Schramm, K. and Skinner, D. (2014) 'Topologies of race: Doing territory, population and identity in Europe', *Science, Technology, & Human Values*, 39(4): 468–487.

Magnet, S.A. (2011) *When Biometrics Fail: Gender, Race, and the Technology of Identity*, Durham: Duke University Press.

Møhl, P. (2019) 'Border control and blurred responsibilities at the airport', in T. Diphoorn and E. Grassiani (eds) *Security Blurs: The Politics of Plural Security Provision*, London: Routledge, pp 118–135.

Olwig, K.F., Grünenberg, K., Møhl, P. and Simonsen, A. (2020) *The Biometric Border World: Technologies, Bodies and Identities on the Move*, London: Routledge.

Palsson, G. (1994) 'Enskilment at sea', *Man (N.S.)*, 29(4): 901–927.

Rao, U. (2018) 'Biometric bodies, or how to make fingerprinting work in India', *Body & Society*, 24(3): 68–94.

Rao, U. and Vijayankar, N. (2019) 'Aadhaar: Governing with biometrics', *South Asia: Journal of South Asian Studies*, 42(3): 469–481.

Selvaggio, L. (2018) *URME Surveillance* and *Leonardo Selvaggio*. Available at: http://www.urmesurveillance.com and http://leoselvaggio.com/ [Accessed 24 June 2020].

Stone, N. (2018) 'Imperial mimesis', *American Ethnologist*, 45: 533–545.

Taussig, M. (1999) *Defacement: Public Secrecy and the Labor of the Negative*, Stanford: Stanford University Press.

Taussig, M. (2018 [1993]) *Mimesis and Alterity: A Particular History of the Senses*, London: Routledge.

Van der Ploeg, I. (1999) 'The illegal body: "Eurodac" and the politics of biometric identification', *Ethics and Information Technology*, 1(4): 295–302.

Von Schnitzler, A. (2013) 'Travelling technologies: Infrastructures, ethical regimes and the materiality of politics in South Africa', *Cultural Anthropology*, 28(4): 670–693.

10

Faking Spirit Possession: Creating 'Epistemic Murk' in Bahian Candomblé

Mattijs van de Port

I was driving Victor, Otávio and his boyfriend Jorge to a celebration of Candomblé in Alaketu, one of Salvador's old and famous *terreiros* (temples). Four gays in a rental car, all dressed in immaculate white, as Candomblé ceremonies require. The initiative for the outing had been mine. As always, Victor had been hesitant to accompany me. He dislikes the ceremonies, with their long hours of just sitting, watching the initiates dance to the drumming, and waiting for the orixá spirits to arrive and possess the initiates' bodies. This time, however, he had figured this was an opportunity for me to meet his friend Otávio, who was 'from Candomblé', and might be of use for my research. I do not know what Victor had told Otávio about me, but I dare speculate that the prestige of a European-who-works-at-a-university-and-writes-about-Candomblé had piqued Otávio's interest in joining.

While driving, I explained that the terreiro we were going to visit was a very old one, of great repute and on its way to be recognized by the Bahian state as 'cultural heritage'. Otávio, a chubby adolescent with straightened black hair wearing a rather risqué lace blouse, exclaimed "*chiquéééérima*", which may be translated as "how fabulous". Otávio lived in a distant suburb of Salvador, and although he was in the early stages of initiation into Candomblé, he had never been to this terreiro. He had not even heard of it. I was reminded, once more, that the Candomblé of anthropologists and the Candomblé of local practitioners pertained to different, only partially overlapping circuits.

Otávio informed us he had just taken a bath, to cleanse himself of the couple of beers he had drank earlier in the day. "With beer in my body, the orixá won't come," he explained. When we arrived at the terreiro he said that the coming of a spirit called Iansá was certain for his body was restless. "It is as if waves of heat and cold go through my body." The eyes of his boyfriend Jorge - a silent kid with the fluffy beginning of a moustache, who followed Otávio like a shadow - shone with excitement. Victor responded by saying that it would be a good thing if he got possessed, because *that* was the kind of stuff I was interested in. Coming from Victor's mouth, the remark sounded much like a command. Entering the terreiro, Otávio told me he was nervous. "I feel like I need to pee all the time!"

The possession came early in the evening. The *xiré*, the opening dance sequence, had only just begun when Iansá arrived. Otávio started jumping on his feet. He leant over to one side, then to the other. He rolled his eyeballs. And then Otávio was gone. He had metamorphosed into the warrior goddess, lady of winds and storms, who announced her arrival with a loud, grunting howl.

With tender care, Jorge took off the goddess' sneakers and rolled up the legs of her trousers, after which a temple dignitary came to take Iansá to the room behind the ceremonial hall. After some ten minutes Otávio returned. The expression on his face was both apologetic and proud. He took my hand to let me feel his. It felt cold. "See? It feels cold!"

I remember that I made a mental note saying: "See? It is always the body that is used to prove the truth of the matter ...!" And indeed, when a young man standing close to us became possessed by a spirit called Xangô and turned up his eyeballs, Otávio hissed at me not to miss out on it: "Look at his eyes! Look at his eyes!"

In Bahia, spirits are widely recognized as a real phenomenon in the world. Likewise, spirit possessions are firmly located in the realm of the possible. However, actual manifestations of spirits and their presence in human bodies are frequently subject to doubt and discussion. The actual presence of spirits in a given ceremony is ambiguous for the spectators, and sometimes even for spirit mediums themselves. The spirits exist, but the authenticity of a particular possession is not given. In an interview, my friend Marcelo gave me an eloquent explanation of the doubt that was troubling him as an initiate into the religion. Here is how he phrased it:

> 'My orixá is very rare. He is of a kind that does not appear very much. He doesn't appear the full hundred percent.

I know that he is there. But others need to be really attentive to perceive his presence. He doesn't dance, he doesn't sing or move. He doesn't do a thing. I know he is in my body. But he is only there to observe. To be present … In some rituals he is more present, that's when I am the focus of the ritual. He then needs to be there, otherwise I could not handle the pain that some rituals inflict on you. I need to be in trance for that to happen … My relation with the orixás is very intuitive. I don't hear them, I don't see them. I have friends, they hear them, they see. They have conversations with the orixás. With me that has not happened until now. Take my brother, who can see the spirits. When we offer food to the spirits, on the street, in the river, he immediately says "okay let's go, they have arrived." Or: "let's stay, they have not yet arrived." In my case, it is more complicated. I have to believe in my intuitions. I'm very skeptical. I do not believe in the things I do not feel. So I have to shake, to cry, to be paralyzed. Otherwise I do not believe it is the real thing. I also test things. I do things to see whether it really is a sign from the spirit. Sometimes you know you should not be having sex [in preparation of a ritual]. But then you are horny, you want it. But for the ritual, that is unacceptable. I already experienced moments of impotence. And over time I came to understand that this was because I was breaking a sacred rule.'

Part of the reasons for doubting the authenticity of a case of possession is that faking possession during ceremonies is a well-known practice and recurrent topic in conversations. It is referred to as '*fingir*' ('to fake' in Portuguese) or to 'give *ekê*' ('to fake' in Yoruba, the ceremonial language used in Candomblé). The practice of 'faking it' helps to explain why time and again I was provided with 'evidence' to support the reality of a spirit being present: bodily signals such as the ones reported by Otávio; miraculous occurrences; or examples of shocking behaviour, such as biting off the head of a living chicken and drinking its blood – something that ordinary human beings would obviously never do.

The uncertainties surrounding the real of spirit possession makes Candomblé an interesting case to see to what effect the figure of the imposter plays itself out in the particular situation of a public religious ceremony.

Candomblé is a maddeningly complex religion, but for the purpose of this chapter it is important to keep two things in mind. First, most

Figure 10.1: Staging the mysteries in a Candomblé ceremony[1]

Candomblé temples have a dual role. They take care of the initiation of orixá worshippers, helping them to regulate their relations with supernatural beings, and organize the yearly cycles of rituals and ceremonies. But to survive economically, terreiros also function as a kind of supernatural service centre, where clients can buy supernatural support in all kinds of matters: health, protection, the problems of family life, the difficulties in love affairs, work, and so on (cf Van de Port, 2011). In relation to this latter function, public ceremonies can be seen as occasions where Candomblé temples seek to display the quality of their connection with the forces of the supernatural, and seek to impress an audience with the power of their *fundamentos*, the mysterious, secret source that feeds the prowess of a priest(ess) as spiritual consultant and operator-of-the-supernatural.

Second, it needs to be pointed out that although the phenomenon of 'giving *ekè*' has been recognized and called into existence by giving it a name, there is no single story to be told as to what faking spirit possession is or means. When starting one's analysis at grassroot level, as befits an anthropologist, one is immediately struck by the plurality of perspectives and interpretative frames that are present in each and every situation. The vignette of the trip to the temple makes this clear: it involved two initiates in Candomblé (Otavio and Jorge), a local skeptic

---

[1] It is the policy of the publisher to blur the faces of individuals in photographs where direct permission has not been sought.

(Victor) and an anthropologist (myself). What 'faking it' is, and how it is understood and evaluated, can only be discussed relative to the different perspectives and different interpretative frames that came together in this particular situation. I will honour this plurality of perspectives by discussing the different takes on the event by the different actors: this is in order to acknowledge the idiosyncrasies that go into the making of these frames, and address the 'discursive heterogeneity' on this topic (Law, 2009: 491). Yet in my conclusions I will also try to arrive at a more general statement as to *what becomes of the situation* (i.e. a ceremony of Candomblé) in this particular assemblage.

Otávio (and Jorge)

In an interview I organized after our visit to the temple, Otávio told me that both his mother and grandmother entertained relations with a spirit called Omulu. Otávio had inherited this spirit. But if his relatives had had to deal somewhat unwillingly with Omulu's presence in their lives – "*pela dor, não pelo amor*", as they say in Bahia: "because of trouble, not because of the love for it" – that was not the case with Otávio. When I met him, he was in the early stages of his initiation, and a real Candomblé enthusiast. Like many lower-class gay men in Salvador, who cannot afford the bus ride or entrance fees to gay clubs, Otávio loved visiting neighbourhood terreiros with friends. "*Correr macumbas*", is what he called it, "running from one temple to another". There are many reasons to engage in this activity, but they are not all strictly religiously motivated. Next to the presence of spirits, who might be consulted on how to handle the difficulties in one's life, Candomblé ceremonies offer their audiences free food, music, singing and the spectacle of the possessed dancing the night away. One of the biggest draws for young gay men like Otávio, however, is the fact that Candomblé ceremonies always have a large presence of other gay men, and the long hours of the rituals give ample opportunity for socializing, flirting and gossiping.

The prominence of homosexuals in the Candomblé lifeworld has long been noted and commented upon in the anthropological literature (Landes, 1940, 1947; Ramos, 1942; Fry, 1986; Birman, 1995). In her magnificent study of gender categories in Candomblé, Birman points out that the Candomblé lifeworld is permeated with a whole spectrum of homo-erotic options, some veiled and coded to the extent that only insiders would know how to read the signs, others more open. Young men like Otávio are called *adé* (an Africanism) or *bicha* (a Portuguese term) in Candomblé: terms that refer to 'effeminate' or 'passive' homosexuals. With their fashionable hairdos, flamboyant outfits and penchant for theatrical gesticulations, adés are a highly visible presence in the Candomblé circuits. Birman notes that they have a reputation for becoming easily possessed: 'at the first beating of the drums, the adés are already reporting to "feel dizzy spells", and gesticulate dramatically, trying not to give in to the force that will make them "fall to the spirit" (*cair no santo*)' (1995: 114). But they fall sooner rather than later. Possession, Birman argues, is the only way for the adé to make it to the dancefloor, be dressed in the spectacular outfit of the female spirits, and dance for the assembled audience.

This is where the 'faking' comes in. In our interview, Otávio had this to say about it:

> 'There are people who fake being possessed by the orixá. Because they think it is pretty (*acham bonito*) when another person manifests the spirit. So when they don't receive the spirit, when they don't have that gift, that mediumship, they fake being possessed. And then there are people who go to the ceremonies of the caboclos and the exus [spirits that are offered large quantities of alcoholic drinks], and they only fake possession so that they'll be offered beer, or champagne, or wine. But they are not possessed. Not the slightest bit. They are faking it!'

Birman elaborates on what Otávio vaguely described as "they think it is pretty". She argues that for adés, the desire to become the female warrior Iansá or the beautiful golden mermaid Oxum is at the root of giving *ekê* or faking. Becoming possessed by female spirits allows for a kind of gender bending: 'the significance of the orixá is first and foremost the fabulous opportunity to embody (*encarnar*) a "Divine Lady", with all the jewelry, flowers and glitter that comes with it' (1995: 114). I wouldn't go as far as to say that gender bending is *all* the adés are interested in, but I did observe that the joy of being dressed in the garments of female spirits, and to dance in front of an audience, is very much what is being discussed among them. It is equally noticeable that the flamboyant female spirits Iansá and Oxum (and a gender ambivalent and 'handsome' spirit called Logum-Ede) are the ones that fuel the adés' imagination: other female spirits, such as the motherly Iemanjá, or the ancient, rheumatic Nanã, rarely figure in their conversations.

Whether on the night of our outing Otávio was faking possession or actually being possessed (or involved in an altogether different act,

Figure 10.2: Glitter and glamour: an adé dressed up for a public ceremony

which escaped my attention) I obviously don't know. The point I want to highlight is that we – spectators to the spectacle – will never know. All I can say is that the local gossip circuit might want to point out that Otávio is clearly an adé (who does these things); that he was eager for Iansá to arrive; and that Victor had put some pressure on him to deliver an occasion that would make the visiting anthropologist happy. It is also interesting to note that Otávio claimed it was Iansá who had possessed him – whereas Otávio was being initiated to Omulu, the male orixá of pestilence and contagious diseases, whose dance is more like a kind of humping, and nowhere close to the flamboyance adés aspire to. Let's say there are reasons to raise some suspicion as to the authenticity of Otávio's possession.

And yet, such suspicions follow a religiously informed frame. Candomblé constructs possession by orixás as an invasion of otherness: orixás are a force from beyond the horizon of man-made worlds, glacially indifferent to the likes and dislikes of spirit mediums. Wanting to be possessed by a particular spirit, which resonates with one's particular human desires, casts doubt on the authenticity of the possession when it happens. However, we ought to take into account that there are other interpretative frames operative in the situation, and I follow Birman's suggestion that for the adés, the gender frames may be equally relevant and appealing as the religious ones. So whereas adés may be 'faking it' in religious terms, there is nothing fake about the adé giving in to his desire to become a power-woman who delivers a smashing performance on the dancefloor to a deeply impressed audience.

Victor

My friend Victor is someone who likes to put himself out there as a sceptic where it concerns Candomblé. He constantly pronounces that there is a lot of lying and cheating in Candomblé, and that he does not believe "in those things". Here is how he talks about it in an interview:

Mattijs:	I would like you to talk about the spirit called 'that one'.
Victor:	But there is no spirit called 'that one'.
Mattijs:	I know, but for you to tell the story!
Victor:	Ah, the story that my sister and I invented?
Mattijs:	And how you would dress up ...
Victor:	No, the clothes were just normal.
Mattijs:	Didn't you tell me?
Victor:	No, we cooked *pirão* and pretended it was food for offering.
Mattijs:	That's the story. About the boy who would beat his aunt. Tell me that story you once told me. That you faked being possessed.
Victor:	Ah, about the neighbour's son. Who always abused his aunt. Beating her. Running after her with a shotgun.
Mattijs:	Tell me that story.
Victor:	So in front of our house stood a São Gonçalinho tree. And there me and my sister would play to be priests from Candomblé. I was the priest, she was the novice. So we took that boy there. And we lit two candles. And I faked to be possessed. I told him that I was possessed by a spirit called 'that one'.
Mattijs:	What was the name of that spirit?
Victor:	'That one'.
Mattijs:	'That one'?
Victor:	And that if he would not stop maltreating his aunt the spirit would catch him and beat him up. The boy was terrified and started to cry, for fear of being punished. After this day he never beat his aunt again.
Mattijs:	But wait, explain that there is no spirit called 'that one'.

| Victor: | There is not. 'That one' is a name I came up with. We invented it. Because we couldn't come up with the name of a spirit. So we said, let's call the spirit 'that one'. |

Victor then continued giving another example of the cheating going on in Candomblé circles.

Victor:	The name of my godmother was Francsica Teles, she was a priestess. When her husband died, I was 13 years of age. I went to live with her, because she needed a man in the house. To cut wood, fetch water. She also needed me because I went to school. I could write, she could not. So when people came to her for a consultation, she had me sitting next to her, with a notebook. I would write down everything she needed for her *trabalhos*. I would write: pumpkin, chicken – you know, balloon-chicken, these really fat ones – okra. All of these vegetables. I also put fish on the list. And some candles. That was all. I would give the list to her clients. When the clients handed in all this stuff, she ordered me to take out the pumpkin, the chicken and the fish. For the offering she only used the candles. We would eat the fish on Friday. And we would eat the chicken on Saturday!
Mattijs:	So the spirit would not see any of the food?
Victor:	She wouldn't use any of this for Candomblé. She would use it to cook us a meal, we would have vegetables, pumpkin … That is when I started doubting Candomblé. They lie a lot, to get the money from the clients.
Mattijs:	You always told me you had difficulties to believe in the reality of the orixás and spirits.
Victor:	I did not believe, because I always saw her lying. Only when I feel an orixá in my body will I believe in the reality of them. But that has never happened.

Victor grounds his scepticism in experiences such as these. As a child, playing 'priest and initiate', he learned that if you fake things they still may have powerful effects. He still knows how to fake possession, with all the right movements and sounds, and this makes for a convincing performance. And then, having been raised in a terreiro de Candomblé, he witnessed the cheating of his stepmother, and the credulence of her clients.

Yet, Victor's take on spirit possession is deeply ambivalent. First of all, the fact that Victor was accompanying me to a temple was highly unusual, as he tends to avoid these places, which make him feel uncomfortable. My fieldnotes from the night with Otávio, for instance, report that Victor was 'trembling all over' when Otávio became possessed. Moreover, knowing Victor quite well, I know that his life was deeply marked by the extreme poverty that followed an episode in his youth when his father lost his house, land and cattle (which is why he was put into the care of his *madrinha*, the priestess). Up until today, the family – including Victor – is convinced that this loss was due to *macumba*, the black magic practiced by Candomblé priests. So if he knows there to be a lot of cheating and make-believe in Candomblé, he also knows the powers of Candomblé to be frighteningly real. Victor's attitude, then, clearly illustrates the aforementioned observation that in Bahian ontologies, the reality of spirits is not questioned: what may be questioned are specific instances of their appearance.

The people from the temple

Recall that when Otávio became possessed by Iansá, the spirit was led away by members from the temple community, and it was Otávio who returned after ten minutes. Backstage, Iansá had been sent back to where she came from (*despachar*). In old and prestigious temples such as Alaketu, this is usually what happens when members from the audience become possessed. Only rarely will the visiting spirit be dressed in a splendid costume, and honoured with three songs to which (s)he may dance. For that is the privilege of spirits pertaining to the initiates of the house, or of invited guests of honour. In the neighbourhood temples, however, all spirits are allowed on the dance-floor. The fact that Otávio's Iansá wasn't allowed to dance (in fact, she reappeared later in the evening, but was sent away again) was much criticized by Victor. In the temple where he grew up, in the Bahian interior, spirits could not be sent away like that, he said. They were *entitled* to their songs and their round of dancing.

On the whole, responses to adés faking possession are ambivalent. Community members and audiences know that adés are all too eager to become possessed, so as to impersonate female spirits, for reasons that are improper with respect to the event. When I asked about faking possession, it was usually condemned as sacrilegious and disrespectful. However, in cases such as the one described above, the temple dignitaries responsible for an orderly ceremony have no other option then to give the possession the benefit of the doubt, because a possession *might* be authentic, and no one wants to risk treating the orixá disrespectfully (which is considered to be very dangerous). I've never witnessed, or heard of stories, where an adé (or other person) faking possession was openly accused of being an imposter by priests or temple dignitaries, and told to stop the impersonation. The one story I came across was of the adé who had faked possession and was allowed to dance, but when he saw from the dancefloor that his friends were all leaving, he called out to them to wait for him. Normally, however, the temple dignitaries may communicate their doubts with facial expressions, but the adé is always led away backstage in the most respectful manner.

Backstage, out of sight, there are several scenarios. One is that the spirit will be sent off. Yet there are also ways to test the authenticity of the possession. In our interview, Otávio mentioned several testing techniques, including asking questions to see whether the answer revealed clairvoyance (a quality associated with spirits), offering certain food items spirits cannot eat, and *banhos* (herbal baths). Frankly I have no idea what these would do: the backstage of public ceremonies is off limits for anyone not associated with the house, and I have never witnessed any such scenes. Zé Adário, a blacksmith and Candomblé priest from Salvador, told me about the testing he knew of.

Zé: In the temple where I was initiated the orixás would come down on Sundays, Mondays and Tuesdays. So what happened? The priestess would lift the skirts of the initiates, take a safety-pin and stick it up their bottom. This she did with the initiated of her own house. Not with the ones that came from elsewhere!

Mattijs: And this with the safety-pin was for what?

Zé: To see whether the spirit was really there! What do you think she would stick a safety-pin in their buttocks for? Wouldn't they go like 'aaaaaai'? A needle would have hurt, let alone a safety pin!

Mattijs: But are you saying there were a lot of people faking it?

Zé: Exactly! But this procedure she did only with her own initiates. Not with those who came from elsewhere.

I guess that adés unmasked as imposters would be reprimanded for their behavior, but this, again, is mere speculation.

Figure 10.3: Adding to the splendour of the ceremony

Another way that faking is dealt with is simply to allow the performance, and appreciate the artistry, beauty and *brille* dancing adés bring to the temple, adding to the splendour of the ceremony, for their talents in performance are well recognized by all. Moreover, priests would tell me, it could be left to the orixá to punish an adé who would be taking things too far.

Me

Last but not least there was me, the so-called participant observer, in my multiple capacity as anthropologist, *gringo*, friend and fellow gay man, bringing yet another perspective on possession and 'faking it' to the situation.

Recall that Otávio tried to give me proof of the authenticity of his possession by making me feel the strange coldness of his body, and then pointed out other bodily states. Obviously, he figured that I, the *gringo* (of whom there are many in the Candomblé circuit, partly because public ceremonies are very much part of the tourist programs Salvador has on offer) was a non-believer. And up to a point, he was right. My academic background, and middle-class upbringing in the radically secularised Netherlands of the 1970s, has inserted me in worlds which leave little to no room for the reality of spirits. As Stefan Palmié, in his study of Cuban Santeria, has commented: 'among academics, nothing brings out positivism more quickly than the subject of ghosts':

> We cannot seem to resist transcribing spirits, gods, or the work of witchcraft into codes that satisfy our deeply held beliefs that stories in which they figure are really about something else: category mistakes, faulty reasoning, forms of ideological misrepresentation, projections of mental states, and so on, figments of the individual or collective imagination that may be profitably analyzed in terms of their psychological or social functions but that cannot be taken literally as referents to a reality that is *really* 'out there'. (Palmié, 2008: 600)

The uncomfortable awareness that this is indeed a perceptual mould I find myself in, made me susceptible to Latour's suggestion to give up on a priori categories of thought and 'follow the natives, no matter what metaphysical imbroglio's they lead us into' (Latour, 2005: 62). So in my research I have tried to keep a kind of agnostic stance, keeping open the possibility of spirits in other world-makings, sensitizing me to that possibility, while acknowledging how my world leaves little space to embrace their reality.

Elsewhere, I have lengthily discussed the moments of bafflement in which I grounded my openness to alternative worldmakings which allow for spiritual presence (Van de Port, 2011). Disbelief, however, is not easily discarded. Although I have been studying Candomblé for two decades, the reality of spirits keeps escaping from my consciousness. Thus, many years ago, when sitting on a terrace in Salvador to have a beer with a friend, a young, homeless woman passed by and asked for a cigarette. Irritated by the demanding tone of her voice, ordering me to give her this cigarette, I refused. She started to make a scene and declined the cigarette my friend offered her, because that one was of a cheaper brand, and she only smoked "the European ones". Soon the waiter came running to our table with a package of Marlboro's, which he gave to her as he sent her away. Everyone on that terrace had recognized the presence of an Exu spirit, except the Dutch anthropologist studying Candomblé.

A more recent example may further illustrate my moments of doubt. It concerns Green Feather, the Indian *caboclo* spirit of a long-time friend, Rogério, who is a priest of Candomblé. Every Thursday night, this caboclo comes down to possess Rogério, an event whereby he dances and drinks for hours and hours to the beating of the drums, and then gives his blessings to the assembled congregation. On this particular night, however, there could be no drumming out of respect for the passing away of a community member earlier that week. So

Green Feather, after having arrived, was a bit at a loss how to entertain his audience. He started up a public conversation with me, asking all kinds of questions about the Netherlands – how long it took to get there, what itinerary the flight would take, what people did for a living, what the cost of living was, how far I lived from the Eiffel Tower, whether indeed it was the case that in summer, the light of day would only disappear around ten o'clock in the evening, and so on – the exact same questions Rogério had been asking me some days before. A profound sense of doubt descended upon me: these were queries that made sense coming from Rogério, who had recently suggested he would like to visit Europe one day. But how could these issues be on the mind of an ancient warrior spirit, a descendant of the Tupi-Guaraní? The immediate, anthropologically correct self-critique that I mobilized in the situation – "well obviously, *you* misconstrued what spirits are, or are not, interested in" – did not diminish the feeling of finding myself to be co-opted in an absurdist theatre production.

A final example concerns an occurrence last summer, when I had embarked on a small boat to film the offerings made to the sea-goddess Iemanjá. The boat was packed with locals from a fishing village, drumming and singing as the boat navigated towards open sea, but it also included a young couple of visitors from Salvador. While nearing the open sea, the girl from Salvador all of a sudden passed out and lost consciousness. Initially there was a concern that she might have had a stroke of some sort, and nearby canoes were called to take her back to land for treatment. Then, however, the crowd learned from her boyfriend that a spirit called Oxum was the master of her head, and all the panic disappeared. The canoes were send back, and the journey continued with drumming and singing. Clearly, Oxum had inhabited the girl in order to join the happy occasion, and would take care of her. I was the only one on board who remained deeply disquieted, thinking "well, this is all very nice, this talk of Oxum, but it is irresponsible to keep the girl from receiving proper medical treatment". I was relieved when, after half an hour or so, she regained consciousness.

Given my incapacity to fully embrace the reality of spirits, there is something oddly impossible about me discussing 'faking it' in Candomblé. After all, I am part of a world which, in a certain way, instructs me to see *all* spirit possessions as fake, independent as to whether locals consider them to be real or instances of 'giving *ekê*'. My way of dealing with this in my writings and filmmaking has been to adopt an interpretative framework that reassembles elements from the baroque, an omnipresent aesthetic in Bahia, and from camp, the aesthetic of gay subcultures in Europe and the Americas (cf Van de

Figure 10.4: They may not be precious stones, but they shine nonetheless

Port, 2012). Both these aesthetics move against the idea that 'the natural', or 'naturalness', is the truth of life and being, and replace this idea with the proposition that human truths are always fabricated and made (up). These aesthetics are rooted in deeply felt experiences of disruption, following a radical break-up of inherited worldviews: the break-up of the sacred canopy during the Reformation, the break-up of naturalized gender patterns when one's sexual desires turn out to be 'out of sync'. Unmasking the natural as yet another pose, both the baroque and camp celebrate the real truth of artifice. They explore the sentiments evoked by plastic roses, neon-colours, over-the-top-theatricality and lilac poodles. One might say that these aesthetics highlight that we are all imposters, that all we ever do, as social and cultural beings, is 'faking it' (cf Miller, 2005). Such sensibilities, needless to say, are helpful for anthropologists who seek to take serious the idea that what is considered to be real differs from place to place, and from time to time: that 'reality' always needs to be thought in the plural.

Imposters as producers of 'epistemic murk'

In these descriptions I have pointed out some of the multiple interpretative frames that came together in the outing with Otávio, Jorge, Victor and myself. I have sought to highlight the co-presence of different perspectives and understandings in the situation, and thus block the production of single-minded academic statements as to what giving *ekê* is about. I thus joined the growing number of anthropologists who have argued that facing the complexities, incongruities and instabilities that are present in concrete situations should not be smoothened, or ironed-out, so as to reduce the messiness of life and produce clarity (cf Law, 2004). And yet, there are questions that I would like to ask about the theme of impostering that transcend the immediacy and particularity of the situation. One such question that is on my mind concerning public ceremonies of Candomblé is: given the presence of these multiple frames, what becomes of the situation? How is the particular assemblage of bodies, objects, ideas, feelings, memories and desires of a situation remembered, discussed, restaged and re-enacted? Could it be that a situation lacking clear, well-delineated and shared understandings comes to be recognized as such? Appreciated as such? Staged as such?

What I find interesting is that public ceremonies of Candomblé produce what Taussig once called 'epistemic murk' (1986: 121), an uncertainty about the real of things, a blurring of clear-cut boundaries between fiction, fact, performance, imagination, reality. Candomblé ceremonies do not shy away from epistemic murk. To the contrary. Adés are not barred from joining the ceremonies, but welcomed, and frequently allowed to dance their rounds. The practice of 'impostering' is eagerly discussed among visitors, and the fake of a possession is always kept up in the air as a possibility. Members of the temple community inform anthropologists about adés, the practice of 'giving ekê', and the modes of testing the reality of spiritual presence. And even clueless tourists, who never cease to express their doubts as to whether what that they see happening in front of their eyes is a 'mere folklore show' or 'really real', are welcomed in to adorn the ceremonial hall with their puzzled faces. 'We really don't know, do we?', seems to be a state of mind that the Candomblé community actively produces.

From an academic perspective, it is difficult to imagine that 'epistemic murk' is actually something one would want. Our methods, as John Law has argued, are primarily about 'distorting reality into clarity' (2004: 4). Now I'm not saying that the people from Candomblé are not interested in this clarity, as the testing to certify the presence of a spirit already suggests. Yet it is interesting to note where, for whom and by whom, 'testing' takes place, and where does it not. The instances that have passed are Marcello's 'private self-testing' ("I wanted to have sex, but couldn't"); Otávio's and Victor's mentioning of bodily states in a tête-à-tête with me, the anthropologist; and tests such as sticking needles in the buttocks of possessed spirit mediums to see if they are not faking it in the backstage of the temple, away from the audiences. Strikingly, testing is *not* part of the public performance of spirit possessions during celebrations. Indeed, other types of 'proof' of the real of spirit presence (for instance the extreme acts which no human would ever do, such as biting off the head of a chicken and drinking its blood) are zealously screened off from public view, and take place in *matança* ceremonies where only the intimate circle of the temple community can be present. What this suggests is that the production of uncertainty, the highlighting of doubt is a quality of *public* ceremonies. The spectacularisation of epistemic murk may be part of what I have elsewhere (Van de Port, 2011) called Candomblé's 'politics of bafflement', as it contributes to the heightening of an awareness that we humans are not the masters of the universe; that our capacities to grasp the world we live in are limited; that there is a beyond to our knowing. As stated, public ceremonies in Candomblé seek to impress audiences with the mysteries that are at the core of their religious and spiritual practices, as well as the spiritual services they sell. For what magic could be produced in a temple where, hypothetically, everything 'is what it is, and not another thing'?

References

Birman, P. (1995) *Fazer estilo criando gêneros: possessão e diferenças de gênero em terreiros de Umbanda e Candomblé no Rio de Janeiro*, Rio de Janeiro: EdUERJ.

Fry, P. (1986) 'Male homosexuality and spirit possesion in Brazil', *Journal of Homosexuality*, 11(3): 137–153.

Landes, R. (1940) 'A cult matriarchate and male homosexuality', *Journal of Abnormal and Social Psychology*, 35: 386–397.

Landes, R. (1947) *The City of Women*, New York: Macmillan.

Latour, B. (2005) *Reassembling the Social: An Introduction to Actor-Network Theory*, Oxford: Oxford University Press.

Law, J. (2004) *After Method: Mess in Social Science Research*, London: Routledge.

Law, J. (2009) 'Actor Network Theory and material semiotics', in B.S. Turner (ed) *The New Blackwell Companion to Social Theory*, Hoboken: Wiley-Blackwell, pp 141–158.

Miller, W.I. (2005) *Faking It*, Cambridge: Cambridge University Press.

Palmié, S. (2008) 'Evidence and presence, spectral and other', in M. Lambek (ed) *A Reader in the Anthropology of Religion*, London: Blackwell, pp 598–611.

Ramos, A. (1942) *A aculturação negra no Brasil*, Rio de Janeiro: Biblioteca Pedagógica Brasileira.

Taussig, M.T. (1986) *Shamanism, Colonialism, and the Wild Man: A Study in Terror and Healing*, Chicago: University of Chicago Press.

Van de Port, M. (2011) *Ecstatic Encounters: Bahian Candomblé and the Quest for the Really Real*, Amsterdam: Amsterdam University Press.

Van de Port, M. (2012) 'Genuinely made up: Camp, baroque and the production of the really real', *Journal of the Royal Anthropological Institute (N.S.)*, 18: 864–883.

11

The Guerrilla's ID Card: Flatland against Fatland in Colombia

Olga Restrepo Forero and Malcolm Ashmore

> He walked through the Amazon jungle in the middle
> of the guerrilla zone with a backpack on his shoulder,
> filled with aguardiente and marijuana and no *cédula*, can
> you imagine? Nobody can exist, in Colombia, without a
> *cédula*. In Colombia even the dead have a *cédula*, and vote
> with it [...]
> - Why the hell don't you have your *cédula* Darío, what's
> the problem?
> - I don't have one, it was stolen.
> - Idiot! Letting someone steal your *cédula* in Colombia
> is worse than killing your mother.
> (Fernando Vallejo, El Desbarrancadero)[1]

Once upon a time,[2] there was a bricklayer and/or carpenter and/or
taxi driver living in the city of Santa Marta on the Caribbean coast.

---

[1] Our translation (the novel has not been translated into English). In his excellent
work on the *cédula de ciudadanía* (Colombian national ID card) Sebastián Guerra
Sánchez (2011) used this fragment as his epigraph. We copy him to honour him.

[2] And several times, we have presented versions of this tale. We thank the organizers
and audiences at the Department of Social Sciences, Loughborough University,
UK; the School of Management, University of Leicester, UK; the Department
of Sociology, Lancaster University, UK (all December 2011); and, of course, the
'Imposters and Gatecrashers' conference, Linköping University, 13 June 2018. An
early version was published in Spanish as Restrepo Forero and Ashmore (2013).

His name was Jorge Enrique Briceño Suárez.[3] At the same time, there was a guerrilla living 'somewhere in Colombia'. And he, too, had the name Jorge Enrique Briceño Suárez.

Just a coincidence, then. A case of random 'homonymy', a namesake with no consequences. But of course not. If that was the case, we would have no story to tell.

So let's start our story with its *dramatis personae*. Conventionally, the characters populating a drama are persons; and indeed some of our characters, like the ones we have already mentioned, and will repeat here, are **people**; some even citizens:

- Jorge Enrique Briceño Suárez, working man of Santa Marta
- Jorge Iván Restrepo Llano[4]

And then there is:

- Mono Jojoy, guerrilla (Jorge Briceño) (Jorge Briceño Suárez) (Jorge Suárez Briceño) (Jorge Enrique Briceño Suárez) (Luis Suárez) (Luis Suárez Rojas) (Víctor Julio Suárez Rojas) (Víctor Suárez Rojas) (Oscar Riaño).[5]

---

[3] Colombian personal names are organized like this: (first given name [*nombre*]) (second given name) (first surname [*apellido*] from father) (second surname from mother). Anglo-Saxon personal names are organized differently, which creates all sorts of misunderstandings in translation, usually avoided by the *apellido-apellido* hyphen. But the hyphen, for the Spanish name system (that does not use it as part of its own internal practices) seems to be the creation of a superimposed, quasi-imposter, identity.

[4] Señor Restrepo (JIRL) is not another namesake, he is a 'numbersake'. His consequential coincidence is numerical: he has the same *cédula* number (70.753.211) as our guerrilla. And because of this he has suffered in a similar fashion to Jorge Enrique Briceño Suárez (JEBS); as we will see as our story proceeds. However, he has an extra disability: his *cédula* number is also attributed to *another* leading FARC (see note 6) guerrilla, another Very Bad Person in not only the Colombian state's files, but also the United States' so-called 'Clinton List' detailing global Badnesses of all kinds. Thus JIRL was not relieved of his burden by what happened to (SPOILER ALERT) Mono Jojoy (our guerrilla) in September 2010; as, at least in principle, JEBS was. Though the details of JIRL's case are interesting (they are set out fully in Restrepo Forero and Ashmore, 2013) we have no space for them here, save the factoids of Box 11.3; and of course, this note.

[5] With this quantity of names, we frequently resort to initials: JB, JBS, JSB, JEBS, LS, LSR, VJSR, VSR, OR.

Yes, that's right. The bracketed names are 'alternatives'. The character, the guerrilla, Mono Jojoy (now, there's an odd name; surely false!) has these other nine 'to his name', as it were. The taking of alternative names is very common among the more (in)famous leaders in the world of Colombian left-wing guerrillas, the FARC-EP, the ELN, and of the right-wing paramilitaries, the AUC and their successors, the so-called Bacrims, and of course the drug gangs.[6] Often, as in the case of our character, these are of two types. A 'nom de guerre', formed on the model of a standard name ('Jorge Enrique Briceño Suárez', for example) and a nickname (like Mono Jojoy).

There are other non-human characters that play even more important roles in this tale. Our second category of actors are **agencies of the state**. These include the police, the army, several judicial agencies and courts, the offices of the attorney general and of the prosecutor, the state intelligence agency, and the *Registraduría,* the organization which issues the national ID card and conducts elections. All of these organizations play their parts, as they will ...

But they cannot do so without the third and final category of actants, the **documents**:

- Firstly, the *cédula*. We detail the history and social role of this identity card elsewhere (Restrepo Forero et al, 2013). Here we need to stress two interlinked features of this *documento*: its relation with impostering and its utter ubiquity – as our epigraph illustrates by showing, dramatically, how the *cédula* is needed, at least by some, even in the jungle and even by dead voters! The *cédula* was introduced originally, in 1935, as a voting card with the stated intention of combating what were understood to be endemic levels of electoral fraud. As its functionality rapidly expanded, official concerns with the population's tendency to fraudulence and impostering grew in tandem, with the *cédula* itself becoming a major focus of anxiety. In 1948, as the first example of a recurring theme in official discourse, it was claimed that there were 1,800,000

---

[6] FARC, or FARC-EP – Revolutionary Armed Forces of Colombia-Popular Army; ELN – National Liberation Army; AUC – United Self-Defence Forces of Colombia, and their successors, the so-called Bacrims – criminal bandits. For a concise description of all these groups, see BBC News (2012). Of course, the situation has changed since the 2016 Peace Accords between the Colombian state and the FARC – the FARC disbanded as an army and is now an as yet unsuccessful political party.

false *cédulas* which was exactly half of all the *cédulas* issued! Such extravagant understandings of the extent of citizen delinquency helps directly to produce a nation of 'fraudsters' as nobody knows who is hiding behind their *cédula*. Thus, *the cédula produces what it is supposed to prevent*: fraud, fear of fraud, and state agencies with a fraud reduction/prevention role, resulting in the mutual distrust of state and citizen. This distrust is maintained and enhanced by standardized patterns of use: the citizen is required to show or quote the ID document for nearly all transactions, however mundane. A person without it *cannot* be trusted. In effect, trust can only be invested, if anywhere, in the *cédula* itself. And as our story will evidence, this distribution of trust has unfortunate consequences.

- Hypertext of the decisions of the Constitutional Court (rights of appeal, *tutela* [writ for protection of the applicant's constitutional rights and for the redress of official decisions], legal suits, judgments, memoranda, reports, arrest warrants, notes, appointments, certificates, and so on).
- Newspapers and magazines, broadcast radio and television.
- Hyperhypertext: the online world.

As another preliminary, we need to introduce you to 'Fatland' and 'Flatland', two domains of reality, two orderings of/in the world:

> *Fatland (el mundo pleno)*: the world of local realities, relations and recognitions, of face-to-face life; but also the world of flexibility, interpretations, uncertainties, and searches for explanations. The life world of Schutz (1967), the interaction order of Goffman (1983), the ethnomethods of Garfinkel (1996 [1967]); or indeed the unformatted 'proliferation of incommensurate entities' that might (or not) 'make up a provisional whole' (Latour, 2005: 243).

> *Flatland (el mundo plano)*:[7] the world of bureaucracy, of documentary realities, texts, and artefacts, paper and

---

[7] Notice that the material style of this text has similarities to that commonly used in presentations; courtesy, of course, of PowerPoint software. We believe that adopting this unusual form in a piece of academic prose ('paper') allows us to visually evidence something of the (anti) style of Flatland where so much of this story is situated. It also allows us to present our story synthetically and to display, in our ANT-like description, the step-by-step connections being made and traces being followed which are the standardized modes of action of many of our actors (Latour, 2005).

electronic, of files and databases and archives, of maps, classifications, orderings, numberings, and rankings. The world in which the synoptic/panoptic/myopic gaze of the state and its officials, rules.

Yes, synoptic, panoptic and myopic, all three. The *synoptic gaze* is, for James C. Scott, a solution to a problem of statecraft, in which the state passes from a 'partially blind' phase with very little knowledge about its subjects, their possessions and localities; and from a lack of a language or a metric to '"translate" what it knew into a common standard necessary for a synoptic view'; to the stage of having simplified and standardized forms of ordering and legibility (Scott, 1998: 2).

Though a fascinating historical account, Scott's thesis suffers from the inappropriate assumption of the pre-existence of an actor, 'the state' with (lack of) sight and (lack of) knowledge. We prefer an account that has 'the state' being an effect of 'its' processes, coming into being alongside 'its' production of docile, legible, simplified 'subjects and their environment'.

The *panoptic gaze* is more familiar and stems of course from Bentham's and Foucault's panopticon, an efficient architectural means of control through an all-seeing machine (Bentham, 1995 [1787]) or an efficient production of the nightmare ('visibility is a trap') of the all-seeing disciplinary state (Foucault, 1977). We see the panoptic as more a totalitarian fantasy than a realizable reality; CCTV systems and other surveillance technologies are never as total or as effective as the image of the panopticon suggests. For a more realistic characterization of the empirical possibilities of 'attempts to see everything at once', we recommend Latour's concept of the 'oligopticon': only a few powerful sights/sites are ever actually covered 'in full' (Latour and Hermant, 1998).

Finally, the *myopic gaze*. What we have in mind, is the ability (and preference?) of officials to narrow their focus to a single document, or a single element of a document, to the exclusion of all else, and to make that one feature 'trump' any alternatives or even collections of alternatives; for example, in the identification of persons, the major task and project of the specifically 'modern' *Flatland*. As Craig Robertson writes, in an analysis of the emergence of the passport in America, the late 19th century witnessed the development of a series of identification practices (replete with well established procedures for classifying, archiving and retrieving) that ended up collapsing all the various personal and legal identities into a single new identity, that, while useful for governing, 'did not fit into an existing relationship of identity and identification based on a logic of trust and practices of self-identification and reputation' (Robertson, 2009: 330). What these new practices of identification, characteristic of *Flatland*, 'did not fit into' was (and still is) the form of life prevalent in *Fatland*.

The opposed pair of *Fatland* and *Flatland* is similar to those that found and sustain modern sociology, professional and lay:

- *Gemeinschaft und Gesellschaft* (*Community and Society*) (Tönnies, 1957);
- the life-world versus bureaucratically/contractually mediated relations (Habermas, 1987);
- Tradition and Modernity;
- *or* in its lay version, Then and Now;
- 'them and *Us*' (them Others, the peasants, the savages, and *Us* civilized moderns);
- *or* 'us and *Them*' ('us' as powerless subjects and the *Them* of '*They* ought to do something about it').

'We' live in both worlds simultaneously. But this does not mean that these two realms are always at peace with each other. Sometimes, in some places, under some circumstances, Flatland can short-circuit the apparently smooth procedures of Fatland. Our story is of one such time, place and circumstance.

The events with reference to the citizen Jorge Enrique Briceño Suárez

In 1997, the citizen (see Figure 11.1) takes a bus ride. He is 'recognized' and identified at a police checkpoint. Here is his account of this incident:

> I remember that time travelling from Valledupar to Pailitas, when the bus was stopped at a police checkpoint, and us passengers were asked to show our papers, as is routine. After seeing mine, the officer went to the lieutenant in charge to inform him that 'Mono Jojoy' was in the vehicle. The lieutenant came over to me and he jokingly told the

Figure 11.1: Jorge Enrique Briceño Suárez

Source: *El Informador*, 2010b.

officers that I could not be that guerrilla, and amid laughter
they let me go. Then I did not think it was important and
I continued my life as normal. (*El Informador* 2010a)[8]

This is his first awareness of the existence of the 'problem' of his
'connection' to Mono Jojoy. Note that he was let go. Note that there
was laughter. Note that an identity built from the documents and
numbers of Flatland was overruled by a non-identity based on a Fatland
recognition that 'I could not be that guerrilla'.

But this victory of Fatland over Flatland would be fleeting …. For
there is an older legal process in which his name and *cédula* number
(CC. 12.536.519) are referenced (which we assume alerted the police
on the bus), dated 26 March 1996:

Regional Attorney's Office – Regional Directorate of Public
Attorney's Unit of Bogotá. In Office 258, 26 March 1996.
Communicates arrest warrant, Attorney's Office before
the 11th Criminal Court, Specialized Circuit of Bogotá
in Office 244 of 27.04.2001 Rad-208059-01. Clarifies
numerical quota, in process 26044: Article 467 C.P. *aggravated*

---

[8]   *El Informador* is JEBS' local paper. The national press also ran this story.

murder, rebellion; Article 169 C.P. *kidnapping for ransom;* Article 343 C.P. *terrorism.* (T-578-10.rtf: 7)[9]

The citizen is arrested and held for several hours after presenting himself, in order, ironically, to clear his name, at the Office of the Regional Attorney in Santa Marta on 23 April 2005. He is harassed on many further occasions, including one extravagant security operation designed to capture Mono Jojoy. As a reporter later told the story:

> he was arrested and handcuffed in the [Santa Marta] Transportation Terminal ... Military Intelligence, the DAS, the police, and other agencies of the security forces came to the terminal to catch the alleged guerrilla leader. 'I could not believe that I was handcuffed in the terminal, as members of the security forces checked the guerrilla's fingerprints against mine to assuage their doubts.' (*El Informador,* 2010a)[10]

As a result, the citizen decides to turn to the law. From this point on, he, as a legal case, lives in Flatland.

The events in/on Flatland

Citizen Jorge Enrique Briceño Suárez (JEBS) issues a 'Petition for information' [*Derecho de petición*] to discover his exact legal status, on 2 June 2009.

He receives responses from the Prosecutor General's Office; from an official of the Department of Administrative Security; from an official of the *Registraduría;* from the Director of Criminal Investigation, National Police; and from the Office of the Attorney General (T-578-10.rtf: 9, 25–33). From these responses to the petition for information made by citizen JEBS, he learns (and so do we) that:

---

[9] For the identity of this reference, which we use repeatedly, see note 12. But please wait until you get there. Translations and paraphrases from this reference are ours. Emphases (of all kinds) here and throughout, are also ours unless otherwise indicated.

[10] One wonders where the comparator prints from 'the guerrilla' were stored; as we will see ....

- JEBS (CC 12.536.519) has accumulated *prison sentences totalling approximately 116 years.*
- JEBS (CC 12.536.519), JBS (CC 12.536.519) and JBS (no CC) have a total of *86 criminal records* including one with INTERPOL.
- His *cédula* (CC 12.536.519) has been *annulled*, and his political rights have been *suspended*. He *cannot vote*. He *cannot access state services*. He *cannot get credit* from any bank. He *cannot leave the country*. He is *subject to arrest or harassment* at any moment.

And all this despite the Office of the Attorney General certifying that 'the true name of the alias MONO JOJOY is LUÍS SUÁREZ identified by CC No. 17.708.695' (capitalization in original). But unfortunately, this specification partly conflicts with that provided by the Director of Criminal Investigation, National Police. While agreeing that the 'true name' of 'the terrorist' and 'the alias' Mono Jojoy is 'Luís Suárez', these two agencies disagree on the most important identifier, the *cédula* number; the police quoting JEBS' number. So it seems that whoever 'Luís Suárez' might be, his Flatland IDs are insecure, because he is attributed with two *cédula* numbers, one of which 'also' belongs to the citizen Jorge Enrique Briceño Suárez. And yet it seems standard and unremarkable practice in Colombian Flatland to treat the *cédula* number as a 'higher level' identifier than, for example, a name. It is treated this way, because it is supposed to be unique. But, in this case not only was the name more important than the number, it was the number(s) that proliferated.

So now JEBS knows the depth of what the state has attached to his name and number. To that extent, he knows 'why' his life has been made impossible. Of course he is puzzled that he, as a citizen of Fatland, could ever be mistaken for Mono Jojoy. But then, he is no longer, relevantly, living in Fatland with all its taken-for-granted ways. His life in that realm has become so thoroughly corrupted by one dimension of Flatland as to be unliveable. Also, his other Flatland connections can no longer run in the smooth, unnoticed way that they used to. And so JEBS makes his next move within Flatland: he applies for a *tutela*, the constitutional guarantee established in 1991, which allows citizens to request a court to grant the immediate protection of their constitutional rights, where these have been violated.

The history of JEBS' application for a *tutela*

In December 2009, the Chamber of the Superior Criminal Court, Judicial District of Santa Marta, admitted our citizen's application for

Box 11.1 Who is Mono Jojoy?

So far you know that he is a guerrilla. He was the major military commander in the FARC-EP, and second in the hierarchy of the whole organization. The agencies of the state (the operators of Flatland) say he is a 'terrorist', and they also say that the name (or names) that he is known by is an 'alias', and that his 'true name' is 'Luís Suárez'. Another (true or false) name by which he is known, is Jorge Enrique Briceño Suárez (or Jorge Briceño Suárez). Under this guise he seems to be the proper referent of all those terrible crimes for which those names (JEBS and JBS) have accumulated so very many sentences, and for which 'he' should be serving so very many years in jail. And the citizen of Santa Marta is terminally '(mis)taken' for him. But this 'mistake' is of a very specific kind, only conceivable in a Flatland that succeeds in ignoring another type of evidence, with its credibility firmly based in Fatland: the citizen looks, behaves and lives, *nothing at all like* the guerrilla.

The citizen   The guerrilla

Source: El Tiempo, 2010c.

a *tutela*, which demanded the reversal of decisions taken by several state agencies, and *if successful* would order the annulment of all his criminal records; the restoration of his *cédula* (12.526.519) and thus his civil and political rights; the deletion of all records linking him with Mono Jojoy; and all this within 72 hours (T-578-10.rtf: 6–7).

Selected responses from the agencies

1. **Second Court of Criminal Circuit, Pasto:** 'BRICEÑO SUAREZ, alias Mono Jojoy who is […] identified by cédula number 70.753.211, which belongs to señor JORE [sic] IVAN RESTREPO LLANO, [a mistake later corrected to] "JORGE ENRIQUE BRICEÑO SUÁREZ, identified by cédula number 12.536.519, of Santa Marta, which accounts for the initial mistake. Since *there are no other irregularities in the case* reviewed, we believe that the tutela cannot succeed'".

2. **First Court of Criminal Circuit, Florencia:** '[3] convictions against señor Jorge BRICEÑO SUÁREZ alias "Oscar Riaño" or "Mono Jojoy"'. 'The court notes that in two of the procedures he is listed as Jorge Briceño Suarez and in the other as JORGE ENRIQUE Briceño Suarez, *but what is constant in the three cases is that he is the popular and widely known "Mono Jojoy"'*.

3. **Supreme Judicial Council:** 'requests that the application for tutela be denied', since there is no violation of fundamental rights because the claim is based on '*subjective judgments that cannot be taken into account* by this court'.

4. **Administrative Department of Security (DAS)**: has information about the [86] criminal records, but 'DAS is the trustee and not the owner of the information that it receives and *is forbidden to modify anything* on its own initiative'.

5. **National Police**: 34 arrest warrants in force: 'However, the judicial authorities who issued the orders have not cancelled them, and the Criminal Investigation Department *does not have the option* to proceed with such cancellations itself'. (T-578-10.rtf: 7– 11, all capitalization in original, emphasis added)

Thus, on 20 January 2010, the Superior Criminal Court of Santa Marta rejected JEBS' petition for a *tutela*. But JEBS didn't give up. He applied to a higher court, the Criminal Chamber of the Supreme Court of Justice. And, on 9 March 2010, this court issued its verdict, and once again it was *negative*. The court said, 'The citizen should show greater diligence and should repeatedly make his case throughout [Flatland]' (T-578-10.rtf: 13). In other words, he should seek redress from every court and relevant agency, individually and serially. A tall order for a citizen of very modest means!

This seems the end of the line, as the citizen has nowhere else to turn. But then, out of the blue, the country's highest court, the

Constitutional Court, decides, in April, to review his case.[11] On 21 July 2010, the court hands down Sentencia [Judgment] T-578/10, in which it *reverses the judgments* of the two courts that had denied our citizen protection and *grants the tutela* providing those fundamental rights violated by the errors of all the agencies previously mentioned.[12] So JEBS with his *tutela*, is finally protected against the ravages of Flatland.[13]

And the press takes notice, as it frequently does with similar judgments of the Court:

'Constitutional Court grants *tutela* to the Mono Jojoy namesake to clear his name'. (*Radio Caracol*, 2010)[14]

'I am a prisoner in my home and in my country': Jorge Enrique Briceño Suárez. (*El Informador*, 2010a)

After 10 years of stigma, Jorge Briceño Suárez regains the name 'stolen' by 'Mono Jojoy'. (*El Tiempo*, 2010c)

How did the Constitutional Court reach its decision, and why was it so different to the previous, negative, verdicts? One very significant reason was that it took into account the type of evidence – Fatland evidence, though of course filtered through the mechanisms of Flatland – that is concerned with *recognition* rather than *identification*.[15] In one of its many reviews of previous occurrences, the Court notes the following conclusion to the swoop on our citizen in the Santa Marta bus station, as recounted earlier:

---

[11] The Constitutional Court, on its own initiative, reviews decisions of the lower courts and selects significant cases where a positive judgement would protect the fundamental rights granted in the constitution. In fact, the lower courts send their judgments directly to the Constitutional Court to facilitate this process.

[12] For the decision in full (18,000 words, and in Spanish) see http://www.corteconstitucional.gov.co/relatoria/2010/T-578-10.htm [Accessed 23 March 2020]. The web page allows downloading of a page numbered file: T-578-10.rtf. All our specific references to material contained in this document are to this file.

[13] Or maybe not. See the section 'Another turn of the screw'.

[14] It was by hearing this broadcast that we first learned of the case and decided to follow it up.

[15] People's practices of *recognition* in face-to-face interactions are based on memory, are socially structured, and work by selectively retaining certain physical traits of the recognized. The anthropometric system involving detailed physical measurements of bodily features devised by Alphonse Bertillon in the 1870s was the first *identification* system used by the police in the West (Cole, 2001). This

Prosecutor 29 ... of the Attorney General's Office issued a resolution on 6 May 2005. The investigation was suspended because 'the arrested person ... does not correspond to the one specified in the investigation, namely JORGE SUAREZ BRICEÑO, alias Mono Jojoy; who is approximately fifty three years old, with fair skin, brown eyes, medium shaped mouth, straight nose, and brown hair, approximately 90 kilos in weight and 1.79 meters tall ... a person PUBLICLY RECOGNIZABLE through photographs and television, who is fully individualised'. (T-578-10.rtf: 23–24, capitalization in original)

This is not the language of names and numbers; it is the language of *anthropometry* (as it is known in Flatland) or of *likeness and recognition* (as it is known in Fatland) (see note 15). This theme reappears later on in the judgment of the Court, where the citizen's 'features' are similarly described and compared to those of his namesake:

Since 2003 the CTI [*Cuerpo Técnico de Investigación*: the forensic service of the Office of the Attorney General] has known that the only Colombian registered with the name of Jorge Enrique Briceño Suárez, holds a *cédula* issued in Santa Marta, and is thin, brown-skinned, with fat lips and a broad nose, that is to say, who has features markedly

---

system, despite its sophistication, systematized elements of recognition coming from other sources. However, a different system of identification, fingerprinting, subsequently appeared more robust. 'Anthropometry emerged in the cities of Europe, while fingerprinting developed in the colonies and on the frontiers of the Western imperial states' (Cole, 2001: 32). On the colonial origins of the 'Henry system' of fingerprinting, see Cole (2001), Henry (2004 [1900]), Kaluszinski (2001) and Sengoopta (2003). On the alternative system developed in Argentina, see Rodriguez (2004) and Ruggiero (2001). Though 'scientific anthropometry' has a decidedly antique feel, it has never really disappeared. The development of ID card systems globally has continued to employ elements of such techniques (Bennett and Lyon, 2008; Lyon, 2009). Rather, anthropometrics have been overlaid and overshadowed by biometrics (massively fingerprinting, though nowadays also by the incipient technologies of iris, facial, and voice recognition, hand geometry and the rest; see Lyon, 2001; van der Ploeg, 2005). Anthropometry has its specific uses, some of its old spirit surviving in 'expert morphological comparisons', such as the one performed in our case in 2003; and even more so in its lay manifestations such as in the decision of the Constitutional Court, acting as it does as a kind of bridge between Flatland and Fatland.

different from those of the military leader of the FARC. (T-578-10.rtf: 32–33)

But it is important to notice that this seemingly more 'humane' system of *likeness and recognition* also enacts a racialized Fatland, where alternative descriptors like 'fair skin'/'brown skinned', 'medium shaped mouth'/'fat lips', 'straight nose'/'broad nose' are doing the work previously done in the standardized practices of anthropometry, that translated and crystallized racial hierarchies (see Box 11.2). As can be seen in both the elements that are specified and those that are not, these descriptors inscribe racialized modes of seeing.

Box 11.2: The 'features' of the two JEBS compared

Names: One	Ten
Jorge Enrique Briceño Suárez	Mono Jojoy
	(Jorge Briceño)
	(Jorge Briceño Suárez)
	(Jorge Suárez Briceño)
	(Jorge Enrique Briceño Suárez)
	(Luis Suárez)
	(Luis Suárez Rojas)
	(Víctor Julio Suárez Rojas)
	(Víctor Suárez Rojas)
	(Oscar Riaño)
Skin colour: Dark (*piel morena*) (African descent)	Fair (*tez blanca; piel blanca*)
Eye colour: [not stated]	Brown (*color cafés*)
Mouth shape/size: [not stated]	Medium/small (*boca mediana; boca pequeña*)
Lips: Thick (*labios gruesos*)	Thin (*labios delgados*)
Nose: Broad (*nariz ancha*)	Thin (*nariz delgada*)
Hair colour: [not stated]	Brown (*castaños*)
Stature/weight: Thin (*delgado*)	Approx. 90kg., almost obese
Height: 1.62 m	1.79 m
Blood type: RH AB+	[not known]

Note: These descriptors appear in the sources referred to throughout the text. They are not ours. [Not stated] means unspecified in these sources.

First expert identification (2003)

The event that happened in 2003 that produced all this knowledge was the first official attempt to answer that irresistible question, Who is Mono Jojoy? What is his 'true identity'? According to a report published in *El Tiempo* (2003), this expert identification was initiated by a group of researchers from the CTI (*Cuerpo Técnico de Investigación*) based at Villavicencio. Motivated by the number of *cédula* numbers that were attributed to Jojoy, all of which were false, except for the one shared with JEBS,[16] and after establishing that the physical characteristics of the man from Santa Marta made it impossible that he could be Jojoy, 'the researchers proceeded by searching through the files [of the *Registraduría*] and found the record of a person identified as Luis Suarez'. Although this is not specifically stated in the *El Tiempo* report, at this point the only method they had was a search for surnames. After drawing a blank with 'Briceño' they had more success with 'Suárez'. They then carried out a 'morphological comparison' together with other specialists, said to be from the (non-existent) 'Department of Physical Anthropology' of the National University:

> ... to verify their first impressions of identity, the researchers undertook a morphological comparison, comparing the image from the *cédula* of Luis Suárez and photographs of Mono Jojoy published in newspapers ... [see Figure 11.2]. The specialists, after an evaluation of nine key features of physiognomy, concluded that there was great similarity

---

[16] This is interesting because it shows a strange degree of carelessness in these seemingly thorough investigations. Let us see exactly what is said in the published report: 'The agents of the [CTI] assigned to one of the several criminal proceedings prosecuted against Mono Jojoy by the Attorney General's Office noticed that [he] appeared to have been attributed with five different *cédulas*. The numbers were *70.753.211*, 70.723.211, 70.732.211, 4.466.970 and 70.309.211 ... The detectives investigated and found that none of these numbers corresponded to a *cédula* issued to anybody (*El Tiempo*, 2003).' Our point is this: *the number 70.753.211 really does belong to at least one citizen*: JIRL. We thank the thorough reading by Jack Yuri Gomez, for having brought this important detail to our attention. Moreover, through a later search for these numbers on the *Registraduría* website, in the way we describe in 'Another turn of the screw', later in the chapter, we are able to state that the number 70.723.211 belongs to a real citizen (whose potential troubles we have not investigated); and also that the number 4.466.970 'belongs to' Oscar Riaño, another of Mono Jojoy's alternate names.

between the two images ... On a scale of 1 to 5, five
morphological features scored 4 (good resemblance) and
the other four features scored 5 (very close resemblance).
(*El Tiempo*, 2003)

Figure 11.2: A morphological comparison

Source: *El Tiempo*, 2003.

Let us dwell briefly on the morphological comparison, noting that these
results ('good resemblance' and 'very close resemblance') are based on
well-prepared materials, such as the grid which allows us to see the
details of the comparison for ourselves. Then there are the two specific
photographs, one a given, reproduced from the *cédula* of Luis Suárez,
the other 'selected', presumably for the purposes of this comparative
exercise, from the multitude of available photographs of Jojoy. The
work of the selection and display of these two images, enabling the
active search for elements that can be highlighted as constituting 'the
likeness', is not described in the article.[17] However, we can make an

---

[17] We, however, are able to do so with the help of Charles Goodwin's 'professional
vision'. Engaging in professional vision involves three practices, all of which are
present in this process of morphological comparison: '1) *coding schemes* used to

Figure 11.3: Luis Suárez' *cédula*

Source: *La Patria,* 2011.

Note: The first expert identification, using 'expert and specialized methods', concluded in 2003 that Mono Jojoy was Luis Suárez.

alternative reading if we wish. We could notice, for example, the striking difference in the eyebrows – one of the elements that would account for the '*mono*' (fair skinned) descriptor of the Mono namesake.

Now, please note the doubly odd label for the image on the left: '**Jorge Suárez Briceño**'. First, this very official document is one of only two places, to our knowledge, where this particular name ordering exists (making it a completely different name).[18] Second (and disregarding the first), this name's 'truth' is undermined by the very process in which it appears; a process that results in its displacement by the label for the right-hand image: '**Luis Suárez**'. Also, the official refusal to give any status whatever to the name '**Mono Jojoy**' is evident here, as it is throughout. This refusal is the source of endless difficulties for the agents of Flatland (as well, of course, for our citizen). It leads them into fruitless searches, like this

---

transform the materials being attended to in a specific setting into the objects of knowledge that animate the discourse of a profession; 2) *highlighting*, making specific phenomena in a complex perceptual field salient by marking them in some fashion; and 3) *the production and articulation of material representations*' (Goodwin, 1994: 606).

18 The other is in the quotation from the judgment of the Constitutional Court ('Prosecutor 29...', T-578-10.rtf: 23–24) encountered earlier. Interestingly, the 'source' of both JSBs is the same organization: the Attorney General's Office.

one, for the 'truth' of someone who already has a fully adequate identity; as the Constitutional Court puts it, citing the statement of prosecutor 29, quoted previously: 'a person PUBLICLY RECOGNIZABLE ... who is fully individualised' (capitalization in original).

Which is how and why Mono Jojoy is (or was – see the following section), for Flatland, 'really Luis Suárez'. And 'Luis Suárez' is registered. He has, or had, the *cédula* (see Figure 11.3) from which his image in Figure 11.2 was taken.

But these things can become confusing. A report on the judgment of the Court in *El Espectador* had this to say about the CTI's conclusions 'The court ... found that even though the ... (CTI) had done a study in which they concluded that the true name of alias "Mono Jojoy" was **Victor** [sic] **Julio Suárez Rojas**, the situation has not been clarified.' (*El Espectador,* 2010).[19] Indeed it hasn't! Interestingly, this (new real) name does not appear anywhere in the document of the Court's decision. Its appearance on 15 September 2010 in *El Espectador* seems strangely prescient, however, as it appears all over the media eight days later.

The death of Mono Jojoy

Two months after the Court's decision, something else happened that revived the press stories of the guerrilla's namesake. Indeed, according to these reports, this event revived his life as much, if not more, than the decision of the Constitutional Court. On 23 September 2010 Mono Jojoy was killed by the Colombian military in 'Operation Sodoma'.

> 'Now I am free', says Jorge Briceño Suárez, namesake of the shot and killed 'Mono Jojoy'. (*El Tiempo*, 2010a)

> The death of Mono Jojoy gives life to Jorge Briceño Suárez, a Santa Marta taxi driver. (Cabrera, 2010)

Immediately upon his death, another operation, 'Operation Identification', was launched. Once again, who is (but now *was*) Mono Jojoy?

---

[19] An earlier report, broadcast on 24 August (*Radio Caracol*, 2010) is similarly, but even more confusingly, prescient. It states that the 'true name' of Mono Jojoy, as declared by the Court, is 'Luis Suárez Rojas'; needless to say, this is not the case. So perhaps this invention marks the earliest point in the strange transition in 'true-name-ship', taken up in the following section, from 'Luis Suárez' to 'Víctor Julio Suárez Rojas'.

Second expert identification (2010)

Minister: the body of Mono Jojoy 'completely identified'

> The body of Víctor Suárez Rojas, 'Mono Jojoy', military chief of the Revolutionary Armed Forces of Colombia (FARC), has been 'completely identified', stated the Colombian Minister of Defence, Rodrigo Rivera, today.
>
> 'We can confirm, not only the morphological, but the scientific identity of the leader,' said Rivera, who reported a 'ten-print fingerprint confirmation' of the guerrilla.
>
> The minister added that alongside the body of the guerrilla leader, the first images of which were released at dawn by the Ministry of Defence, the military found drugs used for diabetes ....
>
> He further stated that the guerrilla was wearing a 'very fine' Rolex watch. (Terra.com Argentina, 2010)

The state of decomposition of the body hinders legal identification of Mono Jojoy

> 'The head of the agency [Forensic Medicine (*Medicina Legal*)] Juan Isaac Llanos said that they had taken his fingerprints and had sent samples to the *Registraduría* to compare them and establish to whom they belonged.
>
> "Officially and reliably it has not been identified" he said in reference to the body claimed to be the military leader of the FARC.
>
> However, the director of the National Police ... said that Mono Jojoy was fully identified through different aspects, including his physical body and the overall context as well as his belongings, which included medicines for diabetes and an original and expensive Rolex watch.' (Radio Santa Fe 2010)

These two reports disagree about the adequacy of the methods used. Yet they share a similar ambivalence about what is meant by 'identification'. On the one hand, there is a concern to establish that the body is/was Mono Jojoy. This is indicated by the references to his 'physical body', the 'overall context' and his belongings, all of which are treated as identifiers ('everyone knows' that Mono Jojoy had diabetes, that he sported an expensive watch, and even that he wore special boots). This eagerness is also depressingly evident in the publication by the newspapers of macabre photos of the dead guerrilla: a production of hyper-reality and hyper-recognition that ends up like the medieval and colonial exhibition of the punished bodies of 'infamous men'.[20]

On the other hand, there is the effort, as in the earlier identification exercise undertaken in 2003, to establish the 'true identity' of this person who is merely an alias. And the method used is said to be fingerprinting the corpse, and the comparison of these prints with those held by the *Registraduría*. Though we are given no details, the result, we are led to believe, is the name (presented as a given referent; as such an oddly unvarnished fact) in the first sentence of the first report: **Víctor Suárez Rojas**. This textual oddity makes us wonder if what is described in these reports can be an 'honest' identification. Because the assumption, of course, is that there *must* be an available match, because every Colombian citizen *must* have a *cédula*, which of course incorporates sets of prints taken at the time of issue. Well, this may indeed be the case for *citizens*. As we have already mentioned, the introduction of the *cédula* in the 1930s made citizenship and *cédula*-ownership mutually defining, as it was progressively required for voting, applying for jobs, market transactions, social insurance, and any instance of life that could possibly be said to require legal identification (Restrepo Forero et al, 2013).

But in what sense can Mono Jojoy be understood as a citizen? According to most of the many biographical articles published in newspapers and the internet, he has always been 'in the FARC', joining the 'Popular Army' at the age of 22 (RFI, 2010) – or in one report, 12 (InSightCrime, 2010). He is said to have been 'born in the FARC' (Calderon, 2010) or 'suckled by the FARC' (RFI, 2010). The statement, 'Jojoy's homeland is the FARC' should be taken quite seriously, indicating that he is not, in any meaningful sense, a citizen, registered as such by the Colombian state. What then is there to suggest that out there in his jungle camp, he has a need of a *cédula*? Or, indeed, that he has ever had one?[21] This argument can be strengthened by the details of the *cédulas* of the two official candidates for Mono Jojoy's 'true identity' (see Figures 11.3 and 11.4). Luis Suárez (whose candidature lasted from 2003 to 2010) was born on 5 February 1953, and was issued with his *cédula* on 28 November 1985, at the age of 32; which is an unusually advanced age, the *cédula* being mandatory at age 21. Víctor Julio Suárez Rojas (overwhelming favourite since September 2010, Luis

---

[20] Thanks to Yuri Jack Gómez for this observation.

[21] Throughout the history of issuing identity cards in Colombia there have always been pockets of people who, for many reasons, manage to escape the control of the state.

Box 11.3: Different forms of the 'crushing embrace' of the Flatland state

Embracing the citizen	Embracing the outlaw
Names: One	Ten
Jorge Enrique Briceño Suárez	Mono Jojoy
Jorge Iván Restrepo Llano	(Jorge Briceño)
	(Jorge Briceño Suárez)
	(Jorge Suárez Briceño)
	(Jorge Enrique Briceño Suárez)
	(Luis Suárez)
	(Luis Suárez Rojas)
	(Víctor Julio Suárez Rojas)
	(Víctor Suárez Rojas)
	(Oscar Riaño)
Cédulas: One	Eight
12.536.519 (JEBS)	12.536.519
70.753.211 (JIRL)	70.753.211 (also attributed to FARC leader Milton de Jesús Toncel Rendón/Joaquín Gómez/Usuriaga)
	70.723.211, 70.732.211, 70.309.211
	4.466.970, 19.208.210, 17.708.695
One name/one official identity/ identified/registered	Many names/no official identity/ fully individualized/not registered (multiple biographies)
Legible	Illegible (and/but completely recognizable)
Excluded/loss of rights (identity, political citizenship, good name, mobility, credit, work)	Excluded/loss of rights/but has work and an alternative life
Localized, immobile	Not localized, mobile
'Suplantados'/(mis)taken identities	'Suplantado'/mistaken identity?

Note: The identity of 'Mono Jojoy' was ruled out by the state, which engaged in a fantasy chase after a 'real ID' from contingently available ghosts – our citizens.

Figure 11.4: Víctor Julio Suárez Rojas' *cédula*

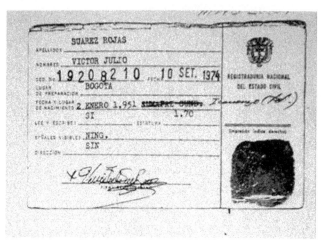

Source: *La Patria,* 2011.

Note: The second expert identification, undertaken following the killing in 2010, using 'not only morphological but also scientific' methods, concluded that Mono Jojoy was Víctor Julio Suárez Rojas.

Suárez having been all but eclipsed) was born 2 January 1951, and gained his *cédula* on 10 September 1974 at the age of 23 (see Figure 11.4). Both of these 'ages of issue' postdate the reported age(s) when Mono Jojoy entered the FARC as a full-time fighter, and, by doing so, either abandoned his citizenship, or, as we are arguing, never gained it.

Another turn of the screw

To close this account, we took advantage of a valuable resource in the *Registraduría* website, which allows anyone to view and obtain a 'certificate of validity' (*certificado de vigencia*) of a *cédula* number. So, at the last minute we decided to search for the four most important numbers for our story. Here's what we found:

- The JEBS *cédula* (12.536.519) had been invalidated on 6 September 2012 'because of loss or suspension of rights' [*baja por pérdida o suspensión de los derechos*]. The agency accredited as the 'informant' for this development is the Ministry of Justice. But how can this be? Did the Constitutional Court not free this citizen from the shadow of Mono Jojoy? After the whole process of the Court granting his *tutela*, Jojoy's death and the (unclear) procedures that led to the current establishment of the 'true identity' of the guerrilla as VJSR, how is it that two years later citizen JEBS receives once again the cruel news that his rights as a citizen have been lost or 'suspended'? We have two possible explanations for this turn of events. We will return to those later.

- The JIRL *cédula* (70.753.211) is valid. Perhaps, then, señor Restrepo's troubles are now over.
- The LS *cédula* (17.708.695) was 'cancelled for double *cédulacion*' (the holder being found to have two *cédula* numbers) on 12 January 2005, by the agency of the *Registraduría* itself. One curious thing about the case is that this 'certificate of validity' varies, in both the date and the place of issue, from the image of the *cédula* which we reproduce in Figure 11.3. The certificate has 9 December 1985 and 'Cartagena de Chaira-Caquetá' while in Figure 11.3 these data are respectively 28 November 1985 and 'Florencia-El Remolino'. So is the certificate 'correcting' mistakes or is its origin an oversight or 'clerical error' leading to new mistakes?[22]
- The VJSR *cédula* (19.208.210) was 'cancelled by death' on 3 May 2011. If the death in question is that of Mono Jojoy, as would appear to be the case according to the second expert identification, this occurred, as we have seen, on 24 September 2010. So we can only note that it took more than seven months to register this event. But then, the tardiness of Flatland is legendary ....

Finally, we return to the cancellation of JEBS' *cédula* in September 2012. It turns out that on 6 July 2012, the President of the Administrative Chamber of the Supreme Judicial Council issued an official communiqué (PSA 12–2546) addressed to all judges of the republic,

---

[22] In a new twist in this endless case, so apparently precise and so confusing, it happens that after the death of Jojoy, the *Registraduría* (2010) posted a news item on their 'Daily News' webpage. The item, originally from the left-leaning internet journal *La Silla Vacía* (2010) was entitled 'Doubly identified'. It refers to Resolution 61 of 12 January 2005, in which the *Registraduría* itself had cancelled the *cédula* of LS, and 'established' that 'LS' and VJSR were one and the same. The strange thing, apart from the circuitous route taken by this information, is that the 'identity' of LS and VJSR which was established in this Resolution had gone completely unnoticed by the authorities, including the *Registraduría*(!) and the Attorney General's Office who fail to mention it in the case before the Constitutional Court, while continuing to identify Jojoy as LS.

Seven years after the death of Jojoy, in a commentary on his honouring by the FARC, we learn that 'His name was Víctor Julio Suarez Rojas ... He was born in Cabrera (Cundinamarca) on February 5, 1953' (*Enfoque*, 2017). Here, the merging of VJSR and LS is complete. The figure now has VJSR's name and LS's birth place and date (see Figure 11.3)! Multiplicity in singularity!

which requested that they proceed to remove JEBS' identity data from the records of proceedings against Jojoy. We also learn that in the same year, the citizen's lawyer had brought a petition for *tutela*, once again in Santa Marta, which led to the opening of a case for contempt of court, because the orders given two years before by the Constitutional Court had remained unfulfilled! It was even necessary to plead:

> Restore to this good citizen the right to walk freely through the streets of the country, and restore to him his civil rights, by correcting the entries in the databases of the agencies of the state, such as the Attorney General's Office, the National Registry of Civil Status, the National Police, the Army, and the DAS.[23]

Here are our alternative explanations for the (re)cancellation of JEBS' *cédula* on 6 September 2012:

> *The combination of inaction and hyperactivity* of the same judicial authorities leads to JEBS losing his rights once more. On the one hand, there is the lengthy delay in relaying and acting on the order of the Constitutional Court and, two years later, the slowness in responding to the circular of the Supreme Judicial Council. On the other hand, these authorities act 'efficiently' in continuing to mount legal proceedings against Mono Jojoy (two years after his death) leaving JEBS, still tied to the guerrilla's shadow by the Flatland link of his *cédula*, to lose his rights. And his drama, his nightmare, starts again.
>
> *The charitable explanation.* The order to cancel JEBS' *cédula*, which comes from the Ministry of Justice, recognizes the dynamic of tardiness mentioned previously, and decides to cut the Gordian knot (described in Box 11. 3). Cancelling the old *cédula* and issuing a new one with another number unties JEBS from his shadow, thus ending his nightmare and restarting his life with a fresh, new, uncontaminated Flatland identity.

---

23 'Difusión de circulares', in Boletín Judicial 006-12, Manizales, 16 July 2012. Available at: http://www.ramajudicial.gov.co/csj/downloads/UserFiles/File/CALDAS/INFORMACION_GENERAL/Boletines/Bolet% C3%ADn%20 Judicial%20006–12.pdf [Article unavailable 2020; Accessed 10 February 2013].

Conclusion: the cracks are how the light gets in

In Colombia, we often say "we have never been modern" (as if we all were latter-day-Latourian converts [Latour, 1993]), meaning that we haven't achieved the goal of producing the ideal-type of the ordered society of the Enlightenment. We have not transformed nature and society according to our own designs, nor have we acquired a well-organized industrial and market economy, or a proper nation state and fully functioning flawless democracy, and we haven't managed to extirpate traditions, and produce a secularized society. These are all Big Topics of debate, but the modern impulse is still there. The obsessive-compulsive history of the *cédula* (along with many other instruments of Flatland) the persistent attempts to make it more secure, more universal, more perfect and less falsifiable is as modern as can be.[24] Weber's 'iron cage', Foucault's panopticon, Scott's high modernity, Bauman's (1991) gardening state, and Gilliam's *Brazil* (1985). All these theoretical prisms have been attempted in efforts to produce some semblance of order in what has been persistently understood as a 'divided society' (Safford and Palacios, 2002), as 'a nation in spite of itself' (Bushnell, 1993), as a society with a 'thin layer of civilisation' (Palacios, 1986), or, finally, when every other explanation bores us, as an unruly Macondo.[25] But modernity's heroic or macabre aims have failed here and everywhere (differently in various locations), overwhelmed by the weight of their impossible promises. The more persistent the work of purification, the more perfect the offering, the more the margins, the hybrids, the cracks, appear and multiply.[26]

---

[24] For example, the *Registraduría* proudly announces that Colombia was the first country in the Americas to adopt the AFIS (Automated Fingerprint Identification System) for civil purposes, to identify the entire population – and not only the criminals, as in other, less advanced, countries ... What an honour!

[25] Macondo is the fictional Colombian town that is the setting for *One Hundred Years of Solitude* (*Cien años de soledad*), the famous 1967 novel by Gabriel García Márquez.

[26] Cracks in the fabric of control should be celebrated. Ahmed Rashid (2016), in his appreciation of the life of Leonard Cohen, quotes a 1993 interview with Thom Jurek for the *Metro Times Detroit* in which Cohen quotes 'the line in "Anthem" that says [*it is only through the material imperfections of all our constructions that illumination arrives* (authors' paraphrase of the actual lyric)]. That sums it up: it's as close to a credo as I've come.' (Rashid, 2016). We too, Leonard. Us too. *Footnote to the footnote:* This reluctant paraphrase, this imposed imposterization, this final Flatland interference: the hole in the heart of the quote of the quote in the quote.

What is the wider relevance of our story of the tragicomic results of mistaken/unstable/impostering identities in Colombian Flatland?

Susan Leigh Star wrote a beautiful article (1991) about the problems of focusing only on the process of standardization without thinking about what is left out, of what is marginalized as the network expands and standards are imposed. She exemplifies these problems with her own case of 'being allergic to onions' in relation to McDonald's. She questioned the logic of network analysis and standards, engaged as she was in understanding their workings. What about people allergic to onions? (MacBurgers *cannot* be served without them.) But also, what about the people that are entangled in the obsessive workings of a network that removes their access to other possible connections?

Of course, we should seek to understand the workings of the state's embrace (Torpey, 2000), and all its networks of rationalization, standardization and bureaucratization. And there is not much in this story that belongs only to the *Colombian* Flatland. The embrace of *any* state can be crushing when it ceases to connect, or enable the smooth transition from one mode of action to the next. However, if we attend only to the story of Flatland and its tools and products (like the *cédula*) through a narrative of monitoring and control, as in much of the governmentality and surveillance studies literatures, we will miss the multiple stories of cracks and hybrids, we will be unable to focus our attention on how the networks crush, flatten and proliferate *their own logics*, producing all the marginalized, the excluded, the possible imposters that defy an impossible modern ordering. In this story, Mono Jojoy manages to achieve self-exclusion from the arrogance of Flatland by means of his own arrogant counter-power (his own Flat networks of guerrillas, drugs and arms trafficking, money laundering, and his Fat powers of persuasion and intimidation); which was only ended by a much more conventional, and brutal, application of state power.[27]

We dedicate this story to **Jorge Enrique Briceño Suárez**, to **Jorge Iván Restrepo Llano** and to the thousands of others allergic to onions (namesakes, numbersakes, clerical errors, all the accidental hybrids, the

---

[27] Perhaps what is most 'macondian' in this story is the fantastic combination of postmodern/pre-modern in the figure of Mono Jojoy and his public image as a celebrity. This, and also the perverse actions of the agents of Flatland tenaciously trying to inscribe him as a citizen, to find a register for him, with a proper number and a proper name: their permanent defence of 'formality, not legality'; as is also the case in the master institution of Colombian Flatland, the notary (*notaría*); on this, see Ashmore and Restrepo Forero (2013).

possible imposters) who get stuck in a singular Flatland. We want their stories not only to produce a smile, and/or an expletive, but that they also teach, enlighten and correct our ways of understanding the density, flexibility, unpredictability, open-endedness; *the Fatness of the world*.

References

Ashmore, M. and Restrepo Forero, O. (2013) 'El documento en su paso por la notaría: confianza, formalidad y credibilidad en Colombia', in O. Restrepo Forero (ed) *Proyecto Ensamblado en Colombia, Tomo 2, Ensamblando Heteroglosias*, Bogotá: CES, Universidad Nacional de Colombia, pp 406–428.

Bauman, Z. (1991) *Modernity and Ambivalence*, Cambridge: Polity Press.

BBC News (2012) 'Profiles: Colombia's armed groups', 29 August. Available at: http://www.bbc.co.uk/news/world-latin-america-11400950 [Accessed 23 March 2020].

Bennett, C.J. and Lyon, D. (eds) (2008) *Playing the Identity Card: Surveillance, Security and Identification in Global Perspective*, London: Routledge.

Bentham, J. (1995 [1787]) *Panopticon Letters*, edited by Miran Bozovic, London: Verso.

Bushnell, D. (1993) *The Making of Modern Colombia: A Nation in Spite of Itself*, Berkeley: University of California Press.

Cabrera, A. (2010) 'Con muerte del Mono Jojoy revivió Jorge Briceño Suárez, un taxista samario', 24 September, 7.02 am, *Radio Santa Fe*.

Cole, S.A. (2001) *Suspect Identities: A History of Fingerprinting and Criminal Identification*, Cambridge: Harvard University Press.

El Espectador (2010) 'Mal tocayo acude a la Corte para demostrar que no es el "Mono Jojoy"' ['Unfortunate namesake goes to Court to show that he is not "Mono Jojoy"'], 15 September. Available at: http://www.elespectador.com/noticias/judicial/articulo-224550-mal-tocayo-acude-corte-demostrar-no-el-mono-jojoy [Accessed 23 March 2020].

El Informador (2010a) '"Me encuentro preso en mi casa y en mi tierra": Jorge Enrique Briceño Suárez' ['"I am a prisoner in my home and in my country": Jorge Enrique Briceño Suárez'], 17 September. Available at: http://www.elinformador.com.co/index.php?option=com_content&view=article&id=4241:me-encuentro-preso-en-mi-casa-y-en-mi-tierra-jorge-enrique-briceno-suarez&catid=71:judiciales&Itemid=415 [Accessed 5 January 2012].

El Informador (2010b) 'Reconocen derechos al verdadero Jorge Enrique Briceño Suárez', 27 September. Available at: http://www. elinformador.com.co/index.php?option=com_content&view=article &id=5781:reconocen-derechos-al-verdadero-jorge-enrique-briceno-suarez&catid=71:judiciales&Itemid=541 [Accessed 5 January 2012].

El Tiempo (2003) '"Jojoy" no es Jorge Briceno', 20 December. Available at: http://www.eltiempo.com/archivo/documento/MAM-1046293 [Accessed 5 February 2021].

El Tiempo (2010a) '"Ahora sí soy libre", dice Jorge Briceño Suárez, homónimo del abatido "Mono Jojoy"', 23 September. Available at: https://www.eltiempo.com/archivo/documento/CMS-7967341 [Accessed 23 March 2020].

El Tiempo (2010b) 'Cuerpo del "Mono Jojoy" permanece en el Instituto de Medicina Legal de Bogotá', 24 September. Available at: https:// www.eltiempo.com/archivo/documento/CMS-7976062. [Accessed 1 January 2021].

El Tiempo (2010c) 'Tras 10 años de estigma, Jorge Briceño Suárez recuperó el nombre que "robó" el "Mono Jojoy"', 18 September. Available at: http://www.eltiempo.com/archivo/documento/CMS-7924722 [Accessed 1 January 2021].

Enfoque (2017) '¿Quién fue el "Mono Jojoy", hoy homenajeado por las Farc?', 22 September. Available at: https://old.periodicoenfoque. com.mx/2017/09/quien-fue-el-mono-jojoy-hoy-homenajeado-por-las-farc/ [Accessed 1 January 2021].

Foucault, M. (1977) *Discipline and Punish: The Birth of the Prison*, translated by A. Sheridan, London: Peregrine Books.

Garfinkel, H. (1996 [1967]) *Studies in Ethnomethodology*, Cambridge: Polity Press.

Gilliam, T. (1985) *Brazil*, Embassy International Pictures.

Goffman, E. (1983) 'The interaction order', *American Sociological Review*, 48(1): 1–17.

Goodwin, C. (1994) 'Professional vision', *American Anthropologist*, 96(3): 606–633.

Guerra Sánchez, S. (2011) 'Cédula y ciudadanía en Colombia: tecnologías documentales y gobierno de la identidad', Tesis Sociología, Departamento de Sociología, Universidad Nacional de Colombia.

Habermas, J. (1987) *The Theory of Communicative Action, Volume 2: Life-World and System: A Critique of Functionalist Reason*, translated by Thomas McCarthy, Boston: Beacon Press.

Henry, E. (2004 [1900]) *The Classification and Uses of Fingerprints*, London: George Routledge and Sons. Digital edition prepared for http://galton.org by Gavan Tredoux.

InSightCrime (2010) 'Victor Julio Suarez Rojas, alias "Mono Jojoy"'. Available at: http://www.insightcrime.org/personalities-colombia/jorge-briceno-suarez-mono-jojoy [Accessed 23 March 2020].

Kaluszinski, M. (2001) 'Republican identity: Bertillonage as government technique', in J. Caplan and J. Torpey (eds) *Documenting Individual Identity: The Development of State Practices in the Modern World*, Princeton: Princeton University Press, pp 123–138.

La Patria (2011) 'Terminaron 35 años de terror', 8 May. Available at: http://www.lapatria.com/story/terminaron-35-años-de-terror [Accessed 5 January 2013].

La Silla Vacía (2012) 'Doblemente Identificado' ['Doubly Identified'], 22 September. Available at: http://www.lasillavacia.com/queridodiario/18280/doblemente-identificado [Accessed 1 December 2012].

Latour, B. (1993) *We Have Never Been Modern*, translated by Catherine Porter, London: Harvester Wheatsheaf.

Latour, B. (2005) *Reassembling the Social: An Introduction to Actor-Network Theory*, Oxford: Oxford University Press.

Latour, B. and Hermant, E. (1998) *Paris Ville Invisible*, Paris: La Découverte Les Empêcheurs de penser en rond.

Lyon, D. (2001) 'Under my skin: From identification papers to body surveillance', in J. Caplan and J. Torpey (eds) *Documenting Individual Identity: The Development of State Practices in the Modern World*, Princeton: Princeton University Press, pp 291–310.

Lyon, D. (2009) *Identifying Citizens: ID Cards as Surveillance*, Cambridge: Polity.

Palacios, M. (1986) *La delgada corteza de nuestra civilización*, Bogotá: Procultura.

Radio Caracol (2010) 'Corte Constitucional concedió tutela a homónimo de lo Mono Jojoy para limpian su nombre', 24 August. Available at: https://caracol.com.co/radio/2010/08/24/judicial/1282649460_348169.html [Accessed 15 March 2021].

Radio Santa Fe (2010) '24 September, 12:07 pm'. Available at: http://www.radiosantafe.com/2010/09/24/estado-de-descomposicion-dificulta-identificacion-legal-del-mono-jojoy/ [Accessed 23 March 2020].

Rashid, A. (2016) 'Death of Leonard Cohen - how the light gets in', *The Spectator*, 12 November. Available at: https://www.spectator.co.uk/article/death-of-leonard-cohen—how-the-light-gets-in [Accessed 5 February 2021].

Registraduría (2010) 'Noticias al día', 27 September. Available at: http://www.registraduria.gov.co/?ano=2010&mes=septiembre&dia=27&page=noticias_aldia [Accessed 12 January 2012].

Restrepo Forero, O. and Ashmore, M. (2013) 'La cédula de ciudadadanía del guerrillero: el mundo plano contra el mundo pleno en Colombia', in O. Restrepo Forero (ed) *Proyecto Ensamblado en Colombia, Tomo 2, Ensamblando Heteroglosias*, Bogotá: CES, Universidad Nacional de Colombia, pp 462–484.

Restrepo Forero, O., Guerra, S. and Ashmore, M. (2013) 'La ciudadanía de papel: ensamblando la cédula y el estado', in O. Restrepo Forero (ed) *Proyecto Ensamblado en Colombia, Tomo 1, Ensamblando Estados*, Bogotá: CES, Universidad Nacional de Colombia, pp 277–327.

RFI (2010) '"Mono Jojoy", el niño amamantado por las FARC'. Available at: http://www.espanol.rfi.fr/americas/20100924-mono-jojoy-el-nino-amamantado-por-las-farc [Accessed 23 March 2020].

Robertson, C. (2009) 'A documentary regime of verification: The emergence of the US passport and the archival problematization of identity', *Cultural Studies*, 23(3): 329–354.

Rodriguez, J. (2004) 'South Atlantic crossing: Fingerprints, science, and the state in turn-of-the-century Argentina', *The American Historical Review*, 109(2): 387–416.

Ruggiero, K. (2001) 'Fingerprinting and the Argentine plan for universal identification in the late nineteenth and early twentieth centuries', in J. Caplan and J. Torpey (eds) *Documenting Individual Identity: The Development of State Practices in the Modern World*, Princeton: Princeton University Press, pp 184–196.

Safford, F. and Palacios, M. (2002) *Colombia: Fragmented Land, Divided Society*, Oxford: Oxford University Press.

Schutz, A. (1967) *The Phenomenology of the Social World*, Evanston: Northwestern University Press.

Scott, J.C. (1998) *Seeing Like a State; How Certain Schemes to Improve the Human Condition Have Failed*, New Haven: Yale University Press.

Sengoopta, C. (2003) *Imprint of the Raj: How Fingerprinting Was Born in Colonial India*, London: Macmillan.

Star, S.L. (1991) 'Power, technology and the phenomenology of conventions: On being allergic to onions', in J. Law (ed) *A Sociology of Monsters: Essays On Power, Technology and Domination*, London: Routledge, pp 26–56.

Terra.com Argentina (2010) 'Ministro: Esta "plenamente identificado" cadaver de Mono Jojoy' ['Ministry: The "fully identified" body of Mono Jojoy'], 24 September. Available at: http://noticias.terra.com.ar/internacionales/ministro-esta-plenamente-identificado-cadaver-de-mono-jojoy,d9beb5502a34b210VgnVCM20000099f154d0RCRD.html [Accessed 21 June 2011].

Tönnies, F. (1957) *Community and Society: Gemeinschaft und Gesellschaft*, translated and edited by C.P. Loomis, East Lansing: Michigan State University Press.

Torpey, J. (2000) *The Invention of the Passport: Surveillance, Citizenship and the State*, Cambridge: Cambridge University Press.

van der Ploeg, I. (2005) *The Machine-Readable Body: Essays on Biometrics and the Informatization of the Body*, Maastricht: Shaker Publishing.

12

Good Enough Imposters: The Market for Instagram Followers in Indonesia and Beyond

Johan Lindquist

Introduction

Instagram is a social media platform founded in 2010 and owned by Facebook (Leaver et al, 2020). It allows users to share 'posts' (photos and videos with accompanying text) which are then seen by their 'followers' (users that have chosen to be notified when a particular account posts new content). With more than one billion users, Instagram has become a major commercial force, not least as the main platform for the rise of 'social media influencers' (Khamis et al, 2017) – a term for popular users who are often paid by brands to feature their products in posts – and increasingly important for politicians (Lalancette and Raynauld, 2019). One influential journalist writes that the Instagram app 'has become a celebrity-making machine the likes of which the world has never seen' (Frier, 2020: xvii). In line with this, Instagram is at the heart of what has been called the 'demotic turn' across the global media landscape, as 'ordinary' people have become increasingly visible – initially through reality television shows such as *Big Brother*, and more recently through social media (Khamis et al, 2017). This has come to form the basis for what has been termed 'microcelebrity', a 'mind-set and a collection of self-presentation practices endemic in social media, in which users strategically formulate a profile, reach out to followers, and reveal personal information to increase attention and thus improve their online status' (Marwick, 2015: 138; see also Senft, 2008: 25).

Instagram has shaped a more specific form of microcelebrity, which Marwick calls 'Instafame', precisely defined as 'the condition of having a relatively great number of followers on the app' (2015: 137). Through Instagram and related social media platforms, such metrics – which are publicly listed – have become critical to contemporary forms of celebrity and the broader attention economy in which what is valued is the ability to gather and maintain an interested audience (Fairchild, 2007). Indeed, these metrics, along with the monetary rewards and perks associated with them, 'encourage people to actively foster an audience' (Marwick, 2015: 140, see also Gerlitz and Helmond, 2013). The generalized ability to engage with a public has thus dramatically expanded to include not only traditional celebrities (film stars and athletes), but also the broader field of microcelebrities (who may only be famous for being on Instagram). In order to access income through sponsored posts or other forms of influence, there are widespread attempts by Instagram users to gain greater visibility and increase their follower counts. While this ideally is related to the account's content, creating posts that their potential audiences want to see, there are many other strategies for inflating follower counts.

To give a few examples: many Instagram users engage in practices such as 'follow-backs' in which they informally agree to follow each other (Abidin, 2016: 3); groups of users – so-called 'pods' – agree to like one another's posts; some internet sites offer free followers in exchange for the right to use user accounts for commercial purposes; and followers from bot-generated[1] accounts are widely available for purchase on internet marketplaces. In particular, highly publicized revelations about 'click farms' and 'follower factories' – digital sweatshops usually located in the global south that produce bot-generated accounts and sell likes, comments and followers on social media platforms – have captured the public imagination in recent years (Clark, 2015; Confessore et al, 2018). In this process, social media appear increasingly compromised, populated by 'fake followers' – imposters.

This chapter approaches the market for Instagram followers as a form of impostering. In line with the classic definition, Miller (2018: 1) claims that imposters 'pose as people they are not'. It follows that impostering is the practice of posing as a person one is not. In relation to Instagram, however, impostering must be considered in relation to

---

[1]   A bot is a software application that runs automated tasks (scripts) over the internet.

user accounts which present themselves as authentic and worthy of attention precisely because of the size and engagement of that audience. More specifically, impostering on Instagram cannot be reduced to a 'real' individual who performs as another, but must rather be understood through a wide range of practices that involve programmers, bots, sellers, platforms, algorithms and individual users. Rather than approaching Instagram impostering through individual actors who attempt to 'game the system', this chapter thus shifts attention to the assemblage of actors and technologies that are involved in regulating, shaping and playing the 'visibility game' that organizes influence on Instagram (Cotter, 2018: 896). With regard to the theme of this book, this highlights an important methodological shift of attention, as the focus on impostering reveals how a range of different practices and technologies are involved in temporarily assembling the 'fake follower' as imposter. This moves us beyond dichotomies such as 'real' and 'fake' to questions concerning how the boundaries between the two are shaped, constituted and contested in the current 'post-truth era' of digital capitalism and politics.

Instagram itself distinguishes between 'authentic' and 'inauthentic' activity, the latter of which is often invoked in blog posts by the company. Though never clearly defined, Instagram often refers to practices such as excessive liking, the mass-production of accounts, or the purchase or exchange of followers as 'inauthentic'.[2] 'Authenticity' is not only critical to Instagram's attempts to describe and regulate activities on the platform, however; it is at the heart of the performance of microcelebrity and Instafame, as seemingly genuine interactions with followers and the revelation of personal experiences are key practices. More generally, 'authenticity's slipperiness is part of what makes it useful: it can satisfy many objectives, and can be interpreted widely' (Marwick, 2013: 121). As will become clear, impostering practices on Instagram are shaped at the highly ambiguous boundary between the authentic and inauthentic.

The follower as imposter takes shape in relation to these distinctions and the practices associated with them. As the demand for followers has increased, so have attempts to regulate them. In the wake of the 2016 presidential election in the United States, in which Russian bots on social media were used to spread misinformation and polarize voters, and increasing concerns with 'Insta fraud,' when influencers are

---

[2] See, for instance, https://instagram-press.com/blog/2018/11/19/reducing-inauthentic-activity-on-instagram/ [Accessed 25 March 2020].

paid to market products based on artificially inflated follower counts, Instagram has gradually developed algorithms that can identity and ban bot-generated accounts: for example, accounts with no posts of their own who like or comment on posts in mechanical ways. In response, providers who produce bot-generated accounts have been compelled to create 'higher quality' accounts – with photos and posts, for instance – that can pass these security measures. An 'arms race' has thus developed over time (Ellis, 2019).

In light of this, the chapter asks: what passes as a 'good-enough' follower within the Instagram economy and the rapidly changing landscape of social media? In his work on brain scans as evidence for explaining socio-medical disorders, Dumit (2000: 210) understands evidence as functioning as '*temporary* resting places for explanations', thus shifting attention away from a concern with scientific truth per se to the different practices, judgements and struggles that come to shape 'good enough' scientific facts. In similar terms, this chapter moves beyond a priori dichotomies between 'fake' and 'real' or 'authentic' and 'inauthentic' followers and focuses on how these and related distinctions are temporarily established and constantly shifting. The chapter thus recognizes the wide range of ways in which Instagram followers are not only produced and valorized, but also surveilled and regulated across a shifting performative landscape. It is in this field of negotiation that good enough imposters find temporary resting places.

The methodological starting point for the chapter is ethnographic encounters with actors who are engaged in the market for Instagram followers, primarily in Indonesia. Indonesians are among the most prolific mobile internet users globally, and Jakarta is arguably the Instagram capital of the world (Google Temasek, 2017; Taylor, 2017; Lim, 2018). Between 2017 and 2019, I made five trips to Indonesia where I contacted approximately 130 sellers identified through internet searches. Of these, 30 agreed to meet in person for interviews and of these I met ten two or more times. Most showed me their work process on their cell phones or laptop computers and three allowed me to visit their work spaces. All interviews were conducted in Indonesian, a language I speak fluently. A further ten interviews were conducted via Skype in 2019, all in English, with providers and sellers in as many countries, including India, Egypt and Morocco. Despite the fact that it was clear that my informants were not always entirely straightforward or honest with me – unsurprising considering the nature of the market – the various meetings gave me a unique perspective on the market for followers and on Instagram more generally. While an increasing number of scholars have come to focus on social media platforms such as Instagram (Leaver et al, 2020), there has been very limited ethnographic attention to the market for 'fake' followers and comparable social media marketing services.

Starting with a diverse range of actors who are engaged in the market for Instagram followers thus allows for reflection on the multiplicity of imposters and impostering practices with regard to Instagram. In what follows, I describe these actors, the technologies and infrastructures on which they depend, the range of Instagram followers for sale on the market, some of the customers on the market, and Instagram's (lack of) engagement with these processes in order to show how the imposter is at the heart of Instagram, public culture and digital capitalism.

The market for followers

I am at a Starbucks in Bekasi, one of the major peripheral cities in the greater Jakarta region, in early 2019. Adi and Handaya, two men in their early twenties, are explaining how they 'sell followers' (*jual follower*) on Instagram. The demand for Instagram followers has created an infrastructure and market that individuals such as Adi and Handaya have become part of and are able to capitalize on. Without any programming skills, and with low start-up costs – a reliable internet connection, a laptop and cellphones; knowledge of Instagram customs and markets is the main investment – they have developed a profitable 'reseller' business for Instagram followers, likes, comments and views, as well as comparable services from other digital platforms. In other words, they do not create accounts themselves but function as brokers within this market. For those who focus on reselling, no programming knowledge is necessary – basic technical knowledge is readily available through internet forums or YouTube tutorials. Indeed, even informants with significant programming knowledge that would allow them to produce bot-generated Instagram accounts or other services turn en masse to purchasing services and instead focus on marketing, search engine optimization and customer service in order to maximize profits.

With the growth of the internet platform economy, the range of what are generally called 'social media marketing services' on the English-speaking market is constantly expanding, and include YouTube subscribers, LinkedIn endorsements, IMDB views, Tik Tok likes, and Spotify premium plays. This market has been dominated by Instagram during the past five years. Approximately 90 per cent of Adi and Handaya's income – which has varied between US$1,500 and $5,000 per month during the past few years (for context, the standard minimum wage in Jakarta was around US$300 in 2020) – is based on Instagram, particularly

Instagram followers. Although there is no limit to how many followers an Instagram account can have, it is currently only possible to follow up to 7,500 other accounts. This means that an account – including bot-generated accounts – can be sold as a follower up to 7,500 times. While Instagram accounts themselves can be purchased and sold, their primary value is relational, as they ultimately are used to develop an audience for other accounts by driving attention towards them (accounts and posts with more likes are more likely to be suggested to other users algorithmically).

By logging onto websites providing social media marketing services and adding funds, Adi and Handaya are able to choose from an extensive menu, enter the Instagram user ID of their client, and place an order, of say, 500 'US', 1,000 'Indonesian' or 1,000 'female Indonesian' Instagram followers, which are gradually added to the customer's account after the transfer of payment. They might purchase these followers for anywhere between US$1 and $5 per 1,000 and sell them at a 20–50 per cent mark-up, profiting through arbitrage. While we are sitting there, Handaya is handling customer service via WhatsApp and Adi makes a few purchases and sales on his cellphone – to a Jakarta fashion model, an online shop for hijabs and a social media manager for a large corporation. Each client is concerned with improving their visibility on social media. "Different customers", Adi tells me, "want different kinds of followers". These followers can, for instance, vary depending on pricing, quality, geography, gender, or by the speed of transfer. They can be so-called 'genuine' accounts accessed through exchange sites, but are mainly bot-generated – *aktif* or *pasif*, respectively, in Indonesian terminology depending on the extent to which they actively like and comment on posts rather than just pose as mute followers.

For Adi and Handaya, then, the good enough imposter takes shape at the intersection between customer demands and the available services on the market. In other words, they are not directly engaged in impostering or producing imposters, but rather in selling and distributing different forms of ready-made imposters on the internet. In fact, as brokers they have a limited understanding of the origin of the services they are reselling. In order to create a sustainable business model, it is of critical importance to keep their hundreds of customers satisfied and to find a balance between cost and quality. Instagram's ongoing security updates, which shut down many bot-generated accounts, temporarily affecting global markets, are a major constraint, as they force Adi and Handaya to deal with unhappy customers while they wait for the providers of social media marketing services to outwit the new algorithms that Instagram develops.

In contrast to Adi and Handaya, Indonesian-born Tom, who lives in Malaysia (Skype, 10 February 2017), has developed a business in which he helps clients develop their Instagram accounts without purchasing followers, acting as a social media manager. While he previously focused on the Indonesian market, his English fluency by way of an Australian mother allowed him to change to that market and its higher profits. His clients give him their Instagram account user ID and password and he generates followers by populating their accounts with attractive content that people will want to follow as well as through practices such as 'follow-backs'.[3] For this he charges US$50 and guarantees between 2,000 and 3,000 followers per month. It takes about 30 minutes to set up the account and then maintenance is minimal.

Some people located at the margins of the business do manual work similar to that of Tom. Fitri, for instance, a female university student in Pekanbaru, Indonesia, generates followers for different accounts passed on by her boss, who offers social media marketing services as part of a broader business. They charge about US$10 for 1,000 followers and she gets a 35 per cent cut. Like Tom, Fitri generates followers by following and being followed back by users that she knows will reciprocate, and it takes about five hours to attract 1,000 followers. When I point out that this is very little, about US$0.70 per hour, she laughs and agrees. She says that she would be playing around with her phone anyway, but more generally she is hoping to develop a long-term entrepreneurial relationship with her boss. When I ask if they don't ever resell followers they have purchased from provider websites, like Adi and Handaya, she says "no", that their business is *murni*, "pure" (10 February 2018).

On this part of the market, both Tom and Fitri see themselves as developing accounts "organically" (Tom's term). Indeed, the followers they are able to attract to their customers' accounts appear interested in the content. In contrast to Adi and Handaya, who broker the circulation of imposters, Tom and Fitri can be understood as imposters themselves, not in relation to the followers – which appear authentic – but with regard to the customers who pay them to develop content and an audience in their name. This form of impostering, however, remains largely legitimate with regard to Instagram's concerns with authenticity, as there is a human rather than a bot behind the mask. This offers a contrast to conventional discussions about imposters, in which the mask is seen as concealing a true identity. In this case, what

---

[3]   See, for instance, https://workmacro.com/instagram/truth-behind-follow-unfollow-method-instagram/ [Accessed 25 June 2020].

is at stake is not the mask as such, but rather the perceived humanity of the wearer.

Located between Adi and Handaya, and Tom and Fitri, are a wide range of different actors or practices. One informant in Jakarta, for instance, Theo, a man in his early thirties who works as a reseller alongside his job as a political consultant, was at any one time also developing and curating more than 20 Instagram accounts that reflected varying identities. Like Tom and Fitri, Theo develops these accounts by adding followers through follow-backs and developing targeted content. This is literally a form of economic investment, as he can use the accounts as an influencer to market products, but also sell the actual accounts (Jakarta, 2 February 2018). He showed me one of them, that of a young woman with a hijab who posted occasional videos. This is in line with classic forms of impostering, as Theo clearly appears as the person behind the mask of the young woman. In this case, however, this relationship is easily displaced, as the mask can be transferred to another user, thus once again illuminating the contingency of the relationship between the 'real' and the 'fake'.

More generally, taking the perspective of different types of actors involved in the market for Instagram followers in Indonesia reveals that the position of the imposter varies significantly. Tom and Fitri's form of impostering remains within the realm of Instagram's terms of use.[4] Similarly, Theo's impostering is legitimate with regard to Instagram as long as he is not impersonating an actually existing person. It is in the process of monetization, when he sells accounts, that he breaches the terms of use – although in practice it is difficult for Instagram to detect this. Finally, Adi and Handaya are not so much imposters themselves as distributing imposters. They all operate amidst a growing demand for imposters on Instagram, which has led to the development of an infrastructure and broader market that allows for the accelerated and increasingly automated circulation of Instagram followers. In this process, a heterogeneous group of buyers is revealed in terms of purchasing power, interested in making impressions on diverse forms of publics.

Consuming attention

While my informants claim that companies, celebrities, influencers, online shops and politicians – the latter often by way of brokers – are

---

4   https://help.instagram.com/581066165581870 [Accessed 25 June 2020].

typical customers of Instagram followers; government agencies, non-governmental organizations, dog owners or provincial police officers are other examples, as are high school students who are able to transform local forms of popularity into broader forms of celebrity (Marwick, 2015: 150). This suggests that pretty much anyone with an Instagram account can be drawn into the field of microcelebrity and the vanity metrics associated with it, thus signalling its expansion into a broader public culture. In this process, the social media influencer, microcelebrity and Instafame have increasingly become global phenomena (for example Abidin, 2016). There are real economic incentives for this. Football superstar Cristiano Ronaldo, for instance, with more than 200 million followers, reportedly makes up to US$1 million per advertised post (Leighton, 2020). The field of influence, however, goes far beyond major celebrities, as around 200 million Instagram users have more than 50,000 followers, the level at which it is arguably possible to make a living wage through posting for brands (Frier, 2020: xvii–xviii).

It is therefore not surprising that there are any number of websites that offer advice for increasing the number of followers and account engagement, for instance, by posting more videos, sharing more selfies and promoting your Instagram content on other social media, or shaping a personal style that is seen by followers as authentic but is actually curated and calculated. As mentioned earlier, in more organized settings, partnerships such as 'pods' – groups of people, often social media influencers, who like and comment on each other's posts after being informed by direct messaging – are widely used (Leaver et al, 2020: 141–143). Engaging in these kinds of practices, however, takes time and effort, and does not necessarily deliver the necessary follower metrics, which leads many to purchase services.

Instagram users who choose to purchase followers do so in light of their interests and the forms of publics they hope to shape. An Indonesian woman running a small online hijab business on Instagram, for instance, is likely to want Indonesian female followers – the primary target for the business. A café in California will most likely want local followers, thus shaping an audience based on a certain form of authenticity and community. In both cases, they will probably search for and find such services on the internet. As already described, there are different kinds of services available for purchase to increase follower counts – those that are more expensive and remain clearly within the realm of what Instagram terms authentic, such as the one run by Tom, and the cheaper ones that allow an endless supply of followers from bot-generated accounts. The Indonesian woman will probably not be able to afford someone like Tom, and may turn to Ali and Handaya to request 'real' female Indonesian followers, which have been accessed through exchange sites, while the café owner may be able to purchase bot-generated followers that are made to appear as though they are located in California or the San Francisco Bay Area. In these cases, the question of what constitutes a good enough follower is not straightforward and depends on available services, costs, and on what is considered acceptable to the customer. If bot-generated 'Bay Area' followers are not available will 'California' or 'US' followers do?

On the Indonesian market a key distinction is made between *aktif* and *pasif* followers. The latter are from bot-generated accounts, which are mainly purchased by Indonesians on the international market, while the former are best understood in the negative, as *not* bot-generated. *Aktif* followers are from Indonesian accounts created by individuals and accessed through exchange sites or hijacked through phishing schemes that deceive Instagram users by offering free followers in exchange for handing over user IDs and account passwords. The importance of *aktif* followers on the Indonesian market has depended largely on the fact that 'Indonesian' bot-generated followers are rarely available for purchase. Indonesian *aktif* followers are highly valued on the national market – and more expensive than non-Indonesian *pasif* followers – since many small businesses, such as the hijab store noted earlier, want a following that appears to consist of interested parties, in this case 'Indonesian female' followers.[5]

In countries such as Turkey, by contrast, (bot-generated) 'Turkish' followers are readily available, most likely because the stronger hacker and programming culture there has led to a greater number of providers – the individuals who mass produce Instagram accounts – and are to a certain degree able to tailor them through clever programming according to market demands. The same will certainly be true with regard to California, as there will most likely be more specific bot-generated accounts available that will be able to offer more specific kinds of followers, such as a 'woman based in California'. As will become evident, the availability of these types of specific services is often changing, as some are closed down by Instagram and new ones developed by providers.

---

[5] It should be noted that the types of purchased followers discussed here do not generate comments or likes automatically. There are, however, other types of bots available for purchase or rent that automatically generate likes and comments for specific accounts.

Rather than any inherent issue of truth, the distinction between *aktif* and *pasif* refers to the apparent agency or lack thereof of these types of accounts: the former acts, the latter does not. For Indonesian customers, these *aktif* followers were generally preferred over bot-generated accounts that came from countries such as India, since, as 'Indonesian' accounts, they appeared as good enough imposters with regard to their specific Indonesian audience. Indeed, using the English term, Theo in Jakarta claimed that *aktif* followers were "more real" (*lebih real*) than those produced through bots since they were created by Indonesian individuals (Jakarta, 2 February 2018). *Aktif* accounts were not without problems, however, as they were able to put their agency to use, since the original owners could choose to unfollow accounts or change their password so that access to the account was lost to the sellers.

Karim, who runs a reseller website in Morocco, explained in contrast to the earlier examples: "if we are working with high-profile celebrities, they don't want expensive high-quality, but less expensive normal quality. They just want to see the numbers. It is the same in every country" (1 September 2019). As part of a broader marketing strategy, influencers, celebrities, corporations or politicians who already had a significant audience could purchase bot-generated followers that informants such as Karim termed "low quality": easily identifiable as bot-generated since they lacked posts, comments and photos, but difficult to notice in a large mass of followers. While 'low quality' followers could jeopardize the perceived authenticity of smaller Instagram players, this was not the case with accounts with hundreds of thousands or even millions of followers. Illustrating this point, Aziz, a Jakarta-based Instagram influencer with more than 100,000 followers, told me that half his followers were purchased, since "no one will check anyway" (16 November 2018).

The wide range of followers available for purchase should fulfil certain aims within the attention economy: in the cases of the hijab shop and the local café culturally specific forms of authenticity, in the case of the celebrities improving metrics in order to create an expansive and diverse audience. In each of the cases, different demands are placed on what makes a good enough follower. In the former cases, a particular form of visibility is critical, as the appearance of the follower, or rather groups of followers – Indonesian women rather than Middle Eastern men – must at least in a superficial sense appear interested in the content of the account, while in the latter case the particular quality of the follower is generally not deemed important, as there is an already existing diversity of followers associated with the accounts.

Higher demands are thus placed on followers that must express some form of authenticity or at least an identity that fits the profile of the customer's account. The importance of *aktif* followers in Indonesia adds a further dimension to the discussion as it illustrates that the follower economy takes shape across an uneven digital landscape shaped by national hacker cultures. The good enough imposter is a socio-technical accomplishment established at the interface between culturally specific demands and local technological limitations – in this particular case, the lack of Indonesian bot-generated accounts.

The economy and ethics of authenticity

If impostering is central to Instagram's self-fashioning enterprise, then the question becomes: what forms of impostering are acceptable and which are not? The previous sections described a range of actors who at different points in time could come to engage in different forms of impostering; for instance, acting on behalf of their customers by operating a particular account, facilitating the exchange and circulation of imposters, or advising customers interested in purchasing imposters with regard to the specific audiences they were concerned with, costs and available services. These different practices are shaped through forms of tinkering in relation to Instagram's algorithms. Following from these discussions, this section therefore shifts perspective by considering the question of the imposter from the perspective of Instagram.

Not all accounts that might be deemed 'inauthentic' by Instagram are shut down. On the one hand, this follows from the broader difficulties of regulating an increasingly complex digital ecology and the arms race that characterizes the relation between Instagram and the market for followers. On the other hand, Instagram's business model is based on advertising, of which the revenues and broader valuation in large part depend on evidence of an expanding user base and data traffic. Even though this remains unacknowledged by corporate platforms, the existence of so-called inauthentic accounts and traffic offer support for their business models (so long as they are not discovered). If the size of the audience is critical to economic success in the era of social media then imposters fit perfectly into the attention economy. In light of this, there are clear risks in putting too much effort into regulating inauthentic activities. For instance, many of my older Indonesian informants described how Twitter's security crackdown in 2013 led many to shift to platforms with laxer measures, notably Instagram. It would thus appear that for Instagram, the *discursive* production of boundaries between 'authentic' and 'inauthentic' followers is more critical than strictly enforcing regulations, as the performance of 'community' is at the heart of its brand (Leaver et al, 2020: 20–25).

As the ethical problems surrounding the business models of Instagram and its influencers have been increasingly publicized Instagram has increased security updates in an attempt to limit 'inauthentic' accounts. This began with the so-called 'Instagram purge' in late 2014, when millions of accounts were deleted to counter 'artificial inflation' in follower counts (Leaver et al, 2020: 137–138), and has continued, for instance, with the closing down of the popular service Instagress, which created bots that automatically commented on and liked other Instagram accounts' photos, thus artificially driving up engagement and making accounts appear more popular and valuable than they in fact were. According to my informants, these security efforts have intensified since 2018.

Alongside this, a wide range of fake-follower-detection tools have been created by private developers, which allow for the auditing of Instagram accounts. For instance, accounts that follow many accounts but never post, lack profile photos, or are based in a country in the global south but mainly follow accounts in California are deemed dubious by these tools and thus seen as part of the Insta fraud economy (cf Ellis, 2019). While Instagram is developing algorithms that aim to minimize the number of bot-generated accounts, other private actors are developing tools that allow potential customers to gauge the quality of influencer accounts (that is, what percentage of their followers are 'real') with regard to potential marketing investments.

The answer to what forms of impostering are acceptable must thus be understood as a moving target, not least since Instagram offers limited public information about their security updates. As a result, providers, sellers and users must engage in testing, tinkering and guesswork to establish an understanding of the different forms the good enough imposter can take. In light of this, the next section considers how the diverse forms of Instagram followers available for purchase on the market have changed over time.

Moving targets

A key problem that resellers such as Ali and Handaya faced was finding reliable social media marketing services, in particular different kinds of Instagram followers. Much of their time was spent scouring websites to compare the pricing and quality of a dizzying array of services. It was generally difficult to judge the quality of followers based on the written description on websites. Instagram followers could come from different providers, with sellers often combining different quality accounts in packages of, say, 1,000 followers. Resellers engaged with the quality of services in a few different ways. Many would primarily rely on recommendations from trusted providers. Stefan, a reseller in Serbia, mainly learned through testing and feedback from his customers (3 June 2019). In terms of testing, he would purchase followers for his own accounts and measure the 'drop rate', which refers to how quickly and what proportion of accounts were closed down by Instagram. For instance, if more than half of the purchased followers were lost within 24 hours, this was often considered a poor service, while if more than half survived 30 days it was considered good.

In his tinkering with services, Stefan would be attentive to what kinds of accounts were part of the service package and consider this in relation to pricing. A low-quality account – which meant, for instance, that it lacked a photo, posts or any activity – easily fell prey to Instagram's algorithms. By contrast, a higher-quality account, which might have a photo, posts, comments and likes, would more easily establish itself as part of Instagram's user base. The most expensive would often be country-targeted – for instance, those posing as male or female 'Turkish' or 'Brazilian' followers. There were many variations in between: one of the main provider websites listed nearly 100 different types of Instagram followers available for purchase. Among the different types of followers, it was also possible to choose speeds of transfer – followers could be 'injected' at once, or, for a higher price, be 'dripped' over time in order avoid suspicion.

Most successful resellers attempted to find a balance between profits and establishing relations of trust with buyers, which was the basis for creating a sustainable business on a cut-throat market. This meant delivering reliable services that rarely dropped, at a reasonable cost. In response to problems with services, some sellers would offer guarantees; free refills, for instance, if more than half the followers dropped within 30 days. When prompted, most claimed they attempted to sell their customers 'middle-range' followers, meaning those that had relatively low drop rates and were not too expensive. The types of middle-range followers available at good prices would change over time, however, and depended on the websites and providers they were purchased from. In other words, in this case the good enough imposter was not necessarily the cheapest or the lowest possible quality with regard to Instagram's algorithms, but one that appeared stable and reliable at a higher cost, thus fostering relations of trust with customers.

It is widely acknowledged that many bot-generated accounts that had previously been able to exist within Instagram's user base were quickly being closed down in the wake of increasing security updates. The survival of accounts or the transfer of services depends on the technological constraints set up by Instagram and is constantly changing over time: some services cease to function while new ones are developed. For instance, after what appears to be Instagram's most significant security update yet in June 2019, Karim from Morocco told me:

> 'We have never seen anything like this before. Before we could transfer 100,000–200,000 followers in one day, but since the update the services are much slower … They have created algorithms that see the automated work of our systems. They have also made a new rule, if accounts have many new followers, like 100K per day, they block this action since they know that this is 100 per cent illegal.' (1 September 2019)

In the context of the ongoing arms race with Instagram, providers must thus continuously innovate. Since limitations are never clarified by Instagram, providers and sellers themselves attempt to figure them out through testing and tinkering. Informants agree, for instance, that the first couple of days are the critical period in the creation of an account. It is in this time it faces the greatest risk of being closed down and must be developed through a process that appears as "natural as possible", as Hendra put it (Jakarta, 6 November 2019), or as "authentic as possible", to paraphrase Instagram's terminology. It is critical to post pictures, edit the profile, to follow new accounts, but not too many, perhaps 10–20 per day. Posting slowly is also considered important, as is the timing.

In this process, Hendra, who has sold followers for more than a decade, said that it is increasingly difficult to distinguish bot-generated from human-generated accounts (16 February 2019), as they were becoming "as human as possible"; a humanness that refers to a particular account aesthetic and activity that could perhaps have been created by an individual. In fact, Ahmet, one of the main providers of Instagram accounts in Turkey, said,

> 'the best accounts are bots because they never unfollow. If you can create profile pictures, give them names, and posts that match, it is like they are real. It is very important to have high quality followers. If your accounts don't have

profile photos and posts it is worthless. You cannot sell them.' (24 June 2019)

In these examples, it is possible to gain some insight into the critical *process* involved in shaping the good enough follower, one that appears 'real' or 'human'. Testing, tinkering, and trial and error become critical in developing practical forms of knowledge concerning the border between what Instagram terms authentic and inauthentic practices. The same is true for the transfer of services. Once it is clear that it is not possible to transfer 100,000 followers a day, there will perhaps be attempts to transfer 75,000 and then 60,000 until the process is successful and limits are identified.

The most recent innovations already described are able to convince both Instagram's algorithms and other users – for the time being. Ahmet said that he had developed further technologies to pass Instagram's algorithms; for instance, a direct messaging system between his bot-produced accounts. In this way, his accounts perform as 'real' users that act, communicate, and engage with other users. Importantly, the development of direct messaging – which is not available for public view – transforms the bot into an agent whose agency is performed for the sole benefit Instagram's algorithms, rather than other users. Its stability as a good enough imposter thus follows not from the visual dimension of the account – a photo, posts, comments – but rather because the account acts authentically in technological terms.

Infrastructures of impostering

Most evidence suggests that automated bot-generated accounts dominate the market. The providers who create these accounts – informants suggest that there are around 300 major players worldwide specializing in different platforms – sell them together with passwords. More common, however, as mentioned earlier, is that followers rather than accounts are available for purchase as the latter are directed to follow specific accounts rather than be for sale themselves. The purchased followers can be resold (but not used) repeatedly after purchase, often through reseller websites that are connected through APIs, application program interfaces, a software intermediary that allows two applications to communicate, thus making the automated exchange of data possible (Snodgrass and Soon, 2019).

APIs are the key 'pipelines' across an expanding digital logistical space (cf Cowen, 2014). Gray and Suri (2019: xiv), for instance, highlight the critical importance of APIs on online labour platforms such as Amazon

Mechanical Turk: 'The MTurk API enabled software developers to write programs ... that automatically pay *humans* to do tasks.' APIs are also critical to the broader shift from social media *sites* to *platforms*. As Helmond writes: 'The moment social network sites offer APIs, they turn into social media platforms by enacting their programmability' (2015: 4). APIs, however, also facilitate the market for Instagram followers and other social media marketing services. In other words, an infrastructure has been created that allows for the intensifying, and often automated, circulation of followers on the transnational marketplace that Adi and Handaya are part of. As Harvey and her colleagues explain: 'infrastructures are extended material assemblages that generate effects and structure social relations' (2018: 5). They are thus 'doubly relational due to their simultaneous internal multiplicity and their connective capacities *outwards*' (Harvey et al, 2018: 5). The available infrastructure shapes the practices of programmers and resellers who through their practices in turn come to influence the development of the infrastructure. Notably, the expansion of this infrastructure has taken place in tandem with the rise of Instagram as a site for advertising and commerce. The infrastructure centered on APIs has thus become critical to a process of acceleration and scaling up of impostering on Instagram. Accounts are shaped by programmers alongside Instagram, which allows for the creation of services such as followers that come to circulate on the market.

As noted at the beginning of the chapter, after registering, sellers such as Adi and Handaya can log into any number of websites selling social media marketing services – buyfollowers.id based in Jakarta, for instance – and can choose from a long list of services after adding credit. Buyfollowers.id may in turn be connected by API to a comparable website in India – say, sellfollowers.com – allowing them to purchase and automatically resell those same Instagram followers to Adi and Handaya without any manual labour. The 1,000 followers that Adi and Handaya are purchasing from buyfollowers.id are in fact coming from sellfollowers.com. Furthermore, it is most likely that sellfollowers. com is reselling those same followers from another website or provider. The difference between them is that Adi and Handaya are engaged in manual sales and the websites in automated transactions using APIs, which allows them to offer lower prices and to increase the scale of their sales. The most successful reseller websites can have thousands of members which in turn sell to other resellers or clients below them.

My research shows that sellers and providers tend to work in small groups of young men connected by friendship, kinship or local ties, sometimes in a shared space, but just as often dispersed across an urban area. This is true not only in Indonesia but also among the individuals that I interviewed around the world. These cottage industries then come to be connected and networked through APIs and digital communication. Brunton writes with reference to Nigerian '419' scams – a well-known advance-fee confidence trick that focuses largely on writing and sending emails – that 'these low-level, somewhat educated scammers, in a society largely without opportunity for those without connections by birth or patronage, have ended up as components in a strange kind of writing machine' (Brunton, 2013: 106). To a large extent, this description fits well with regard to Ali, Handaya and my other informants, who have become part of a new kind of machine, reflecting the enduring inequality – but not exclusion – at the centre of the fake follower economy.

Most of my Indonesian informants came from the Jakarta region, many from major cities such as Bandung, Pekanbaru, Yogyakarta, and a few from smaller towns; a pattern replicated in my other Skype interviews with individuals from major cities such as Cairo and Dhaka. The individuals involved in the market are almost exclusively young men, the majority in their late teens and early twenties, with the oldest being in their early thirties. Most had a lower middle class to middle class background – parents who are entrepreneurs, office workers or teachers – and more than half some college education. In other words, these are individuals who are generally part of a broad and expanding middle class that is aspirational but lacks access to the patronage networks that have historically been critical in Indonesia and other countries. They have come of age in an increasingly digital environment in which jobs are not readily available, but in which entrepreneurial skills can be put to use. Most claimed that they were temporarily engaged in an industry that was characterized by increasing competition and inherently unsustainable.

Many came to the business through various forms of play or tinkering, either online games such as World of Warcraft – in which many quickly became aware that there was money to be made in selling accounts – or by spending hours on the internet figuring out ways to make money. In other words, the key actors in this market – the entrepreneurs and programmers – are generally 'geeks' who become experts through their own initiative rather than formal training (Jones, 2011: 19–20). Adi, for instance, calls himself *autodidakt*, as he had initially learned how to engage with the market through surfing the internet alone or with his friends. While making money was certainly a key incentive, ethical issues were generally downplayed, but it is clear that the practices of deception and impostering were generally understood as a form of frivolous play – much like Instagram itself – which they then were able to transform into a business model.

In sum, it is evident that the market for Instagram followers and other social media marketing services depend on a data infrastructure that has been able to match increasing demand following the rise of Instafame, social media influencers, and the growth of an online marketplace centered on Instagram. This infrastructure is integrated with a wide range of cottage industries at the interface of a market run by young men who have grown up playing and tinkering on and with the internet. It is unclear where exactly we should locate the imposter or the practices of impostering in this chain of relations, as it comes to include not only individuals but also technologies.

If, as the editors of this volume suggest in Chapter 1, 'the imposter is presently having a moment' (p 5), then it is important to note that the accelerated process of Instagram impostering relies on an evolving socio-technical assemblage. This moves us beyond a performative understanding of imposters as strictly engaged in 'the staging of social identities' and comes 'to include the staged realities of things and materials' (p 16). This takes on a particular digitized form once we consider that the acceleration of impostering with regard to social media marketing services is made possible by a highly networked socio-technical infrastructure.

Conclusion

Brunton (2013) approaches the history of spam as a shadow history that comes to mirror broader developments on the internet during the past decades. As he puts it, 'spamming is the project of leveraging information technology to exploit existing gatherings of attention' (Brunton, 2013: xvi). More generally, spam is a form of 'technological drama' (Pfaffenberger, 1992, cited in Brunton, 2013: xvi) that takes shape over time, drawing together a wide range of actors and technologies that come to develop practices in response to one another. It is significant that the research Brunton's book is based on ends in 2010, the same year Instagram was founded – a critical shift in the media landscape. Like spam, the Instagram follower economy can be understood as being centered on the exploitation of gatherings of attention, which is increasingly critical with regard to phenomena such as Instafame (Marwick, 2015). As such, the rise of the follower economy comes to mirror the broader rise of social media and the ongoing transformation of public culture and digital capitalism. It is in relation to this particular shadow economy that a technological drama comparable to spam has taken shape at the intersection between Instagram's increasing security measures, the acceleration and infrastructuralization of the market for followers, the innovations of a diverse group of providers, and the demands of a wide range of customers. It is at the centre of this drama that the good enough follower or imposter takes shape.

This chapter, like many of the others in this volume has shifted attention away from an a priori binary between 'fake' and 'real' to take the good enough imposter as a starting point for better understanding the attention economy, Instagram and social media. This market depends on socio-technical assemblages that include data infrastructures that allow for automated transactions and an extensive cottage industry of young men that takes a similar form around the world. It is based on an ongoing arms race in which providers, sellers and users, as well as Instagram itself, are negotiating which forms of impostering are acceptable within the Instagram platform. In this negotiation, the good enough follower emerges as a temporary placeholder (cf Dumit, 2000) – sufficient for the present purposes.

Recognizing that impostering is shaped through a range of relations in the market for Instagram followers allows us to consider not only the position of the imposter in social media, but also to approach impostering as a key dynamic within the attention economy. What is revealed is precisely that impostering is the norm rather than an aberration in this economy. There is no stable position occupied by the imposter in these processes; it takes a variety of different forms and positions, all of which are in the process of taking shape or being discarded. The social media influencer, providers of bot-generated followers, sellers, and Instagram itself are all concerned with purchasing, creating, banning, selling and distributing imposters in an attempt to stake out a position in the attention economy. It is precisely the fact that the imposter is impossible to pin down that allows it to become a force in digital capitalism.

This pushes us to rethink the figure of the imposter itself, which takes shape not in contrast to a 'real' account or individual, but as a socio-technical accomplishment in itself, which sets the limits of, rather than appearing as the inverse of authenticity. If we imagine the imposter, the good enough Instagram follower, as located at the center rather than the margins of the attention economy and digital capitalism, it becomes possible for us to consider Instagram itself as a hoax – 'operating in a constant crossfire between play and truth' (Miller, 2018: 7). This play of identities and the intensification of imitation is in constant tension with the demands of accountability normally placed on the production of economic value (Miller, 2018: 5–7). As social media platforms have come to reshape the contours of economic value as well as of public and political culture, it becomes increasingly critical to take practices of impostering, such as those described in this chapter, as a critical vantage point from which to critically engage with the particular crossfires between play and truth that characterize the contemporary world.

Acknowledgments

Thanks to Else Vogel, Steve Woolgar and David Moats for incisive comments on earlier versions of this chapter. The research upon which the chapter is based was funded by the Swedish Research Council (number 2017–02937).

References

Abidin, C. (2016) 'Visibility labour: Engaging with influencers' fashion brands and #OOTD advertorial campaigns on Instagram', *Media International Australia*, 161(1): 86–100.

Brunton, F. (2013) *Spam: A Shadow History of the Internet*, Cambridge: MIT Press.

Clark, D. (2015) 'The bot bubble: How click farms have inflated social media currency', *New Republic*, 21 April. Available at: https://newrepublic.com/article/121551/bot-bubble-click-farms-have-inflated-social-media-currency [Accessed 12 January 2020].

Confessore, N., Dance, G., Harris, R. and Hansen, M. (2018) 'The follower factory', *The New York Times*, 27 January. Available at: https://www.nytimes.com/interactive/2018/01/27/technology/social-media-bots.html [Accessed 1 June 2020].

Cotter, K. (2019) 'Playing the visibility game: How digital influencers and algorithms negotiate influence on Instagram', *New Media & Society*, 2(14): 895–913.

Cowen, D. (2014) *The Deadly Life of Logistics: Mapping Violence in Global Trade*, Minneapolis: University of Minnesota Press.

Dumit, J. (2000) 'When explanations rest: "Good-enough" brain science and the new socio-medical disorders', in M. Lock, A. Young and A. Cambrosio (eds) *Living and Working with the New Medical Technologies: Intersections of Inquiry*, Cambridge: Cambridge University Press, pp 209–232.

Ellis, E.G. (2019) 'Fighting Instagram's $1.3 billion problem: Fake followers', *Wired*, 10 September. Available at: https://www.wired.com/story/instagram-fake followers/ [Accessed 5 June 2020].

Fairchild, C. (2007) 'Building the authentic celebrity: The "Idol" phenomenon in the attention economy', *Popular Music and Society*, 30(3): 355–375.

Frier, S. (2020) *No Filter: The Inside Story of Instagram*, New York: Simon & Schuster.

Gerlitz, C. and Helmond, A. (2013) 'The like economy: Social buttons and the data-intensive web', *New Media & Society*, 15(8): 1348–1365.

Google Temasek (2017) 'e-Conomy SEA spotlight 2017: Unprecedented growth for southeast Asia's $50B internet economy'. Available from: https://www.thinkwithgoogle.com/intl/en-apac/tools-research/research-studies/e-conomy-sea-spotlight-2017-unprecedented-growth-southeast-asia-50-billion-internet-economy/ [Accessed 1 June 2020].

Gray, M.L. and Suri, S. (2019) *Ghost Work: How to Stop Silicon Valley from Building a New Global Underclass*, New York: Houghton Mifflin Harcourt.

Harvey, P., Jensen, C.B. and Morita, A. (eds) (2018) 'Introduction: Infrastructural complications', in *Infrastructures and Social Complexity: A Companion*, Abingdon: Routledge, pp 1–22.

Helmond, A. (2015) 'The platformization of the web: Making web data platform ready', *Social Media & Society*, 1(2): 1–11.

Jones, G.M. (2011) *Trade of the Tricks: Inside the Magician's Craft*, Berkeley: University of California Press.

Khamis, S., Ang, L. and Welling, R. (2017) 'Self-branding, "micro-celebrity" and the rise of social media influencers', *Celebrity Studies*, 8(2): 191–208.

Lalancette, M. and Raynauld, V. (2019) 'The power of political image: Justin Trudeau, Instagram, and celebrity politics', *American Behavioral Scientist*, 63(7): 888–924.

Leaver, T., Highfield, T. and Abidin, C. (2020) *Instagram: Visual Social Media Cultures*, Cambridge: Polity.

Leighton, H. (2020) 'Cristiano Ronaldo is the first person to reach 200 million followers on Instagram', *Forbes*, 30 January. Available at: https://www.forbes.com/sites/heatherleighton/2020/01/30/cristiano-ronaldo-is-the-first-person-to-reach-200-million-followers-on-instagram/#1a31ecf33ddf [Accessed 9 June 2020].

Lim, M. (2018) 'Dis/connection: The co-evolution of sociocultural and material infrastructure of the internet in Indonesia', *Indonesia*, 105: 155–172.

Marwick, A. (2013) *Status Update: Celebrity, Publicity, and Branding in the Social Media Age*, New Haven: Yale University Press.

Marwick, A. (2015) 'Instafame: Luxury selfies in the attention economy', *Public Culture*, 27(1): 137–160.

Miller, C.L. (2018) *Impostors: Literary Hoaxes and Cultural Authenticity*, Chicago: University of Chicago Press.

Pfaffenberger, B. (1992) 'Technological dramas', *Science, Technology, & Human Values*, 17(3): 282–312.

Senft, T. (2008) *Camgirls: Celebrity and Community in the Age of Social Networks*, New York: Peter Lang.

Snodgrass, E. and Soon, W. (2019) 'API practices and paradigms: Exploring the protocological parameters of APIs as key facilitators of sociotechnical forms of exchange', *First Monday*, 24(2). Available at: https://doi.org/10.5210/fm.v24i2.9553 [Accessed 10 September 2020].

Taylor, E. (2017) 'The most geo-tagged city on Instagram may surprise you', *Vogue*, 2 August. Available at: https://www.vogue.com/article/jakarta-indonesia-instagram-stories [Accessed 1 June 2020].

13

Thinking beyond the Imposter: *Gatecrashing* Un/Welcoming Borders

Fredy Mora-Gámez

Hesitation

The song sounded familiar. I had doubtless heard it before. Its lyrics were fervently performed by its interpreters while standing inside a moving bus. This version of Juan Gianitti's *El Camino del Dolor*[1] – the Path of Pain – was carefully attended to by most of the passengers in the bus. After the song finished, the younger member of the duo saluted the audience and went to every seat to collect coins and food. This was not the first time they had done this, as I could deduce from their skill in managing the movements of the bus while successfully playing the guitar and singing. I was returning from Altos de Cazucá, a slum in a town called Soacha located in the south-west of Bogotá where people moving into the city usually settle after escaping from the still ongoing situation of violence in Colombia. When the duo came off the bus, I decided to join them and initiate a conversation. This impulsive and intuitive act resulted

---

[1] The lyrics of this song describe the story of a farmer who once lived happily in the countryside with his partner, until a tragedy dispossessed him from her and their house, forcing him to continuously walk the path of pain.

in a revealing encounter with Luis and his son who generously shared their story after I explained my interest in their performance and experiences (Figure 13.1).

When asked about their lives before coming to the city, Luis seemed to repeat the lyrics of the song they had just performed in the bus. Like the song's author, they too had once owned a farm. They lived there until 2008, when an armed group that occupied the region assaulted and assassinated Luis' wife and daughter before expelling him and his son from their land. "Even if it was safer to go back, I would not do it", were the words he used to describe their reluctance to revisit a place full of painful memories – their own *Camino del Dolor*. Luis told me how he moved among different towns trying to find a job and a place to stay with his son. They both claimed to be earning enough money to survive in the city through their performances on buses.

Something drew my attention while talking. Luis constantly asked his son to hand him things or inquired what time it was. I initially interpreted this as their merely being in a hurry. A powerful revelation in Luis' story, however, suggested more was going on. Luis had a visual impairment and played the guitar by heart. When walking, he was most of the time guided by his son. Yet he described his visual impairment as "nothing worthy of stopping trying to earn some money to survive in the city". Luis did not want his audience to become aware of his progressive limitation, so he was extremely careful and effective when he moved inside buses. "I don't want to earn this money because of their pity for me", he explained. Luis was pretending that he could see well all the time, so his impostering act was flawless. I was struck not only by his ability to move around effectively, but also by the reason he claimed for doing it.

Since 2011, following the passing of the Law of Victims (LV), which first publicly recognized the existence of the armed conflict, the Colombian state formalized its legal obligation both to compensate victims and to assist in the restoration of land to its displaced occupants. There were already bureaucratic procedures in place before the passing of the LV, however, following the guidelines of previous similar regulations. Luis explains how since their arrival in the city in 2009 they unsuccessfully applied several times for registration as official victims of the armed conflict. Some of the reasons for their rejection to be registered in the official record included presumed inconsistencies in the information they provided, the lack of evidence about their forced displacement, and an existing registration in the public health

Figure 13.1: Luis and his son in Bogotá

system which supposedly prevented them qualifying for assistance and benefits as victims of the armed conflict redundant. The long waiting times within bureaucratic systems in Luis' story were unsurprising to me. I had encountered them before in my research on the LV. Luis and his son had been advised by several friends, mostly experts in matters of administrative procedures, to use a legal resource to appeal their rejection from the Registro Único de Víctimas (henceforth

RUV).[2] Through their friends, mainly people living in Soacha, they found out that official communications take a long time, so they had been exploring legal options such as actions of protection (*tutelas*) and rights to petition. In the meantime, there were not many jobs Luis could take given his visual impairment.

Luis and his son went through unthinkable experiences before arriving in Bogotá. They had to leave behind their family, homes, jobs, friends and everything they knew. People in similar situations of forced displacement due to the armed conflict share common challenges: they have to engage in administrative procedures to be granted some sort of governmental aid and they have to make use of their own skills to find forms of employment available for them in the city. Stories like Luis' have made me wonder what exactly is granted as 'reparation' and what it takes to be given the official status of victim, which seems to be bestowed only after the successful completion of standardized administrative procedures validated by state representatives.[3] Back then, while looking at him relying on his son and still performing his magnificent act during our encounter, Luis generously revealed that he had never reported his impairment in the applications for registration. Because of his omission, his claims for compensation were based exclusively on his forced displacement. From his point of view, this made them "fair claims". By contrast, asking for compensation on the grounds of his impairment would simply make him feel like a "beggar".

Luis could have been registered in the RUV sooner. All he had to do was ask the evaluators to tick the disability box in the forms, so his condition could have been rapidly certified by medical experts. Instead, he confabulated with his son to avoid this label, to obtain benefits due to the violence he experienced and nothing else. In the assistance offices where they applied for registration, Luis asked his son to drive his fingers across the form to sign it and fingerprint it. Although evaluators in reception offices might have noticed his condition, this information was omitted from Luis' claims. He diverted the attention of the system and exposed its inability to determine applicants' physical impairments – a feature that such infrastructure should usually capture as it constitutes a crucial vulnerability condition within population assessment. When 'infrastructures meet people' (Star, 1999: 92), fitting stories of migration and violence into standards of classification unfolds

---

[2]   Official Record of Victims administered by the Unit for Assistance and Reparation of Victims (UARIV).

[3]   For an extended study of these bureaucratic procedures see Mora-Gámez (2016a).

in complicated ways. A multiplicity of (dis)orders are produced within and around migration management infrastructures. The case of Luis is a preliminary inspiration to reflect on this multiplicity.

In line with the general theme of this volume, Luis' revelation was unique and provocative. Although the notion of 'imposter' might to some extent be useful analytically to expose the cracks and failures of the RUV as an example of migration management infrastructures, I hesitate to describe Luis' act exclusively in those terms. My hesitation in the analytic use of the imposter to describe the experiences of my interlocutors in Colombia and Greece is what inspires this chapter. Recent research at the intersection between migration studies and science and technology studies has analysed the production of evidence and narratives in bureaucratic and procedural standards within migration management infrastructures (Papada et al, 2019; Pelizza, 2019; Pollozek and Passoth, 2019). An important portion of the literature on migration studies has documented cases of impostering in asylum applications (Bohmer and Shuman, 2010; Terretta and Derrida, 2015), spousal migration (Charsley and Benson, 2012) and migration based on faith-based claims (Madziva and Lowndes, 2018). Other studies have documented the challenges of determining the veracity of asylum requests in vulnerable populations (Lelliott, 2019), family reunion (Olwig, 2019) and cases involving LGBTQ populations (Hertoghs and Schinkel, 2018). These studies explain how impostering is problematic for migration management infrastructures and how it exposes their fractures. Luis' strategies, however, point in a different direction. Luis was not omitting information about his condition so as to avoid and overcome evaluators' suspicions. Neither was his intent to pass as able-bodied. He was doing it to obtain aid that was not related to his impairment, but to the violent displacement and dispossession he had experienced in the past.

The contributions to this volume explore different kinds of activity that can count as impostering and explore the *imposter* as a way of rethinking social relations. Some of these activities involve playing within the rules of a specific game, and being capable of using, translating and betraying those rules in different ways. In the case of migration management infrastructures, practices such as rural migrants' exchange of information about bureaucratic procedures were geared towards conforming to the rules and standards of official registration. However, other practices of my interlocutors worked to escape (Papadopolous et al, 2008) management infrastructures and to exceed state projects (Ranciere, 1999). My hesitation to employ the imposter as an analytical device is deeply grounded in my experience of these latter practices: arrangements like street memorials which make visible state crimes;[4] empathy chairs displaying stories of the armed conflict; and weaving communities embroidering textile pieces which tell stories about war violence and its affective aftermath (Sánchez-Aldana et al, 2019; among others). These practices are not geared towards playing by the rules of capture embedded in population management systems. Instead, they pose strategies either to challenge state-infrastructures or to produce alternatives to it (Papadopoulos, 2018). Luis did not try to conform to the standards of disability in the application procedure for compensation, nor did he try to pass as a victim deserving assistance. Instead, his efforts were about altogether evading such classification and capture, about avoiding representation by governmental systems as a disabled body.

My hesitation in engaging with the *imposter* as a theoretical tool is also informed by a reflexive concern about the marginalized situation of those who I accompanied in their bureaucratic journeys and everyday struggles. Using 'imposter' to designate those who have shared their everyday life with me as a researcher, a Colombian national, a person also experiencing migration, a volunteer and, in some cases, a friend, generates strong feelings of unease, arising from reflections on the significance of caring and knowing as relational processes (Puig de la Bellacasa, 2012). In the paragraphs that you are about to read, I try to unfold these feelings by 'thinking-with' my interlocutors and their marginalized statuses, 'dissenting-within' the imposter as a provocation, and 'thinking-for' the effects that the use of the imposter can convey to the audiences of this text (Puig de la Bellacasa, 2012). This chapter seeks to move the study of migration away from the 'world that comes with' the imposter.

So, my central concern is: How can we reframe the notion of impostering and think beyond it to reflect on this kind of practices and social orders? What alternatives can we propose to study what exceeds the limits of impostering as an analytical category? In this chapter, I suggest that 'the imposter' does not fully equip us to understand the politics at play in many of the practices around bureaucratic procedures. I draw on these questions to develop an alternative take on the matter in two steps. First, I analyse empirical instances where evaluators, rather than applicants, engage in faking practices. Thus, I advocate an analytical shift, or relocation, of the locus where impostering might take place, thus broadening the range

---

[4]   For a detailed ethnographic account of these memorials see Mora-Gámez (2020).

of actors involved. Instead of considering impostering as an applicants' activity, I reflect on its relationality and distribution within migration management infrastructures. Second, I describe applicants' efforts to overcome extensive and prolonged bureaucratic procedures in order to achieve asylum and relocation. Drawing on the notion of *gatecrashing*, I describe how engaging in everyday crafting practices has allowed some of my interlocutors unexpectedly to trespass 'gates' or socio-technical boundaries of containment.

Relocating: the relationality of impostering

Luis and his son were not the only ones applying for registration. Since 2012, the LV deployed technological bureaucratic tools like the RUV to assess the conditions of applicants and decide on their access to monetary compensation, psychosocial assistance, their participation in financial projects funded by the state, and their access to job databases and land restitution (Ibáñez and Velásquez, 2006; Franco-Gamboa, 2016). Every applicant must provide their statements in assistance centres located across the country. The information is collected using the Formato Unico de Declaración (FUD) a paper-based form designed for this purpose. After being signed and fingerprinted, these forms are sent to the assessment office in Bogotá, where their consistency will be determined using the information available in governmental databases. A formal communication with an official response is expected three months later.

After becoming familiar with the application procedures experienced by Luis and many other informants, I visited the assessment offices in Bogotá. On many occasions, I was surrounded by an uncountable number of boxes containing FUDs from different regions. I frequently felt overwhelmed when thinking of the stories behind those boxes, of the years of experiences of pain, perpetrations and assaults that inevitably reminded me of people like Luis. I imagined the necessary chains of translation between those stories and these forms, there in a single place waiting for evaluation. This pile of boxes and the database produced after their digitalization are still important means by which the Colombian state captures the *Camino del Dolor* of applicants for registration. Elsewhere, I have documented how the objectivity of the assessment is not a pre-existing feature of the procedure, but a relational achievement (Mora-Gámez, 2016b). As part of the procedure, every acceptance or rejection of an application strengthens and corroborates the official account of the conflict promoted by one of its actors: the Colombian state. The veracity of RUV applications is produced while simultaneously establishing the objectivity of the army and police databases used in the assessment. Hence, the RUV and the FUD, among other objects and entities, enact a population management infrastructure which is the cornerstone of current reparation policies in post-conflict Colombia.

By the time of my visits to the assessment offices, however, a few well known cases of 'fake victims' had seized the attention of the local media. I managed to engage a civil servant in a short interview. An important portion of the victims were registered through the FUD, whereas the remaining percentage is included by court rulings and judicial sentences of land restitution. My interviewee also asserted that the cases of rejection are specifically for situations such as domestic violence, common crimes and traffic of narcotics. These cases should be treated by other legal mechanisms and regulations that are different from post-conflict reparation and the LV. Another reason for the rejection of application was 'fraud'.

My interviewee said that the office had identified several 'fake' applications made by *tramitadores*.[5] These are people, usually civil servants within the bureaucratic system, who use their expertise in the application procedures to guide applicants so as to increase their chances of registration. According to official accounts, some *tramitadores* charge applicants fees for this service, based on a percentage of the compensation awarded by the government. UARIV recently established an anti-fraud committee to identify cases in which false statements might have been provided, and fake documents were produced as evidence during the applications.

My interviewee at the office shows me a poster designed for their latest anti-fraud campaign (Figure 13.2). The poster alludes to recent complaints about tricksters, represented as wolves, having involved victims of the armed conflict in fraudulent practices while charging a portion of the received compensations. My interviewee further explains that following anonymous complaints, they found out that applicants already included in the RUV had apparently provided false statements during their applications. A number of investigations were carried out by the District Attorney. "Making an application with a fake *cédula* (ID), superseding another person, or pretending to be a victim to receive compensations is a crime and must be punished",

---

5   'Trámites' are administrative procedures; *tramitadores* may be understood as people with some expertise in those administrative procedures, in this case, agents processing the applications within the system.

Figure 13.2: Pamphlet from campaign 'Ojo con el fraude'

Source: With permission from Unidad para la Atención y Reparación Integral a las Víctimas (UARIV).[6]

---

[6]   The text in the image reads: 'Beware of scams ... Do not let yourself be fooled. Learn here how to prevent fraud ....'

says my interviewee, arguing that public resources must be preserved and delivered in a responsible way to "the people who deserve them". I can tell that my interviewee finds my question about fraud uncomfortable. He downplays the point with a final statement: "It has happened before, when the assessments were done in the local offices throughout the country, and there are people in prison because of that, but it is not happening anymore." In direct contradiction to that statement, however, local newspapers at that time suggested that at least 400 cases were under investigation for fraudulent applications (*El Nuevo Siglo*, 2014).

In a more convincing account about cases of fraud in the RUV, one of the evaluators in the assessment centre tells me that at least 40 cases that he had to assess were under investigation because the statements had been apparently copy-pasted by civil servants in assistance centres filling out the forms of several applicants. The only differences between the applications were the names, places, dates and some other details.

> 'It was frustrating that the first encounter of those applicants with the state had been conducted in that way, the civil servants did not even take the time to listen to them and transcribe what they said. They just copy-pasted the same text for every single applicant. ... I was annoyed, the only contact we have with those people is through the narration of events imprinted in those forms, so I felt it was a mere production of documents.'

The words of the evaluator are revealing for a variety of reasons. First, the person that makes the decision about the registration has no personal contact with the actual applicant, but only with their written narratives, which are codifications and translations of the original narratives offered in the assistance centres. Second, what the evaluator claims about the declarations under investigation as "mere production of documents" is true for every single FUD under assessment. Every narration is produced as a result of highly structured questions specifying dates, places and events. The standardized narratives are also productions of documents and elaborated versions that fulfil the technical criteria for assessment. The main difference between fraudulent and legitimate FUDs here rests in their *degree* of homogeneity. Therefore, the homogeneity of the documents produced must not be so evident or explicit, because this would threaten the central ethos of the evaluation system, its alleged objectivity and capacity to detect inconsistencies. But an evident homogeneity among FUDs also threatens the capacity of RUV to differentiate good narratives, those fitting, from deficient ones. The practices of security and the narratives against fraud and impostering also promote the idea of objectivity, transparency, rigour and responsible management of the information by state representatives. Objectivity, rigour and transparency, then, are values performed together with the status of official victim within an information infrastructure. But this cannot be achieved by applicants by themselves. My interviewees at the assessment centre claim that fake applications are the outcome of applicants' lack of information, education and technological literacy. They also identify civil servants acting as *tramitadores* as key players involved in impostering, filling the forms or accessing the databases.

A few months after my visits to the assessment centre, the media followed a case where seven civil servants were arrested for having accessed the RUV databases in order to modify stored information and use it to register people, supplant applicants to claim their benefits, speed up claims of monetary compensations, and divulge information about the perpetrators in applicants' statements (Bargent, 2014; Redacción Judicial, 2014). At the time I learned about such developments, I was visiting a rural town in north-west Colombia. There I met a social movement of women already registered in the RUV, who were offered the possibility of speeding up compensations in exchange for a proportional fee. Similarly to Luis, these women had been forcedly displaced from their farms and were now part of a collective seeking the restoration of their communities and lands. Due to the absence of titles over their land, however, most of their applications were initially rejected. But then civil servants offered them the possibility to be registered in exchange for a percentage of their compensations as official victims.

It was not clear what the civil servant was able to do to obtain such an outcome. Whether it was a specific way to fill the FUD, or a direct intervention in the database: it was a mystery to me as to the women they targeted. Yet, they were offering people effectively displaced by the violence of the armed conflict the possibility of being registered as victims within an information infrastructure, despite their previous rejection. In other words, a group of applicants, who reasonably should qualify for compensation, might be recognized as victims only after deceiving the system by producing evidence and statements that were not necessarily consistent with their situation. Their accounts had to be shaped in particular ways by civil servants with access to the system. The outcome of such action is the production of applicants, whose everyday life was undoubtedly affected by war, as 'certified victims' – but at the same time, as imposters. Crucially, this involved the participation of other deceiving actors, and access to and transformation of objects like forms and databases.

This experience exemplifies how the misuse of migration management infrastructures can be carried out by evaluators and other people with access to databases. It is not exclusively the practices of applicants that deserve scrutiny. Producing fake statements is a relational achievement involving different actors, instances, objects and procedures. The official status of victim in the RUV is not the mere outcome of applications and assessments. It involves inscriptions, transformations, assessments and validations mediated by databases, forms and platforms. Whereas applicants can produce statements and forge evidence to navigate the system, evaluators in different instances can also alter the trajectories of forms and evidence, producing fake victims as outcomes. But there is an additional form of deceiving enacted as part of this arrangement. As mentioned before, the assessment process of every application takes three months, as the national guidelines dictate. However, the actual contact of civil servants with FUD is restricted to the moment the document is filled in. Evaluators take an additional 20–40 minutes for its assessment and production of the official communication. In the remaining time, the FUD merely travels to the assessment office, waits for evaluation, and travels back to the original assistance centre where applicants will be notified about the outcome. Nonetheless, the length of the whole procedure has a performative effect on how people imagine the assessment. Luis, the women from the collective, and many other of my informants, often expressed their belief that the length of the process was an indicator of its thoroughness, objectivity and rigour.

This effect might not be intentional on the part of civil servants and evaluators. It is perhaps the way in which bureaucracy works when dealing with marginalized populations like people on the move fleeing war. Yet, because they are rejected by the RUV or perhaps because they do not even engage with it, the stories of people affected by violence lack legitimacy. In any case, by circulating a narrative of objectivity and rigour associated with the long waiting times, the migration management infrastructure remakes the state while violently dispossessing applicants' accounts from their truth and veracity. Thinking with the *imposter* in these cases would focus on whether actors (applicants or evaluators) do or do not embrace the standards of the registration system; it would stress the way they transform the rules of migration management infrastructures. Yet, this analytical move is in danger of disregarding what falls outside such order, what people do around the rules of the registration procedures and how those practices challenge the boundaries enacted by information infrastructures. How to address these practices? Is the figure of the imposter useful in understanding how they expose, produce and trespass the cracks of migration management infrastructures?

Gatecrashing boundaries of containment

For a variety of reasons, my own journey as a migrant and a researcher from the global south pursuing a career in the global north led me to the southern European border where Greece is located. Many of my reflections gathered with social movements in Colombia resonated with my experience of squats, reception centres and concentration camps in Athens and Lesbos. There, I saw how the confinement and restriction performed by borders extend beyond their material and concrete limits. The bureaucratic procedures for asylum and relocation requests and residence and work permits are extended forms of surveillance and control (Jacobsen, 2015; Boswell, 2016). These procedures enact infrastructures shaping the everyday life of migrants by creating boundaries of containment. For instance, working becomes illegal, and the extensive waiting times of often almost two years actively discourage migrants from persisting with their applications. Besides the impossibility of engaging in paid labour, the deplorable conditions of refugee camps and the controls imposed by extended bordering practices in Greece, and the south of Europe in general, make the situation of migrants even more precarious. In this section, I explain how engaging in sewing, as a collective crafting practice, allows its participants to trespass bureaucratic gates, overcome waiting times, expand collaborative networks and unexpectedly 'crash' the gates of asylum in the EU.

I had the privilege of coming across many experiences of mobility and endurance in Athens. But one of those stories was particularly insightful and revealing of the bureaucratic challenges that asylum applications pose for people on the move, and the alternatives they engage with to overcome them. This was the case of Ivan, a migrant, an asylum seeker, or, as he described himself, a "constant traveller". Ivan narrated the story of his arrival while he was sewing in a workshop in Athens. As in many other stories of mobility, Ivan frequently mentioned his emotional journey, and the importance of the rubber boats in departing from Turkey and finally arriving on the coast of Mytilene, Lesbos. Instead of focusing on the reasons why he fled his home country, Ivan chose to share his experience in a squat and a sewing workshop in Athens.[7] The restrictive conditions of refugee camps in Athens made Ivan choose to participate in a squat instead. As a member of the squat, he was part of the teams in charge of cooking and cleaning among other tasks. A few months before I met him, he was invited by a group of volunteers to become part of workshop for sewing using recycled materials. Gina, a volunteer I met through Ivan, generously invited me to the workshop.

My visits to the workshop took place in the midst of sewing machines, tables, thread, chairs, drawers and workshop participants. I was initially unaware of the final outcomes of their sewing practices. During my first visit, Ivan and Gina took me to the storage room where I would come to have a better understanding of the materials used at the workshop (see Figure 13.3). Loads of rubber pieces from boats, discarded floating life vests, belts, bags and boxes were spread around the floor of the storage room. It took me a while to make sense of the trajectories of those materials, and their owners. Those materials were the remains of migration across the southern sea borders of the EU; the remnants from crossing the sea seeking safety. These objects had been in contact with hundreds of migrant bodies during their journey towards Lesbos, bodies that succeeded and others that failed in the attempt to reach Europe. I stood in front of these materials trying to comprehend and reflect on their mediations between 'various worlds that tend to be kept apart', such as 'Europe and its others, care and surveillance' (M'charek, 2018: 94).

At the time of my fieldwork, several boats arrived on the Greek coast of Mytilene every week, transporting dozens of migrants from different countries whose final point of departure was Turkey. Volunteers in Mytilene tried in particular to collect complete pieces and functional vests, and sent these materials to Athens by truck or boat. The usable but still dirty pieces were then cut using a variety of cutting instruments. Only at that point were the rubber and the usable belts cleaned for the first time.

---

[7]   Drawing on this same ethnographic case in Athens, I put forward the argument that the material politics of these Recrafting Workshops consists of the enactment of spaces of contestation of citizenship in the EU. I also contrast this approach with the notion of 'acts of citizenship' coined by Isin (2009). See Mora-Gámez (2020).

Figure 13.3: Ivan sewing at the workshop after classifying and cleaning the materials

Workshop participants explained how they try to remove as much dirt as possible. Meanwhile, other volunteers like Gina preferred to remove some of the dirt but not all of it. As Ivan and Gina were somewhat silent about the outcome of all these procedures, I remained puzzled about the goal of the workshop. Ivan allowed me to talk to him about this while he was sewing.

Volunteers and other participants had taught Ivan the basics of sewing. While he was firmly holding the rubber, I sometimes imagined him trying to hold onto the same kind of rubber, but inside the boat on his journey to Europe. When I asked what he thinks of while sewing, Ivan said: "At the beginning [of his time at the workshop], I used to have memories of my arrival all the time, but it is not the case anymore". Because he was aware of the origin of the materials, he was more careful about the results of his work. The same hands were again touching rubber. But this time, the aim was not to cross a geographic border, but, as I would come to understand, to challenge borders and gates of an entirely different nature. Dozens of bags, wallets and keychains were sewed in the workshop and sent to groups of friends (volunteers and migrants) to be sold in Germany, Sweden and the Netherlands.[8]

My experience of researching migration management infrastructures in Colombia and Greece was enriching and revealing, but my interest while visiting this squat was different. I was now tracing material practices of making and transforming objects, and the ways they enact alternative forms of solidarity in self-organized communities. Back in the workshop I asked Ivan about his struggles during the application period and his time in Greece. Most of his friends from his time living in refugee camps had their applications rejected, some were caught working illegally and imprisoned, others were deported, and yet others possibly made their way to other EU countries as stowaways departing from Patras towards Italy. Despite all the concerns he expressed about his time in Athens – hunger, loneliness, worry and boredom – Ivan emphasized how his participation in the workshop and the skills he developed were crucial for enduring such challenging conditions.

> 'You see, I am now kind of an expert at sewing, more like a tailor, my work helps people and collects donations for friends in Germany. They help us in the same way by sending supplies. I eat here, have coffee, sometimes a nap, before going back to the squat to fulfill my duties. I am making something out of all this waste, all this misery, all this rubbish charged with so many bad feelings. I make something useful, nice, clean from all those pieces. This lets me better endure the time waiting for a response'.

I came to realize that the sewing workshop does not consist exclusively of the transformation of objects like rubber and thread. Boats and vests are reconfigured into mobile and transgressive objects that make a part of their journey visible. The material transformation of rubber was also a way of defeating the various forms of confinement that those involved faced. With the donations collected across the EU, the workshop guarantees the transportation of materials and a space where migrants share their crafting skills, while also extending their solidarity networks with already settled migrants living in Athens and other locations of the EU. Some of the migrants participating in

---

[8]   I have described elsewhere how these solidarity networks are also crucial for the exchange of information regarding asylum applications and everyday practicalities (Mora-Gámez, 2020).

the workshop were already trained tailors in their home countries, whereas others like Ivan had never sewed before. This crafting space then, also offers their participants the possibility of learning new skills, exchanging information with already settled migrants about how to deal with bureaucratic procedures, and facing up to other everyday 'gates' by relying on cooperation networks.

While sewing, the journey is remade, and rubber is transformed. Through such work, the confinement posed by the EU migration management infrastructures appears less restrictive. These infrastructures materialize forms of population management embedded in narratives of humanitarian policies towards migrants. By contrast, the sewing workshop encourages community attempts to share literacy among the participants. The cooperation of Ivan and his community was neither accidental nor a product of formal training. Their skills for recrafting rubber were precisely the result of engaging with the materiality of sewing machines while exploring their potentialities.

Crafting in this sewing workshop offers possibilities to workshop participants. The long waiting times, reminiscent of those in the Colombian procedures for victim recognition, discourage applicants from persisting with their applications for asylum. Conversely, this sewing workshop opens new channels of participation and provides means of persistence within extended bordering processes. The distribution of bags and wallets is also led by relocated migrants across the EU who send the donations back in the form of material aid to help ease the general precarity of their conditions. At one level, then, by expanding the alternatives of the participants of the workshop and by using donations, these collaborations and crafting practices transforming rubber and thread enable them to endure the long waiting times of asylum application and relocation.

But the sewing workshop is not only about enduring long waiting periods of precarity. In the form of bags and wallets, these practices of crafting also reach and produce other audiences mobilizing alliances and their accounts of their journey across Europe, thus enacting new social orders. They also transform their audiences. The preference of people like Ivan to clean the materials can be seen as an attempt to materialize the persistence of something good and transparent despite the difficulties of the journey. The desire of other volunteers to keep some of the dirt and rustiness, however, addresses an additional concern. The dirt in the rubber stands as evidence of its authenticity and of the intensity of the experience. It produces an audience for the journey of people fleeing violence and seeking asylum and refuge – a promise made, at least on paper, by Europe. In Gina's words:

'Our purpose is not only to make nice bags to conquer a market of white activism, but to make people in Germany, Sweden and other countries affectively aware of what is going on in the borders, of how painful it is to cross a border for migrants and their families'.

Migrant bodies recraft the objects that transported them across the geographic border. The kind of solidarity materialized in this space of rubber and thread makes visible how migrants may overcome their experiences of their journeys. The sewing workshop reassembles everyday spaces across different EU member states by granting partial visibility to usually neglected stories. This is a relational achievement that I call *gatecrashing* the boundaries posed by socio-technical procedures of extended borders.

Gatecrashing paradoxical 'invitations'

In his ethnography of the French State Council, Latour defines *value objects* as emerging entities making their way through the law (2010: 140). At different points, these value objects are mobilized in the transit of law and its intersections, or 'intertextualities' (Latour, 2010: 256), with other laws and regulations. Using this vocabulary, we might say that the value object 'Human Rights' is predominant in the intertextuality of regulations like the UN Convention, the Dublin Regulation, the International Humanitarian Law and the Colombian Law of Victims, among others. In these regulations, Human Rights, as a modern entity coming 'with its own world' (Puig de la Bellacasa, 2012), shapes the policies and procedures of humanitarian assistance of asylum seekers, refugees and internally displaced people. Elsewhere,[9] I have examined how the world of Human Rights poses a narrative where states openly welcome people on the move seeking refuge from war violence and perpetrations. This narrative constitutes an implicit *invitation* to participate in the humanitarian project of modern states. Nonetheless, this invitation also poses a paradox.

Once the geographic or internal borders are crossed and people on the move settle partially or permanently, they still have to engage with points of bureaucratic control as 'periodic reminders' (Vlachou, 2017: 141) that they are not entirely welcome or that they have no

---

[9]   I analyse this intertextuality in detail for the case of the Law of Victims in Colombia and the International Humanitarian Law (see Mora-Gámez, 2016b).

equal rights as citizens. In this regard, bodies that following equality/diversity policies and global humanitarian laws are supposedly welcome; they are considered and treated as 'space invaders' (Puwar, 2004). The presence of 'space invaders', on the one hand, justifies and further legitimizes the intertextualities and practices that sustain states as the legitimate managers of mobility. But on the other hand, 'space invaders' show that state solidarity and reparation are designed with precisely the purpose of managing and exploiting mobility. For people on the move as potential gatecrashers, this is their living paradox: to be un/welcome only as 'space invaders'.

Despite the un/welcoming invitation and the discouragement enacted by population management infrastructures – posing as 'gates' preventing one from entering un/welcoming states – spaces like the workshop permit the persistence of applicants, the exchange of invaluable information and skills, and the configuration of solidarity networks with already settled communities. This is not to say that by engaging in these practices, people on the move overcome all the difficulties in their journeys and the extensive waiting times on their way to fulfilling the promise of relocation and asylum. Increasing and changing forms of control and confinement pose new challenges, and henceforth the development of other strategies.

Although the status of victim, refugee or asylum seeker is granted by legitimized state infrastructures and border keepers, these statuses imply a paradoxical in/exclusion. The conventional leitmotif of gatecrashing is 'entering a space even though you are not supposed to be there'. *Gatecrashing* un/welcoming gates and spaces through material interconnected practices, then, is not only about resisting bureaucratic categorization or crafting means of psychosocial support. Material practices like sewing that allow participants to survive, build networks, establish connections in and through achieved relationality, challenge institutional gates, boundaries and regimes. Even if these practices do not amount to material transition across borders, these collective material practices permit them to trespass the gates thrown up by a paradoxical un/welcoming invitation. People on the move crash population management infrastructures as gates that discourage their success, pose continuous restrictions, both at the entrance and, once they are inside, in protected sovereign spaces. Among the multiplicity of achievements of the presented cases in Colombia and Greece, I have drawn special attention towards the social orders produced from practices of material transformation like crafting as a way of trespassing un/welcoming 'gates'. This unconventional sense of *gatecrashing*, I suggest, offers a possible way to understand emerging orders and their material conditions of possibility while addressing migrants and applicants as more than (potential/suspected) imposters navigating bureaucratic systems.

By *gatecrashing* un/welcoming gates, people on the move form collaborations and networks and come to be part *of* communities and alternative hybrid alliances in other countries of the EU already, even though physically and bureaucratically they remain in their arrival point in Greece. Although using the notion of imposters serves to reflect on the case of migrants crossing borders, it also presupposes that migrants become politically successful only when they manage to beat the system and get access/recognition, even by deceiving the system. The politics of gatecrashing, by contrast, is not about playing, cheating or even changing the game; it is about finding ways of living within and outside it. These ways of living enact orders that simultaneously crash the 'gates' of in/exclusion by blurring their boundaries, softening their demarcation and forming alternative forms of existence.

In the study of migration and population management infrastructures, thinking with imposters brings our attention to people's attempts at beating the system and passing through by playing within its rules. As I described before, faking and forging are relational achievements within population management infrastructures involving alliances between different actors. However, I have advocated for an alternative lens which does more justice to the political implications of the practices I witnessed. *Gatecrashing*, as used in this chapter, avoids the performative risks of using the imposter to address people and communities in already marginalized and confined situations. Additionally, it points out a different claim for agency by migrants. A claim unfolded in gradual investments and everyday practices that do not necessarily engage but coexist with the rules of management infrastructures and states. It is a claim for agency that becomes visible when migrants 'crash' into bureaucratic and restrictive 'gates' and become part of a space, despite being paradoxically un/welcomed. *Gatecrashing* is then an alternative term to better understand how crafting, and the sociomaterial exchanges it entails, allows migrants to trespass 'gates' mainly comprised of the forms of socio-technical confinement posed by surveillance technologies, bureaucratic requirements and extensive waiting times.

In this chapter I have thought beyond the imposter as an analytical tool. Thus taking distance from a notion like impostering is also an ambitious invitation to my readers to reflect on the categories we use as researchers, and the performative implications they may have for our diverse audiences. Consider this chapter an attempt, then, to also *gatecrash* the boundaries of social concepts, by exposing some of their cracks and limitations, even if this makes us feel un/welcome in certain spaces.

References

Bargent, J. (2014) 'Restitución de tierras en Colombia es socavada por venta de información de las victimas', *InSight Crime*. Available at: https://es.insightcrime.org/noticias/noticias-del-dia/restitucion-tierras-colombia-venta-informacion-victimas/ [Accessed 5 February 2020].

Bohmer, C. and Shuman, A. (2010) 'Contradictory discourses of protection and control in transnational asylum law', *Journal of Legal Anthropology*, 1(2): 212–229.

Boswell, C. (2016) 'The "epistemic turn" in immigration policy analysis', in G. Freeman and N. Mirilovic (eds) *Handbook on Migration and Social Policy*, Cheltenham: Edward Elgar, pp 11–27.

Charsley, K. and Benson, M. (2012) 'Marriages of convenience, and inconvenient marriages: Regulating spousal migration to Britain', *Journal of Immigration, Asylum and Nationality Law*, 26(1): 10–26.

El Nuevo Siglo (2014) 'La Unidad de Víctimas contra los avivatos'. Available at: https://elnuevosiglo.com.co/articulos/4-2014-la-unidad-de-victimas-contra-los-avivatos [Accessed 10 April 2020].

Franco-Gamboa, A. (2016) 'Fronteras simbólicas entre expertos y víctimas de la guerra en Colombia', *Antípoda*, 24: 35–53.

Hertoghs, M.A. and Schinkel, W. (2018) 'The state's sexual desires: The performance of sexuality in the Dutch asylum procedure', *Theory and Society*, 47(6): 691–716. doi:10.1007/s11186-018-9330

Ibáñez, A.M. and Velásquez, A. (2006) *El proceso de identificación de víctimas de los conflictos civiles: una evaluación para la población desplazada en Colombia*, Bogotá: CEDE.

Isin, E.F. (2009) 'Citizenship in flux: The figure of the activist citizen', *Subjectivity*, 29(1): 367–388.

Jacobsen, L. (2015) *The Politics of Humanitarian Technology: Good Intentions, Unintended Consequences and Insecurity*, New York: Routledge.

Latour, B. (2010) *The Making of Law: An Ethnography of the Conseil d'État*, Paris: Polity.

Lelliott, J. (2019) 'Children's rights and crimmigration controls: Examining Australia's treatment of unaccompanied minors', in P. Billings (ed) *Crimmigration in Australia*, Singapore: Springer, pp 275–301.

M'charek, A. (2018) '"Dead bodies at the border": Distributed evidence and emerging forensic infrastructure for identification', in M. Maguire, U. Rao and N. Zurawski (eds) *Bodies as Evidence: Security, Knowledge, and Power*, Durham: Duke University Press, pp 89–109.

Madziva, R. and Lowndes, V. (2018) 'What counts as evidence in adjudicating asylum claims? Locating the monsters in the machine: An investigation of faith-based claims', in B. Nerlich, S. Hartley, S. Raman and A. Smith (eds) *Science and the Politics of Openness*, Manchester: Manchester University Press, pp 75–93.

Mora-Gámez, F. (2016a) 'Reconocimiento de víctimas en Colombia: sobre tecnologías de representación y condiciones de estado', *Universitas Humanística*, 82(1): 75–101.

Mora-Gámez, F. (2016b). *Reparation beyond statehood: Assembling rights restitution in post-conflict Colombia*, doctoral dissertation, University of Leicester, UK.

Mora-Gámez, F. (2020) 'Beyond citizenship: the material politics of alternative infrastructures', *Citizenship Studies*, 24(5): 696–711. doi: 10.1080/13621025.2020.1784648

Olwig, K.F. (2019) 'The right to a family life and the biometric "truth" of family reunification: Somali refugees in Denmark', *Ethnos*. doi: 10.1080/00141844.2019.1648533

Papada, E., Papoutsi, A., Painter, J. and Vradis, A. (2019) 'Pop-up governance: Transforming the management of migrant populations through humanitarian and security practices in Lesbos, Greece 2015–2017', *Environment and Planning D: Society and Space*, 38(6): 1028–1045.

Papadopoulos, D. (2018) *Experimental Practice: Technoscience, Alterontologies, and More-than-Social Movements*, Durham: Duke University Press.

Papadopoulos, D., Stephenson, N. and Tsianos, V. (2008) *Escape Routes: Control and Subversion in the Twenty-First Century*, Ann Arbor: Pluto Press.

Pelizza, A. (2019) 'Processing alterity, enacting Europe: Migrant registration and identification as co-construction of individuals and polities', *Science, Technology, & Human Values*, 45(2): 262–288. doi: 10.1177/0162243919827927

Pollozek, S. and Passoth, J.H. (2019) 'Infrastructuring European migration and border control: The logistics of registration and identification at Moria hotspot', *Environment and Planning D: Society and Space*, 37(4): 606–624. doi: 10.1177/0263775819835819

Puig de la Bellacasa, M. (2012) '"Nothing comes without its world": Thinking with care', *The Sociological Review*, 60(2): 197–216.

Puwar, N. (2004) *Space Invaders: Race, Gender and Bodies Out of Place*, Oxford: Berg.

Ranciere, J. (1999) *Disagreement: Politics and philosophy*, Minneapolis: University of Minnesota Press.

Redacción Judicial (2014) 'Descubren red que vendía información de víctimas del conflicto en Antioquia', *El Espectador*. Available at: https://www.elespectador.com/noticias/judicial/descubren-red-vendia-informacion-de-victimas-del-confli-articulo-508750 [Accessed 5 August 2016].

Sánchez-Aldana, E., Pérez-Bustos, T. and Chocontá-Piraquive, A. (2019) 'What are textile activisms? A view from feminist studies to fourteen cases from Bogota', *Athenea Digital. Revista de Pensamiento e Investigación Social*, 19(3). doi: 10.5565/rev/athenea.2407

Star, S.L. (1999) 'The ethnography of infrastructure', *American Behavioral Scientist*, 43(3): 377–391. doi: 10.1177/00027649921955326

Terretta, M. and Derrida, J. (2015) 'Fraudulent asylum seeking as transnational mobilization', in I. Berger, T. Redeker Hepner, B.N. Lawrance, J. Tague and M. Terretta (eds) *African Asylum at a Crossroads*, Athens: Ohio University Press, pp 58–74.

Vlachou, M. (2017) 'Researching experience in global higher education: A study of international business students in the UK', doctoral dissertation, University of Leicester, UK.

Postscript: Thinking with Imposters – What Were They Thinking?

Agnes, Forrest Carter, Civet Coffee Bean, Cuckoo, Iansá and Oxum, Sarah Jane, Han van Meegeren, David Rosenhahn, Diederik Stapel and Jorge Enrique Briceño Suárez

So, that's it.

Yes, all done.

Great book!

Yes, absolutely.

Full of wonderful, rich examples.

Yes. Which were your favourites?

I loved learning from fakes, and the Ku Klux Klan guy, and epistemic murk …

… the welfare cheats, the good enough Instagram followers, and the conjuring …

Wasn't that cool, how the magician …

… the conjuror …

Ah yes, sorry, the conjuror. How the conjuror drew parallels between the art of conjuring and the craft of academic writing.

Yes. A great demonstration of the effective use of dialogue to bring complex arguments to life.

Clever!

I especially like that the editors commissioned expert imposters to write the Postscript, rather than the usual choice of a prestigious academic.

Yes, nice touch.

A welcome alternative to inviting a Big Name who then largely ignores the preceding chapters and proposes some grand theory based on personal anecdote.[1]

Good to avoid that.

Mind you, I do think the collection seems hardly to scratch the surface of the whole imposter phenomenon. I mean, there are no focused case studies of undercover policing, the imposter syndrome …

It's interesting though isn't it, that as soon as one person mentions the term imposter, someone else feels obliged to provide yet more examples of the category. It's as if the very term 'imposter' has some kind of intrinsic capacity to generate more examples of itself.

… professional actors, trompe l'oeil illusions, fake composers …

Fake composers?

'Albinoni's' Adagio in G minor, for example?

Oh, OK.

… the Australian lyrebird, spies …

Have you already forgotten that the whole volume began with the Russian spies?

With *possible* Russian spies.

OK, yes, *possible* spies. Or tourists.

And what about all those related figures and perspectives, didn't one of the reviewers of the original book proposal mention some of them: the anansi spider (Hylland Eriksen, 2013), the coyote trickster (Haraway, 1991), agnotology (Proctor and Schjebinger, 2008) …?

Yes, and I noticed that despite its catchy subtitle, the book says little specifically about gatecrashers[2] and charlatans.

But they couldn't include every imaginable empirical example of imposters, could they? The book would have been massive!

---

[1] We recognize that certain Big Names are justifiably famous for their celebrated Postscripts, for example Deleuze's (1992) 'postscript on the societies of control', Garfinkel's (1967) postscript about Agnes, as discussed in Chapter 1 of this volume, and Kierkegaard's (1941 [1846]) ironic pseudonymous work.

[2] Except for Mora-Gámez (Chapter 13).

And that's the point isn't it. Imposters are everywhere. The examples in the book are just a taster of an apparently limitless phenomenon. Without wishing to sound like its own back cover endorsement, this book now opens up a whole new field of imposter studies. I for one am inspired to think with imposters in my own research from now on.

An especially important new field because, as the editors argue, the cases selected for this collection help build a set of general insights into impostering which can 'reset the agenda for mainstream social theory'.

A bit over the top, that claim, wouldn't you say?

No. I find it very persuasive. And you have to agree that social theory is badly in need of respecification. Society is no longer the organism that many theorists made it out to be. The functioning whole has become a fiction, an ideal more than anything, often defended by unsavoury types, White supremacists, xenophobes and the rest. We have to think relations differently, and the imposter can help us to do so.

And that's its critical potential isn't it. This new approach departs from the common sense notion which implies (malicious) intent on the part of the imposter. It offers ways of analysing how people, groups or things come to be designated, suspected or treated as imposters.

So we both agree …

A terrific book.

Except I do wonder about the Introduction.

What about it?

Well. Don't you think it a bit strange? The editors give us this excellent introductory chapter which extols the virtues of paying attention to disordering. That the intriguing thing about imposters is where order is not easily resolved …

Yes.

… but then include an ordered description of the actual content of each of the chapters about imposters!

Ahh.

A definitive listing of the actual ways in which imposterings do disorder!

Yes I see. But it's only an ordering device, isn't it. They just wanted a way of organizing the chapters. All collections do this don't they? Are not editors expected to dream up a rationale which connects and holds the different contributions together?

At this point the camera pans back to reveal the actual identities of the discussants.

Harold Garfinkel: That's just typical of conventional social science: one is expected to concoct some kind of artificial narrative arc to help the reader. This is what I mean when I describe the technical practice of sociology as 'shoving words around' (Garfinkel et al, 1981: 133).

Erving Goffman: But that's OK then, isn't it? The editors are just following the rules of the game (Goffman, 1969). I don't see there is any alternative really. You're being a bit harsh.

HG: Well, maybe. But it is disappointing isn't it? They want to celebrate the open endedness of interpretation and the unresolvedness of impostering. They claim that the disordering done in impostering is actually generative of culture and society. They claim this is a new way of generating insights into social theory. But they then give us this same old formulaic ordered representation of the actual contents and their relations to each other.

EG: Wow Hal, you are such a purist!

HG: Wow Erv, you're such a poser!

EG: We are all posers, I've always said so.

HG: Yes but I hate that. You think all we do is go around posing, somehow mysteriously following the dictates of social convention.

EG: Seriously though, how else would you expect them to manage it any better? You can't have an edited collection without a table of contents. The point is they are just acting like good editors.

HG: *Acting* like good editors!? There we have it again! You are so hung up on performance.

EG: But that's the point. We are all acting all of the time.

HG: Yes but 'acting like editors'? Are you suggesting they are impostering as editors?

EG: We now know from this book that it's not that easy. There is no simple way to decide whether or not they are imposters. Indeed, it is much better not to decide. It's much more interesting that way. We need to ask: are we talking about the kind of imposter who has long been exploited by social theorists to make theoretical points about societal order? Or are we talking about the much more intriguing, indeterminate, open ended and uncertain imposter, the figure now so effectively emancipated by this book?

At this point HG and EG are joined by an unexpectedly erudite social scientist.

AS: I wonder if you are both missing the point here. It's pretty obvious isn't it? Disordering and indeterminancy have to be understood in terms of the transformative hermeneutics constitutive of boundary transgression.

A few seconds pause. HG and EG are momentarily taken aback.

HG: I see. Good point.

AS: Ordering requires boundaries and transgressing the boundaries necessarily entails situated hermeneutic transformation.

EG: Yes, I hadn't thought of that.

AS: So the editors are inevitably and thoroughly complicit in their own externalized transgressivity.

HG: Yes, of course!

EG: Yes, that's it!

But just as all parties seem to agree, AS undergoes a sudden transformation. We see him metamorphose from his social scientist persona. He is now bristling with moral outrage, possibly borne of overzealous commitment to ideals of purity and exactitude. The mask slips, and he reveals himself to be, after all, just an angry physicist.

AS: That's enough of all this! What a load of nonsense!

EG and HG: What? Who are you?

AS:	Who am I? The question is: who do you think you are? As usual you have got yourselves in a hopeless mess. Do you really think this load of rubbish is going to pass muster with the Bristol University Press reviewers?
HG:	They seemed pretty sympathetic before, when the editors submitted their proposal.
AS:	Oh this is just typical of cultural studies! No proper standards. No proper criteria of evaluation ...
EG:	... I'm not sure the editors would like you calling this book 'cultural studies' ...
AS:	... any kind of claptrap can get published. All posing as serious academic work when it's actually nothing of the kind.
HG:	... Whoa, hold on! I think these nice people are only trying their best.
AS:	It's so easy! And please don't pretend you've never heard of my definitive *exposé* of intellectual impostering. It's even discussed in the Introduction to this volume. My fake cultural studies research report, which passed scrutiny by the journal reviewers and was published, even though it was meaningless drivel? And even knowing all that, you still failed to spot my little deception just now.
HG:	Ahh, er Alan, Alan something ...
EG:	A deception? Oh dear ...
HG:	Alan, Alan er ...?
EG:	... What a pity. I quite like the sound of 'transgressive hermeneutics' ...
HG:	On the tip of my tongue, Alan, err ...
EG:	... A shame. Makes a lot of sense, I thought. Pretty good social science insight really ...
HG:	... Socalled?
Alan Sokal:	SOKAL!
HG and EG:	Ahh!
Alan Sokal:	Yes! So pathetic! My goodness, it's so easy to imposter as a social scientist (Sokal, 1996)!

EG:	Hmm. Actually you know, CF, I'm not sure Sokal would put it quite like that.
HG:	*Flustered* 'CF'?? You mean Garfinkel?!
EG:	Err ...
Steve:	OK. Let's drop the masks people. I think our reader has got the point by now.
CF [Claes-Fredrik]:	Yes, time to stop pretending, or at least pause it for a bit ...
Else:	Oh thank goodness. I'm really not comfortable impersonating dead White males.
David:	And I am really not sure this will please the Bristol University Press reviewers. Please can we just go back to the normal writing format, as I suggested from the start?

This volume makes clear the importance of researching imposters. This is not only because impostering practices are intriguing in themselves but mainly because, as the contributions show, thinking with imposters provides an alternative way of thinking social and cultural relations. By positioning imposters as quintessential and omnipresent disrupters, a whole new field of research comes into play. More questions arise and, as postscripts often predictably declare, more research is needed.

Among the key questions still remaining: Can anyone or anything become (staged as) an imposter? Can anyone or anything become an imposter researcher? Can any activity get to count as, or be cast as, impostering? Certainly as Garfinkel, Goffman and Sokal just showed, attempts to answer this often face considerable difficulty and mess. The wide range of impostering activities and situations presented in this volume suggest there is no *principled* way of answering this. But what kinds of *pragmatic* solution might be attempted? What, in other words, are possible vehicles for the governance of imposter research? What kinds of social arrangement (organization, network, assemblage) might be appropriate to decide what will and will not count as impostering? One of us (CF) suggested a provocative thought experiment. What if we established a Society for Imposter Studies (SIS, or if royal patronage could be secured, the Royal Society for Imposter Studies RSIS)?[3]

---

[3] It often helps new organizations to have an independent authority figure as a patron. Yet the advantages of royal association are equivocal in present times, given the risk of being tainted by scandal.

By considering how SIS operates, we might learn more about what in practice gets to count as impostering. A learned society has the advantage of lending prestige to the field, handing out prizes (Imposter Researcher of the Year; Lifetime Award for Excellence in Impostering Research, and so on), and more generally create an aura of respectability for the field. Separate categories of membership would be established: members, fellows and employees. Mechanisms would be set up to determine the merits of claims to membership, and to adjudicate in cases where members are asked to leave. Is payment of a membership fee sufficient to allow entry? Should there be assessment of the extent to which an applicant's imposter scholarship has advanced the cause of impostering? How well have they feigned expertise in understanding impostering practices? Establishing appropriate mechanisms and procedures for addressing these questions will help institutionalize what counts as research into imposters. So institutions of future imposter governance like SIS will be busy with the brutish day to day work of sorting things out. They decide as a matter of practical expediency who or what falls under the definition of imposter, and what counts as imposter research. Pragmatic decisions are made. Persons and events are determined acceptable or not.

To see a little more how this might work, consider a bid for membership of SIS by Alan Sokal. On what grounds should he be accepted? Should he be judged on whether or not he succeeded as an imposter (in Sokal, 1996)? Was his effort authentic? Was it sincere? Does it matter that Sokal's impostering was copied by others (Lindsay and Boyle, 2017; Pluckrose et al, 2018)? That it involved considerable scholarly craft?[4] Should he be admitted on the basis of his claim that he and Jean Bricmont exposed imposter scholars in their book *Intellectual Impostures* (Sokal and Bricmont, 1998)? They targeted thinkers such as Jacques Lacan, Gilles Deleuze and Bruno Latour. But are these scholars actually imposters? Or is this merely the upshot of posturing by Sokal and Bricmont? If we accept they are indeed imposters, then Lacan, Deleuze and Latour also merit membership of SIS alongside Sokal and Bricmont.

The only possible conclusion is that Sokal both is and is not a worthy member of SIS. A matter of no small practical inconvenience perhaps, but such awkwardness is a small price to pay for the more important organizational goal of sustaining indeterminacy. For to determine that

---

[4]   One observer pointed out that to succeed as imposters they had to do considerable scholarly work, much like 'real' authors (Soar, 2018).

Sokal both is and is not a member upholds the foundational ontological commitment of impostering to disorder and indeterminancy.[5] Such apparently curious and often self-contradictory determinations of impostering have to be understood as thoroughly (in)consistent.

What then is the value of 'the imposter' for social theory? We suggested in Chapter 1 that the imposter provides a figure of significance to sit alongside other famous figures in social theory. For example, the figure of the parasite alerted us to the close interrelation between order and disorder, how the disorder of one order always implies an order from a different perspective (Serres, 2007). The stranger (Simmel, 1908) drew attention to group formation and processes of inclusion and exclusion. The figure of the imposter alerts us to the omnipresent puzzle of the relations between appearance and reality. Not just as a problem of metaphysics, something to be solved theoretically, or to be used in its disturbance as a methodological probe. But rather to be observed and understood as an engine which is constitutive of social worlds. The collection has shown how the appearance/reality puzzle is actively created and sustained by a variety of actors: by technologies (Grunenberg, Chapter 9; Lindquist, Chapter 12), bureaucracies (Kaufman, Chapter 8; Restrepo and Ashmore, Chapter 11), celebratory performance (Rappert, Chapter 7; van de Port, Chapter 10), in film and literature (Merck, Chapter 5; Abbott and Large, Chapter 6); audiences and communities (Rosenthal, Chapter 2; Derksen, Chapter 3) and so on. It has also surfaced the nature and extent of the complexities involved, for example, with the observation that imposters are themselves sometimes used to 'reveal impostering'.[6] The importance of attending to the dynamics of the appearance/reality puzzle is that its constitution and maintenance has important social and political effects (Rosenthal, Chapter 2; Coopmans, Chapter 4; Mora-Gámez, Chapter 13).

More profoundly, perhaps, we think the chapters in this volume suggest that the figure of the imposter offers a challenge to the very idea of social theory itself. For if social theory is about ordering and imposters

---

[5] Pollner (1987) similarly argues that in making routine decisions and determinations, traffic courts maintain and uphold basic ontological commitments. The day-to-day business of determining whether and by how much a driver had exceeded the speed limit frequently involve 'reality disjunctures' where the police and the driver's version of the speed of the vehicle differ. In determining which version is correct, the significant work of the court is not just to determine the facts of the matter. It also reaffirms that a car cannot be travelling at both 30mph and 60mph at the same time.

[6] As for example is claimed by Sokal and Bricmont (1998).

are about indeterminancy and disordering, it follows that imposters do not merely have the capacity to respecify social theory, but perhaps also to cast doubt on the entire project of theorizing. The figure of the imposter which emerges from this volume requires us to think and see social and cultural forms differently. The indeterminancy of the imposter invites us to think and see *with doubt*. In the face of propositions that the world is a certain way, thinking with imposters leads us to wonder: is it? The broad idea that social theory provides a framework for studying and interpreting social phenomena, is now unsettled by the proposition that both framework and phenomena are indeterminate. They are both themselves part of the appearance/reality puzzle.[7]

Do you think this is enough of a Postscript?

I don't know. What happened to the idea that we would craft a Postscript which charts the way forward for studies of imposters? I suppose, in the end, we are saying there is no way to resolve this?

We are saying we need to avoid a resolution.

So we succeeded then, right?

References

Ashmore, M. (1987) *The Reflexive Thesis: Wrighting the Sociology of Scientific Knowledge*, Chicago: Chicago University Press.

Bloor, D. (1976) *Knowledge and Social Imagery*, Chicago: Chicago University Press.

---

[7] We see the imposter as a contribution to, and pushing forward, a long-standing series of conversations about the relations between appearance and reality, and the frailty of any order. These conversations occur across the social sciences and humanities, and have at times relied on and at times struggled with the very idea of stable social order. In Science and Technology Studies alone they take the form of a succession of different, perhaps increasingly bold, ways of doubting, and of different targets of doubt, stretching from Merton (1973) and Kuhn (1962) to ontological politics and beyond, via the strong programme (Bloor, 1976), the Sociology of Scientific Knowledge (Star, 1991; Collins, 1992), Actor Network Theory (Callon and Latour, 1981; Law, 1994), reflexivity (Ashmore, 1987) and multiplicity (Mol, 2003), to name but a few. The long history of these moves can be understood as a history of different modes of 'otherwising', (Woolgar, forthcoming): successive demonstrations of the indeterminancy of a wide range of increasingly recalcitrant phenomena. The imposter is both a reminder of the need to further this conversation, and a figure which helps us push it forward.

Callon, M. and Latour, B. (1981) 'Unscrewing the big Leviathan: How actors macro-structure reality and how sociologists help them to do so', in K. Knorr-Cetina and A.V. Cicourel (eds) *Advances in Social Theory and Methodology: Towards an Integration of Micro- and Macro-Sociologies*, London: Routledge & Kegan Paul, pp 277–303.

Collins, H.M. (1992) *Changing Order: Replication and Induction in Scientific Practice*, Chicago: University of Chicago Press.

Deleuze, G. (1992) 'Postscript on the societies of control', *October*, 59: 3–7.

Garfinkel, H. (1967) *Studies in Ethnomethodology*, Englewood Cliffs: Prentice Hall.

Garfinkel, H., Lynch, M. and Livingston, E. (1981) 'The work of a discovering science construed with materials from the optically discovered pulsar', *Philosophy of the Social Sciences*, 11(2): 131–158.

Goffman, E. (1969) *Strategic Interaction*, Philadelphia: University of Pennsylvania Press.

Haraway, D. (1991) *Simians, Cyborgs and Women: The Reinvention of Nature*, New York: Routledge.

Hylland Eriksen, T. (2013) 'The anansi position', *Anthropology Today*, 29(6): 16–19.

Kierkegaard, S. (1941 [1846]) *Concluding Unscientific Postscript to the Philosophical Fragments,* Copenhagen: University bookshop Reitzel.

Kuhn, T.S. (1962) *The Structure of Scientific Revolutions*, Chicago: University of Chicago Press.

Law, J. (1994) *Organizing Modernity*, Oxford: Blackwell.

Lindsay, J. and Boyle, P. (2017) 'The conceptual penis as a construct', *Cogent Social Sciences*, 3: 1330439. Since redacted but available at https://web.archive.org/web/20170520194758/https://www.cogentoa.com/article/10.1080/23311886.2017.1330439.pdf [Accessed 24 June 2020].

Merton, R.K. (1973) *The Sociology of Science: Theoretical and Empirical Investigations*, Chicago: University of Chicago Press.

Mol, A. (2003) *The Body Multiple: Ontology in Medical Practice*, Durham: Duke University Press.

Pluckrose, H., Lindsay, J.A. and Boghossian, P. (2018) 'Academic grievance studies and the corruption of scholarship', *Areo*, 2 October.

Pollner, M. (1987) *Mundane Reason: Reality in Sociological and Everyday Discourse*, Cambridge: Cambridge University Press.

Proctor, R.N. and Schjebinger, L. (eds) (2008) *Agnotology: The Making and Unmaking of Ignorance*, Stanford: Stanford University Press.

Serres, M. (2007) *The Parasite*, Minneapolis: Minnesota University Press.

Simmel, G. (1908) *Soziologie: Untersuchungen über die Formen der Vergesellschaftung*, Leipzig: Duncker & Humblot.

Soar, D. (2018) 'Short cuts', *London Review of Books*, 40(20): 20–25.

Sokal, A. (1996) 'Transgressing the boundaries: Toward a transformative hermeneutics of quantum gravity', *Social Text*, 46–47: 217–252.

Sokal, A. and Bricmont, J. (1998) *Intellectual Impostures*, London: Profile Books. [A very similar book, but with a different title was published as Sokal, A. and Bricmont, J. (1998) *Fashionable Nonsense: Postmodern Intellectuals' Abuse of Science*, New York: Picador.]

Star, S.L. (1991) 'Power, technologies and the phenomenology of conventions: On being allergic to onions', in J. Law (ed) *A Sociology of Monsters: Essays on Power, Technology and Domination*, London: Routledge, pp 26–56.

Woolgar, S. (forthcoming) *It Could Be Otherwise: Provocation and its Limits*.

Index

References to footnotes show both the page number and the note number (318n1).